Pore Structure of Cement-Based Materials

Modern Concrete Technology series
A series of books presenting the state-of-the-art in concrete technology

Series Editors
Arnon Bentur
National Building Research Institute
Technion-Israel Institute of Technology
Technion City, Haifa 32 000
Israel

Sydney Mindess
Office of the President
University of British Columbia
6328 Memorial Road, Vancouver, BC
Canada V6T 1Z2

Pore Structure of Cement-Based Materials

Testing, interpretation and requirements

Kalliopi K. Aligizaki

CRC Press
Taylor & Francis Group
Boca Raton London New York

CRC Press is an imprint of the
Taylor & Francis Group, an **informa** business

A TAYLOR & FRANCIS BOOK

CRC Press
Taylor & Francis Group
6000 Broken Sound Parkway NW, Suite 300
Boca Raton, FL 33487-2742

First issued in paperback 2019

Typeset in Sabon by
Newgen Imaging Systems (P) Ltd, Chennai, India

No claim to original U.S. Government works

ISBN-13: 978-0-419-22800-4 (hbk)
ISBN-13: 978-0-367-86383-8 (pbk)

British Library Cataloguing in Publication Data
A catalogue record for this book is available
from the British Library

Library of Congress Cataloging in Publication Data
Aligizaki, Kalliopi K.
     Pore structure of cement-based materials : testing, interpretation
and requirements / Kalliopi K. Aligizaki.
        p. cm. – (Modern concrete technology series ; v. 12)
     Includes bibliographical references and index.
     1. Cement–Testing. 2. Porosity. I. Title. II. Series.
TA435.A48 2005
620.1'35–dc22   2005004165

Visit the Taylor & Francis Web site at
http://www.taylorandfrancis.com

and the CRC Press Web site at
http://www.crcpress.com

Contents

List of figures

List of tables

Foreword

Concrete has been considered to be a load bearing material in the first place and therefore main interest and research were focused essentially on mechanical properties for a long time. Compressive strength used to be and still is the decisive criterion for classifying different types of concrete. Although it has been recognised at an early stage that porosity has a strong influence on strength, interest in pore size distribution remained very limited with one exception. The characteristic pore size and the geometrical distribution of artificially entrained pores in concrete became important parameters in materials testing and in practice in the context of frost resistance after the pioneering work of T.C. Powers. Under these conditions pore size distribution of hardened cement paste and concrete must have seemed to be rather esoteric for many engineers dealing with concrete.

Interest in durability and service life of reinforced concrete structures grew gradually as the cost for maintenance and repair of existing structures became too important. Detailed investigations showed clearly that the infrastructure in many countries can not be maintained in the designed and in a satisfying state with available financial resources. The initially promoted idea to solve all durability problems by increasing the compressive strength of concrete (high performance concrete) has been just as widespread as it is wrong.

Durability of porous materials such as concrete can be understood and improved only by studying the complex interaction between the porous structure and the environment. Most deteriorating mechanisms are tightly linked to migration processes. Absorption of water and salt solutions by capillary suction is one example of a migration process which limits service life of concrete structures in many cases and which is controlled by porosity and pore size distribution. Dissolved ions may also penetrate cement-based materials by diffusion through the pore liquid. In addition, reactive gases such as CO_2 may enter the structure of the material via partially water filled pores and initiate reactions with the solid skeleton. The pore size distribution in cement pastes covers a wide range from nm to mm. Migration mechanisms and as a consequence migration rate vary considerably with the pore size. For this reason the porous structure of concrete is the key issue for durability.

Porosity and pore size distributions of hardened cement paste and concrete have been studied by numerous authors in the past. A multitude of fundamentally different test methods has been applied. If we have a look into the existing literature, however, we will observe considerably differing and sometimes even contradictory results. Results of different test methods have seldom been critically compared. One reason for this unsatisfactory situation certainly is the lack of a comprehensive book on the pore structure of cement-based materials.

This obvious gap will be closed finally with this book. The author describes all relevant experimental methods in detail. Both simplistic methods to determine total porosity and most sophisticated methods applying small-angle neutron scattering are covered. The physical background of each test method is described. Principles of modern apparatus are introduced. Finally all statements are substantiated by numerous relevant references.

Progress of research is slow and the quality of individual contributions varies a lot if in a specific field of science and technology no generally accepted and comprehensive text book exists. With this book a solid common basis is provided. Nobody will have an excuse from now on if he or she does not take the actual state-of-the-art as described in this volume into consideration when planning an experiment or when interpreting results in porology of cement-based materials.

It is my sincere hope that this book will find its way to the research groups active in durability studies and others concerned very quickly.

Prof. Dr Folker H. Wittmann
Aedificat Institute Freiburg and
Qingdao Technological University

Preface

Several techniques have been developed and are currently used to characterize the pore structure of cement-based materials due to our need to understand the hydration process of cement, to develop models that describe the cement paste microstructure, and, consequently, to improve the properties of the binding material. The experimental techniques have been developed based on break through theoretical formulations in Applied Physics and Materials Science. Most of the experimental techniques used and the analysis procedures followed have been adaptations to cement pastes from other materials, e.g. from catalysts or rocks. Some of the assumptions and the limitations of the techniques become less familiar as they continue to be used throughout time.

In recent years, the development of technology, the widespread use of computers, and the development of sophisticated software have made significant steps in analyzing experimental results. Scientists make efforts to analyze and explain new experimental findings and refine the models that describe cement paste microstructure. In addition, as new techniques are developed, criticism and comparison are mentioned often to the more "traditional" techniques. The available information has become so numerous, that it becomes at times difficult to follow the developments, and keep up-to-date with new information.

This book presents most of the techniques currently used to determine the pore structure of cement-based materials. For each technique presented, an effort is made to describe its historical development, the theoretical principles and assumptions, the experimental procedure, the analysis of results as they relate to pore structure parameters, and its limitations. Even though the basic equations are given, their full derivations are not included here, but rather the original references are cited. In addition, some basic chemistry and physics background is required for the reader to understand the methods presented. This volume has only focused on techniques that characterize pore structure and does not extend to techniques that are used to describe or model transportation procedures.

The aim of this book is to educate engineers, students, and scientists, so that they have a clearer understanding of the techniques available and used to characterize the cement paste pore structure, and of the procedures followed to analyze experimental data. The book started from my personal need to collect information from the technical literature, in order to use and understand the results from techniques as applied to cement science. As more information on analysis procedures and comparison of results have become available, I considered it a challenging project to organize the material and make it available to others, who might find reference to these topics useful.

The information in the technical literature has been overwhelming and a selection of references that have been used was inevitable, with preference given to sources that are more widely accessible to the scientific community. I have tried to credit the research groups that have introduced and developed the techniques to cement-based materials; however it is inevitable that some omissions might have been made, for which I sincerely apologize in advance. In addition, I have made the effort to organize the material presented so that repetitions are avoided and similarities are pointed out. For possible errors that exist even when conversions to units and symbols were made, I welcome with great appreciation corrections and suggestions, and I hope that the book can be useful enough to compensate for any errors and omissions.

Kalliopi K. Aligizaki
November 2004
pore-structure@gmx.net

Acknowledgments

This book has become possible through the contribution of individuals from different sources. As a minimal recognition, I would like to extend my gratitude to the following:

I am first and foremost grateful to my publisher, Taylor & Francis for its trust in undertaking publication of this project. I am tremendously thankful to Tony Moore, Senior Editor, who worked with me patiently and tirelessly until the end and oversaw the project from beginning to end. My special thanks are also extended to Monika Faltejskova, Editorial Assistant, for her care during preparation of this publication, and to the other staff at Taylor & Francis and Newgen Imaging Systems who contributed to the preparation of this volume.

The book would have not been possible without the endless research efforts of numerous researchers, who are cited throughout the chapters. Many experimental results are reproduced here, and I am grateful to the following publishers who offered generously their permission to reproduce copyrighted material used here; more specifically: The American Ceramic Society, ASTM International, Elsevier Science, Institute of Physics Publishing, Materials Research Society, National Research Council of Canada, RILEM Publications, Taylor and Francis Inc., and the Transportation Research Board. Also, my very special thanks to the authors who granted us their permissions as well.

I am indebted to Dr Joe Larbi, TNO Building and Construction Research, Delft, The Netherlands, for making microscopic pictures suitable for use in this book.

I would also like to thank the staff of the Libraries at The Pennsylvania State University, USA, and The Delft University of Technology, The Netherlands, for offering their help in obtaining scientific references essential for this book.

This work would have never been possible without the support from Drs Della Roy and Rustum Roy, of the Materials Research Laboratory at The Pennsylvania State University. The knowledge I have acquired carrying out research in their laboratory facilities has been invaluable. Also, my very special thanks to the late Dr Bryant Mather for his encouragement to put the effort on this work; it is regretful that I will not have the opportunity to get his numerous comments and suggestions.

I have been very fortunate to have had the unlimited support from colleagues and friends, especially during the last year of my work on this book. This volume comes as a special thank you to them, and I hope that they will feel it was worth their support and encouragement.

List of symbols

Symbol	Parameter	Unit	Chapter
A_a	total air content	%	1
A_p	cross-sectional area of a pore	mm^2	1
A_{paste}	air content of paste	%	1
A	constant that can be derived theoretically	m·K	5
A	total area analyzed	m^2	8
$A(t)$	normalized magnetization as a function of time t	unitless	6
A_0	constant		7
A_A	area fraction of the plane surface occupied by the particles of interest	unitless	8
A_m	average area occupied by one molecule of adsorbate in the completed monolayer	m^2/molecule	4
A_p	total area of particles	m^2	8
A_s	total surface area of the pores	mm^2	1
B	constant	m	5
B_0	externally applied magnetic field	T	6
B_{inc}	incoherent scattering contribution to the total scattering (background signal)	cm^2	7
C	circularity	unitless	1
D	fractal dimension of the line	unitless	1
$D_V(r)$	pore size distribution function	—	1, 3
E	energy of a photon	J	6
E_{ml}	energy of a level	kcal/mol	6
F_1	force driving mercury out of the pore	N	3
F_2	force driving mercury into the pore	N	3
$F(R)$	size distribution of spheres	—	8
G	shape factor	unitless	1
I	spin number	unitless	6
I_i	intensity from pore i		6
$I_i(r)$	intensity from pore i		6
$I(0)$	incident intensity	counts/s	7
$I(Q)$	SANS scattering intensity	counts/s	7
$I(\lambda,\theta)$	scattered neutron flux	counts/s	7
$I_0(\lambda)$	incident neutron flux	counts/s	7
$\underline{L}$	total lengthy of a line	m	1, 8
$\overline{L}$	spacing factor	mm	1
LF	loss factor	%	2

Symbol	Parameter	Unit	Chapter
$L(r)$	distribution of the total length of pores (whose radii fall between r and $r + dr$)	—	4
L_L	length fraction of a line within a particle of interest	unitless	8
M	mass of the material	g	1
M	molecular weight of the adsorbate	unitless	4
M_S	mass of adsorbate at saturated vapor pressure	g	4
M_0	equilibrium longitudinal magnetization moment	T	6
M_{cap}	magnetization due to the water in the capillary pores	T	6
M_{gel}	magnetization due to the water in the gel pores	T	6
M_i	magnetization of the ith pore	T	6
$M(0)$	magnetization moment at time $t = 0$	T	6
$M(t)$	magnetization moment as a function of time t	T	6
$M(\tau)$	measured magnetization at different delay times τ	T	6
$M_i(t)$	magnetization moment at time t	T	6
$M_{xy}(t)$	transverse magnetization at time t	T	6
$M_z(t)$	longitudinal magnetization at time t	T	6
$M_s(t)$	magnetization after relaxation delay t	T	6
N	anormalization constant	T	6
N_A	Avogadro's number ($= 6.02325 \cdot 10^{23}$)	molecules/ g/mol or molecules/ mol	4
$N_L(l)$	number of intersections per unit length of secant	unitless	8
$N_V(D)$	distribution of spheres with diameter D	—	8
$N_V(l)$	number of spheres in the solid with chord length l	unitless	8
N_P	number of scattering particles	unitless	7
N_α	population of protons in the low energy state	unitless	6
N_β	population of protons in the high energy state	unitless	6
P_p	perimeter of the pore cross section	mm	1
P	pressure	kPa	4
P	pressure applied on mercury to intrude the pore	N/m^2	3
P	net pressure across the mercury meniscus at the time of the cumulative intrusion measurement	N/m^2	3
P_0	saturated vapor pressure of the adsorbed gas at the adsorption temperature MPa	N/m^2	4
P_1	atmospheric pressure	N/m^2	5
P_2	applied pressure	N/m^2	5
P_3	equilibrium pressure after pressurization	N/m^2	5
$P(r)$	volume probability density	unitless	1, 6

Symbol	Parameter	Unit	Chapter
$P(r)$	pore size distribution	—	6
$P(Q)$	a function called the form factor or shape factor for each particle	—	7
P_P	fraction of points falling within the particles of interest	unitless	8
P/P_0	relative pressure of vapor in equilibrium with a meniscus having a radius of curvature	unitless	4
$\bar{Q}$	scattering vector	—	7
Q	modulus of the scattering vector	1/m	7
Q_P	Porod's invariant	1/m⁴/sterad	7
$\bar{R}$	average radius of gyration of a collection of pores or particles	m	7
R_i	radius of gyration of the ith class of pores or particles	m	7
R_g	radius of gyration of a particle	m	7
RF	retention factor	unitless	3
R	roundness	unitless	1
R	gas constant $= 8.314$	J/(Kmol)	4, 5
R_{max}	maximum size of the spheres	m	8
S	surface area of the particles	m²	7
S	specific surface	m⁻¹ or m²/g	1, 3, 4, 6
$\underline{S}$	pore surface area	m²	3
$\bar{S}$	Philleo factor	mm	1
S_{mol}	area occupied by one mole of liquid nitrogen	mm²	4
S_P	surface area of a particle with a radius of gyration R_g	m²	7
$S(Q)$	interparticle structure factor	unitless	7
T	neutron transmission of the sample	unitless	7
T	ambient temperature	K	5
T	absolute temperature	K	4, 6
T_0	temperature at the triple point of water	K	5
T_1	spin–lattice (longitudinal) relaxation time	s	6
$T_{1,b}$	longitudinal relaxation time for water molecules in bulk	s	6
$T_{1,i}$	longitudinal relaxation time of the ith pore	s	6
$T_{1,s}$	surface affected longitudinal relaxation time	s	6
T_2	spin–spin (transverse) relaxation time	s	6
T_2^*	effective transverse relaxation time	s	6
$T_{2,cap}$	transverse relaxation time of the capillary pores	s	6
$T_{2,gel}$	transverse relaxation time of the gel pores	s	6
$T_{2,i}$	transverse relaxation time of the ith pore	s	6
$T_{2,s}$	relaxation time in the surface phase	s	6
T_t	relaxation time	s	6
T_m	normal melting point of the bulk material	K	6
$T_m(x)$	melting point of a crystal of linear dimension x	K	6

Symbol	Parameter	Unit	Chapter
V	volume of gas adsorbed at pressure P	cm^3	4
V	volume of the solid at the D-dried state	m^3	5
V	sample volume illuminated by the neutron beam	m^3	7
V	volume of the scattering particle with radius of gyration R_g	m^3	7
V	bulk volume of concrete	mm^3	1
V	volume of the solid	m^3	8
V_0	volume of micropores that has been filled at relative pressure P/P_0 and at temperature T (current adsorption)	cm^3/g	4
V_M	molecular volume	cm^3/g/mol	4
V_P	volume of scattering particle(s)	m^3	7
V_P	total volume of the particles in the solid	m^3	8
V_S	volume of adsorbate at saturated vapor pressure (equal to the total pore volume)	m^3	4
V_S	volume of the porous sample	m^3	5
V_R	volume of the calibrated reference cell	m^3	5
V_V	volume fraction of the particles of interest	unitless	8
V_a	total volume of air voids	mm^3	1
V_a	volume of adsorbed vapor	cm^3/g or mm^3/g	4
V_b	the rapidly exchanging volume of bulk liquid	m^3	6
V_b	bulk volume of the material	m^3	1
V_c	volume of the empty sample cell	m^3	5
V_{con}	bulk volume of concrete	mm^3	1
V_{fp}	cumulative volume occupied by freezable pore water	m^3	5
V_m	volume of gas adsorbed to form a monomolecular layer	cm^3/g or mm^3/g	4
V_{max}	total volume of mercury intruded at the maximum applied pressure	m^3	3
V_{mol}	volume occupied by one mole of liquid nitrogen (equal to $34.6 \cdot 10^{24}$)	mm^3	4
V_{molar}	molar volume of the condensate	m^3/mol	4
V_p	total pore volume in the bulk material	m^3, mm^3	1, 4
V_p	total volume of the particles in the solid	m^3	8
V_p	volume of the pores	m^3	3, 6
V_{ret}	retained volume of mercury	m^3	3
V_s	rapidly exchanging volume of surface liquid	m^3	6
V_{tot}	total volume of the material	m^3	5
V_{tr}	true volume of the material	m^3	5
a	constant determined by the structure of the sample	unitless	1, 7
c	constant, function of net heat of adsorption $(q_1 - q_L)$ of the monomolecular layer	unitless	4
d	pore width	m	3
d	scattering length	m	7
$dN_L(l)/dl$	slope of the distribution	unitless	8

Symbol	Parameter	Unit	Chapter
dP	change in the applied pressure	N/m^2	3
dS	surface area of the pore wetted by mercury	m^2	3
dV	change in the pore volume	m^3, mm^3	1, 3
dW_1	work required to increase the area wetted by mercury	Nm	3
dW_2	work required to force mercury into the pore	Nm	3
dr	change in the pore radius	mm	1
dq/dt	heat flow		5
dσ/d$\Omega(Q)$	microscopic differential scattering cross section	cm^2	7
dΣ/d$\Omega(Q)$	macroscopic differential scattering cross section	cm^2	7
dγ/dT	temperature coefficient for surface tension	N/m/K	3
$f(r)$	size distribution of the sections	—	8
$f(T_1)$	distribution of relaxation times	—	6
f_b	volume fraction of pore fluid with bulk properties (or proportion of fluid in the bulk phase)	unitless	6
f_s	volume fraction of pore fluid with surface-affected properties (or proportion of fluid in the surface phase)	unitless	6
g	acceleration of gravity (=9.81)	m/s^2	3
h	maximum height of the sessile drop	m	3
$h, \hbar$	Planck's constant (=6.626 × 10^{-34})	Js	6
k	Boltzmann's constant	J/K	6
k	constant, function of the vapor and of experimental conditions	m	4
k_α	characteristic parameter	mole/JK	4
$\underline{k_1}$	empirical constant	m^2/N	4
k_0	wave vector of the incident beam of neutrons	—	7
$\bar{k}_1$	wave vector of the scattered beam of neutrons	—	7
k_f	melting point depression constant	KÅ	6
k_f	constant	KÅ	6
k_p	Porod constant	1/m^5/sterad	7
l	length of the cylindrical pore being wetted	m	3
l	length of the pore	m	3
l	total length of the segments falling in a constituent	m	8
l	chord length	m	8
l_p	length of a cylindrical pore	mm	1
m	mass of the specimen	g	5, 7
m_D	mass of dry material	kg	5
m_1	magnetic quantum number	unitless	6
m_{OD}	mass of specimen after oven-drying	g	2
$m_{f\text{-}d}$	mass of specimen after freeze-drying	g	2
m_s	mass of saturated, surface-dry sample	kg	5
m_{sub}	mass of the saturated sample when it is immersed in water	kg	5

Symbol	Parameter	Unit	Chapter
n	number of pores having radius r_p and length l_p	unitless	1
n	total number of pores	unitless	1
n	amount of gas adsorbed at pressure P on 1 gram of adsorbent	moles	4
n	amount of gas in the sample cell at pressure P_1	moles	5
n	number of scatterers	unitless	7
n_1	amount of gas occupying the remaining volume in the sample cell at pressure P_1	moles	5
n_2	amount of gas occupying the remaining volume in the sample cell at P_2	moles	5
n_V	number of air voids per unit volume of cement paste	voids/mm^3	1
n_R	amount of gas in the reference cell at pressure P_1	moles	5
n_i	number of pores with radius r_i	unitless	1, 7
n_m	amount of gas adsorbed when the entire surface is covered by a monomolecular layer per 1 g of adsorbent (= moles of adsorbate per g of adsorbent) = monolayer capacity	moles	4
p	cement paste content	%	1
p	volume fraction of one of the two phases	unitless	7
r	length of the short segment generating the line	m	1
r	curvature of the liquid–solid interface	m	5
r	distance from the electronic center of gravity	m	7
$\bar{r}$	arithmetic mean of the radius of all pores	m	1
r_1	minor axis of the elliptical cross section of the pore	m	3
r_2	major axis of the elliptical cross section of the pore	m	3
r_K	apparent pore radius (radius into which condensation occurs)	m	4
r_g	geometrical mean of the radius of pores	m	1
r_h	hydraulic radius	m, mm	1
r_i	radius of a given pore i	m	1
r_p	pore radius	m, mm	1, 3, 5, 6
s	standard deviation	—	1
s_g	standard deviation of $\ln(r)$	—	1
t	thickness of the adsorbed film that is built up on the pore wall at pressure P	mm	4
t	time	s	5
t_f	thickness of the nonfreezable pore surface water	Å	6
v	volume of the monolayer of adsorbed water on the internal surface of the pore space	m, mm^3	5
v_m	molar volume of water	m^3/mol	5
x	the fractional inversion	unitless	6
Δ	thickness of the water layer on the pore surface	m	6

Symbol	Parameter	Unit	Chapter
ΔB	variation in the main magnetic field over the region of interest	T	6
ΔE	energy difference between the two levels	cal/mol	6
ΔH_f	the bulk enthalpy of fusion	J/mol	6
ΔH_m	change in the molar heat of the ice-water transition	J/mol	5
$\Delta H_m(t)$	change in the molar heat of the ice-water transition	J/mol	5
ΔT	depression of the freezing or melting temperature T of water	K	5
ΔT	temperature change	K	6
ΔT_m	temperature reduction of the melting point	K	6
ΔV	change in solid volume from a D-dried state to 11% RH	m^3	5
ΔV	volume of an adsorbed water monolayer	m^3	5
$\Delta l/l$	solid particle swelling	unitless	5
$\Delta \Omega$	solid angle element defined by the size of a detector pixel	sterad	7
$\Delta \rho$	difference in neutron scattering length density or contrast electron density	$1/m^2$	7
$(\Delta \rho)^2$	scattering contrast	$1/m^4$	7
ΣS	specific surface area	m^2/g	3
α	specific surface of air voids	mm^{-1}	1
β	affinity (or similarity) coefficient	unitless	4
γ	surface tension of a liquid adsorbate	N/m	4
γ	magnetogyric ratio	MHz/T	6
γ_1	surface tension of mercury at temperature T_1	N/m^2	3
γ_2	surface tension of mercury at temperature T_2	N/m^2	3
γ_m	surface tension between mercury and the pore wall	N/m	3
γ_{sl}	interfacial tension between ice and water	N/m	5
ε	total porosity of the material	%	1, 5
$\eta(\lambda)$	detector efficiency (sometimes also called the response)	—	7
θ	contact angle between liquid, ice and pore wall	degrees	5
θ	contact angle between mercury and the pore wall	degrees	3
θ	contact angle between the liquid and the pore wall (usually assumed to be 0)	degrees	4
2θ	scattering angle	degrees	7
λ	wavelength of the neutron or X-ray beam	m	6, 7
μ_z	component of the nuclear magnetic moment in the z axis	T	6
ν	frequency of the photon	Hz	6
ν_0	Larmor frequency of a nucleus	cycles/s	6
ρ	density of the adsorbate	g/mm^3	4
ρ	surface relaxivity parameter	m/s	6
ρ	density of the solid/sample	g/m^3	6

Symbol	Parameter	Unit	Chapter
ρ_{CSH}	neutron scattering length density of C-S-H gel	$1/m^2$	7
ρ_{H_2O}	neutron scattering length density of water	$1/m^2$	7
ρ_{app}	apparent density of the material	kg/m^3	5
ρ_L	density of liquid nitrogen	g/m^3	4
ρ_l	density of the liquid in the pores of the sample	kg/m^3	5
ρ_m	density of mercury	kg/m^3	3
ρ_{tr}	true density of the material	kg/m^3	5
$\rho(T)$	density of water as a function of temperature T	g/m^3	5
$\rho_{ice}(T)$	density of frozen water	g/cm^3	5
$\rho_{water}(T)$	density of freezable pore water	g/cm^3	5
σ	surface energy at the liquid–solid interface	N/m	6
σ	area occupied by one molecule of nitrogen	mm^2	4
τ	thickness of a monomolecular layer of adsorbate	mm	4
ϕ	shape factor	unitless	3
ω	angular frequency $= 2\pi\nu$	cycles/s	6
ω_0, ω_L	Larmor frequency	cycles/s	6

List of abbreviations

This table lists some abbreviations and acronyms that are mentioned in this book or are usually found in the related technical literature. Some of the lower case abbreviations might be found in upper case, and vice versa.

abs	absolute
ADC	analog-to-digital converter
ads	adsorption
AFM	Atomic Force Microscopy
AFNOR	Association française de normalisation (French Standards)
AS	Australian Standards
ASTM	American Society for Testing and Materials
atm	standard atmosphere
a.u., au	arbitrary units
av	average
BET	Brunauer–Emmett–Teller (method)
BJH	Barrett–Joyner–Halenda
BS	British Standard
BSE	backscattered electron
BSI	British Standards Institute
°C	degree Celcius
ca	approximately
CASRN	Chemical Abstracts Service Registry Number
cc	cubic centimeter
cgs	centimeter–gram–second system
c.m.	center of mass
COE	US Army Corps of Engineers
conc	concentrated, concentration
const	constant
CPMG	Carr–Purcell pulse sequence, Meiboom–Gill modification
cps	cycles per second/counts per second
CRT	cathode-ray tube
C–S–H	calcium silicate hydrate (gel)
cu	cubic
CW	continuous wave
d	day

deg	degree
diam	diameter
DIN	Deutsches Institut für Normung (German Standards)
DR	Dubinin–Radushkevich (method)
DSC	Differential Scanning Calorimetry
DTA	Differential Thermal Analysis
EDS	Energy dispersive X-ray spectroscopy
EDX	Energy dispersive X-ray detector
EN	European Norms (European Standards)
EPMA	Electron probe microanalysis
eq, eqn	equation
erf	error function
e.s.d.	estimated standard deviation
ESEM	Environmental Scanning Electron Microscope
est	estimate, estimated
eV	electrovolt
exp	exponential function
expt	experimental
ext	external
°F	degree Farenheit
FA	fly ash
FEM	field emission microscopy
FID	free induction decay
fl	fluid
FLR	Fluid liquid replacement
fp	freezing point
FT	Fourier transform
ft	foot
FTIR	Fourier transform infrared spectroscopy
fus	fusion (melting)
GGBS	ground granulated blast furnace slag
hr	hour
HREM	high resolution electron microscopy
Hz	Hertz
i	square root of (-1)
imm	immersion
INS	inelastic neutron scattering
ISO	International Standards Organization
ITS	international temperature scale
IU	international unit
JAS	Japanese Standards Association
K	degree Kelvin
keV	kiloelectrovolt
lb	pound
lim	limit
liq	liquid
ln	logarithm (natural)

log	logarithm (common)
MAS NMR	magic angle spinning nuclear magnetic resonance
max	maximum
min	minimum, minute
MIP	mercury intrusion porosimetry
MKS	meter–kilogram–second system
ml, mL	milliliter
mmHg	millimeter of mercury
MP	micropore method
mp	melting point
MRI	magnetic resonance imaging
MRRA	magnetic resonance relaxation analysis
MS	mass spectroscopy
NA	numerical aperture
NMR	nuclear magnetic resonance
NNLS	nonnegative least squares method
NS	Norwegian Standards
NTP	normal temperature and pressure
obs	observed
OM	optical microscope
oz	ounce
pH	negative log of hydrogen ion concentration
pixel	picture element
ppb	parts per billion
ppm	parts per million
psd	pore size distribution
PVT	pressure–volume–temperature
rad	radian
RF	radio frequency
RH	relative humidity
RILEM	International Union of Laboratories and Experts in Construction Materials (Réunion Internationale des Laboratoires d' Essais et de recherché sur les Matériaux et les constructions)
rms	root mean square
rpm	revolutions per minute
SAM	Scanning Acoustic Microscopy (also Scanning Auger Microscopy)
SANS	small-angle neutron scattering
SAS	small-angle scattering
sat, satd	saturated
SAXS	small-angle X-ray scattering
sd	standard deviation
SE	secondary electron
SEM	scanning electron microscope
SF	silica fume
SFS	Finnish Standards Association

SI	International System of Units
sln, soln	solution
S/N	signal-to-noise ratio
sol	soluble, solution
sp gr	specific gravity
SPI	single point imaging
sqrt	square root
std, stnd	standard
STEM	scanning transmission electron microscope
STP	standard temperature and pressure
sub	sublimation
TEM	transmission electron microscopy
temp	temperature
TGA	thermogravimetric analysis
theor	theoretical
trs	transition
TS	transition state
ULV	ultra-low viscosity
UOP	Universal Oil Products
vap	vaporization
WAXS	wide angle X-ray scattering
w/c	water–cement ratio
WMIP	Wood's metal intrusion porosimetry
wt	weight
XRD	X-ray diffraction
y, yr	year

Units and conversions

Since the 1960's a coherent set of units, the SI units (Système Internationale d' Unités) was adopted, based on the definition of five basic units: meter (m), kilogram (kg), second (s), mole (mol), and Ampere (A).

Throughout this book the metric system of units has been used and all units from original papers have been converted to be consistent throughout the text. However, a conversion table follows to facilitate the reader when s/he refers to the original sources of the technical literature. It should be noted that occasionally practicing scientists continue to use older units when the SI units have not become a standard practice for convenience reasons. The most common prefixes are also listed.

More information on other units may be found in the "Manual of Symbols and Terminology for Physicochemical Quantities and Units," Whiffen, 1979, Pergamon Press.

SI prefixes

Symbol	Prefix	Multiple
n	nano	10^{-9}
μ	micro	10^{-6}
m	milli	10^{-3}
c	centi	10^{-2}
k	kilo	10^{3}
M	mega	10^{6}
G	giga	10^{9}

Recommended values of fundamental physical constants

Quantity	Symbol	Value	Units
Acceleration due to gravity	g	9.81	m/s^2
Avogadro constant	N_A	$6.02205 \cdot 10^{23}$	mol^{-1}
Boltzmann constant	k	$1.38066 \cdot 10^{-23}$	J/K
Electronic charge	e	$1.60219 \cdot 10^{-19}$	C

Quantity	Symbol	Value	Units
Faraday constant	F	$9.64846 \cdot 10^4$	C/mol
Gas constant	R	8.314	J/(molK)
Ice point temperature	T_{ice}	273.15 (exactly)	K
Molar volume of ideal gas	V_m	$2.24138 \cdot 10^{-2}$	m³/mol
Planck constant	h	$6.6262 \cdot 10^{-34}$	Js
Standard atmospheric pressure	p	101 325 (exactly)	N/m²
Atomic mass unit	m_u	$1.660566 \cdot 10^{-27}$	kg
Speed of light in a vacuum	c	$2.997925 \cdot 10^8$	m/s

Conversion factors

Multiply	By	To obtain
Ångström (Å)	1×10^{-10}	meter (m)
atmosphere (atm)	760	mm mercury (Hg)
atm, std	1.013×10^5	pascal (Pa)
bar	1×10^5	Pa
Btu	1.055	Joule (J)
cal	4.184	Joule
centimeter (cm)	3.281×10^{-2}	foot (ft)
cm	0.394	inch (in)
cubic foot (ft³)	7.481	gallon
cubic meters (m³)	1,000	liters
electrovolt (eV)	1.602×10^{-19}	joule (J)
foot (ft)	0.3048	m
gram (g)	2.205×10^{-3}	pound (lbm)
inch (in)	2.540	centimeter (cm)
in of Hg	0.0334	atm
kilopascal (kPa)	0.145	lbf/in² (psi)
MPa	145	lb/in²
mm of Hg	1.316×10^{-3}	atm
Newton (N)	0.225	lbf
Pascal (Pa)	9.869×10^{-6}	atmosphere (atm)
Pa	1	Newton/m²(N/m²)
Pa s	10	Poise (P)
Pound (lbm)	0.454	kilogram (kg)
lbf	4.448	N
lbf/in² (psi)	0.068	atm
lb/in²	6896.55	Pa
Torr	133.322	N/m²
Torr	1	mmHg

Temperature conversions

$°F = 1.8(°C) + 32$

$°C = (°F - 32)/1.8$

$°K = °C + 273.15$

Units and derived units in SI

Force	$1\,N = 1\,kgm/s^2$
pressure, stress	$1\,Pa = 1\,N/m^2$
energy, work	$1\,J = 1\,Nm = 1\,kgm^2/s^2$
length	$1\,m$
mass	$1\,g$
time	$1\,s$
temperature	$1\,K = 1°C$
amount of substance	$1\,mole$
density/mass density	$1\,kg/m^3$
specific volume	$1\,m^3/kg$
surface tension	$1\,N/m$

1   Introduction

Pore structure is a very important microstructural characteristic in a porous solid because it influences the physical and mechanical properties, and controls the durability of the material. The physical and mechanical behaviors of a porous material are strongly affected by the way in which the pores of various sizes are distributed within the solid. Materials with the same total pore volume can exhibit entirely different characteristics, depending on whether the material contains a small number of large pores or a great number of small pores. A lot of research has been carried out in the past in order to understand the principles and reactions involved in the formation of the pore structure of a composite material, and to provide models that describe it. The accuracy of a model developed to predict the pore structure in a porous solid can be determined from how well the experimental data fit the outcome predicted by the model. Characterization of the pore structure is complicated by the existence of pores having different shapes and sizes and by the connectivity between pores.

Hardened cement paste, cement mortar, and concrete are porous cement-based materials, into which liquids and gases can penetrate. The properties and behavior of the principal binding material, the hydrated Portland cement paste, control the properties and behavior of the composite material. Pores can exert their influence on the properties of cement-based materials in various ways. *Compressive strength* and *elasticity* are primarily affected by the total volume of pores; however, they can also be influenced by the size and the spatial distribution of pores, maximum pore size, pore shape and connectivity. *Permeability* and *diffusivity* are influenced by the total volume, size, shape, and connectivity of the pores. *Shrinkage* is largely a function of changes in surface energy at the pore walls and, therefore, depends upon the total surface area of the pore system. Durability to *freezing and thawing* and deicer scaling are controlled by the volume and the spacing of entrained air voids. One of the most important challenges in concrete science, and an area of considerable research activity are understanding the pore structure of hardened cement paste and how it affects the properties and behavior of cement-based materials. Pore structure measurements in hardening cement pastes are also used to establish a realistic picture of the hydration process. The understanding of the cement hydration process and the formation of the pore structure may lead to the development of new materials with improved performance, and broaden the range of applicability of cement-based materials. Reviews of the research carried out in the past on the influence of pore structure on the properties of hardened cement pastes have been published in the technical literature [1.1,1.2,1.3].

Accurate characterization of the pore structure in hardened cement pastes is quite difficult because cement paste involves additional complications compared to other porous materials, which include the following characteristics:

- The range of pore sizes in hardened cement pastes is very wide compared to most other porous materials. Hardened cement paste in concrete contains entrapped and also usually deliberately entrained air voids, which are much different in shape, distribution, connectivity, and extent of fluid filling than the "intrinsic" general pores. In concrete, pores and voids can range in size from 1 nm to 1 cm, i.e. seven orders of magnitude.
- The pore structure of hardened cement paste in concrete is extremely tortuous and complex and spatially inhomogeneous; e.g. the cement paste close to aggregates has a different pore structure than the bulk cement paste. In addition, fine and coarse aggregate particles in concrete are porous having their own pore systems, which might be entirely different from the pore system of the cement paste.
- The fluid in the pores of cement paste is not water, but highly concentrated potassium and sodium hydroxide solutions. Drying of the hardened cement paste, which is needed before applying some of the testing methods, leaves deposits of these solids inside the pores, reducing in this way the available pore space and thus underestimating the pore size.
- The fragile microstructure of the cement paste matrix can be damaged during testing by the technique employed for the pore structure characterization.
- Many of the techniques used to determine the pore structure parameters require pretreatment of the specimen, most often drying. There are clear indications that the drying procedure changes the microstructure of hardened cement paste permanently. Therefore, even though the pretreatment methods can be relatively reliable for materials with pore structures that remain stable on removal or addition of water, they might not be as reliable when used in hydrated Portland cement paste.
- Several indirect methods for characterizing pore structure make assumptions about the pore geometry in order to analyze experimental data; therefore, the pore geometry assumed might be very different from the true pore geometry. When only one technique is used for pore structure characterization, interpretation of results is restricted, because each method is applicable for a certain range of sizes and is based on a number of underlying assumptions. Different techniques yield different values for the same pore structure parameters, depending on the size range on which each technique applies. For this reason, several experimental methods are used, sometimes on the same specimen, because then it is possible to compare and contrast results from overlapping of the available information and gain extra insight about the characteristics of the pore structure.

In addition to the intrinsic general pores of the hardened cement paste, air voids can be present in cement paste after using an air-entraining chemical admixture. Air-entrained concrete was discovered during the mid-1930s and has been a major development in concrete science to improve durability of concrete subjected to freezing and thawing. Air voids are minute, spherical pores, well distributed within the cement paste. Practical and theoretical considerations suggest that it is not only the total amount of entrained air voids, but also their size and distribution that are important factors contributing to the observed increased resistance of concrete to freezing and thawing. A tremendous effort has been put in the past to characterize the air void system in hardened cement-based materials and determine the adequacy of air voids to protect concrete against freeze–thaw action.

1.1   Scope of the book

In the following chapters, the methods most commonly used in concrete science to characterize the pore structure and air void parameters are presented. Even though there have been numerous publications describing the various techniques used to determine pore structure in porous materials, there is not much compiled information on how these techniques apply on cement-based materials and what are their limitations. New techniques are developing continuously with new applications, and new models and interpretation of results are introduced. This book focuses on the pore structure characterization techniques that have been developed and are used extensively in concrete science. The methods used most commonly to prepare specimens for microstructural and microscopical analysis are described in Chapter 2. Chapters 3–7 describe different techniques used to characterize the microstructure of hardened cement paste, including mercury intrusion porosimetry, gas adsorption, pycnometry (displacement techniques), thermoporometry, nuclear magnetic resonance, and small-angle scattering. Each technique is presented in a separate chapter, with the theoretical background, experimental procedure, analysis of results obtained, and its advantages and limitations. The direct microscopical methods that have been developed through current sophisticated technology to apply to cement paste microstructure are described in Chapter 8, with emphasis given on stereology and characterization of air voids. Chapter 9 provides comparison of results as they have been obtained by different techniques for the pore structure parameters of interest. Mercury intrusion porosimetry and nitrogen adsorption are two techniques that have been used extensively and for a long time for characterization of the pore structure in hardened cement-based materials; therefore, most often comparison is done with results obtained by other techniques to results obtained by mercury intrusion porosimetry and nitrogen adsorption. Even though characterization of pore structure can also be carried out in actively hydrating cement pastes in order to determine the hardening process of the cement paste, the development of pores in fresh cement pastes will not be dealt here, but emphasis is given rather to pores in hardened cement pastes.

1.2   Pores in cement paste

Hardened cement paste is produced when cement chemically reacts with water. For civil engineering applications, Portland cement is predominantly used and it is the cement considered here. Dry cement consists of four main oxide phases: tricalcium silicate [$3CaO \cdot SiO_2$ or $C_3S(*)$], dicalcium silicate [$2CaO \cdot SiO_2$ or C_2S], tricalcium aluminate [$3CaO \cdot Al_2O_3$ or C_3A], and tetracalcium alumino-ferrite [$4CaO \cdot Al_2O_3 \cdot Fe_2O_3$ or C_4AF]. Cement also contains gypsum that is added to control the setting process. C_3S and C_2S constitute approximately 75% by mass of dry cement. When cement is mixed with water, chemical reactions take place that result in the formation of hydration products. The two silicate phases, C_3S and C_2S, give calcium silicate hydrate (C–S–H) and calcium hydroxide [$Ca(OH)_2$] as hydration products, which constitute approximately 50–60% and 20–25% of the total volume of the hydrated product respectively. Hardened cement paste contains C–S–H gel, crystals of calcium hydroxide, ettringite, minor residues of the original unhydrated cement, and residues of the original water-filled spaces in the fresh paste. Further, details on cement hydration can be obtained elsewhere [1.4,1.5].

When cement paste is combined with fine aggregate (maximum particle size 4.75 mm), it forms a mortar, and when cement paste is combined with fine and coarse aggregate it forms concrete. Mortars and concrete are the most used cement-based materials for engineering applications and their properties are significantly controlled by the properties of the cement paste. Pores in cement mortars and concrete vary in size, shape and origin, and can be subdivided into the following classes:

- pores in the cement paste matrix, which include gel pores, capillary pores, hollow-shell pores, and air voids;
- pores in the aggregates;
- pores associated with the interface between aggregates and cement paste;
- water voids, i.e. voids under aggregate particles or reinforcing bars created by bleeding water of the concrete mixture, and voids due to construction deficiencies, e.g. honeycombing due to inadequate compaction;
- internal discontinuities in the cement paste associated with dimensional instabilities that occur during humidity and temperature changes.

Throughout this book, attention will be given mainly to the pores in the hardened cement paste matrix, i.e. gel pores, capillary pores, hollow-shell pores, and to air voids. Voids like honeycombing will not be considered, because they are the result of poor practice and are not inherent to properly prepared concrete. In addition, no distinction will be made about the location of pores, i.e. at the interface between aggregates and cement paste, or in the cement paste matrix.

---

* In cement chemistry, the following abbreviations are used to shorten the presentation of chemical composition of compounds: $C = CaO$, $S = SiO_2$, $A = Al_2O_3$, $F = Fe_2O_3$, $M = MgO$, $\bar{S} = SO_3$, $H = H_2O$.

1.2.1   Gel pores

The C–S–H gel is the main component of cement paste and is responsible for the strength and microstructure of cement paste. The C–S–H gel is a colloidal amorphous gel, which contains pores of approximately a few nanometers in size that are called gel pores. The C–S–H gel itself has a porosity of about 28%. Studies by S. Diamond, 1976, and H. Jennings *et al.*, 1981, have indicated that the C–S–H gel exists in a variety of forms: fibrous particles, honeycombs, flattened particles, and irregular grains [1.6,1.7]. The C–S–H gel can only be produced in the originally water-filled capillary cavities in fresh cement paste. The bulk volume of C–S–H gel after a cement grain is fully hydrated requires 60% more volume than the original volume of the unhydrated cement grain and the water, and this expansion moves into capillary pores, which are described in Section 1.2.2 [1.8]. As hydration proceeds, the amount and distribution of capillary and gel pores changes considerably: the capillary pore volume is reduced because the capillary pores become filled with hydration products, and the gel pore volume increases as more gel is formed. In addition, there is a net reduction in total porosity.

Theoretically, at complete hydration, 1 g of cement binds chemically approximately 0.23 g of water during the cement hydration reactions, and 0.19 g of strongly physically bound gel water. Complete and unimpeded hydration is therefore only possible at water–cement ratios above 0.42 (0.23 + 0.19). The chemically bound water, also called non-evaporable water, is an integral part of the structure of the solid gel, and is only driven off by heating the specimen at about 1000°C for approximately 15 minutes. The amount of mass loss minus the ignition loss of the original cement gives the non-evaporable water content of the specimen [1.4]. Unbound water in the cement paste is called free water or capillary water, indicating that it is the water present in the capillary pores. Only capillary water is unimpededly accessible for cement hydration.

Gel porosity has a major effect on hydration rates, because gel pores exist inside the hydration products that accumulate between the liquid phase and the anhydrous cement grains [1.1]. Gel pores have a size of about 0.5 nm [1.9]. The selection of this value is based on the assumption that hydration products cannot precipitate in pores having diameters smaller than about 2 nm. Due to the small size of the gel pores, which are only an order of magnitude greater than the size of the water molecules, and due to the great affinity of water molecules to the gel surfaces, the movement of water in gel pores does not contribute much to cement paste permeability [1.2].

Several microstructural models have been proposed in the past, in order to describe the structure of the C–S–H gel and the formation of gel and capillary pores. A measure of how well the microstructure system is understood is the ability of the microstructural models to predict the outcome of a pore structure characterization experiment. Of the models proposed in the past to describe the structure of C–S–H gel and the formation of gel and capillary pores, four have been used extensively, have been analyzed, and continue to provide the basis for further development. These models have been proposed by Powers and Brownyard, Feldman and Sereda, Wittmann, and Jennings and Tennis, and are described briefly in the following paragraphs. The models by Powers and Brownyard, and Feldman and Sereda, describe a layered structure for the C–S–H gel, while the models by Wittmann, and Jennings and Tennis describe a colloid structure.

1. T. Powers and T. Brownyard, 1948, 1958, [1.9,1.10,1.11] proposed a model for the structure of cement gel, based on their comprehensive studies of water vapor sorption isotherms (see Chapter 4) and chemically bound water in hardening cement pastes. They presented cement paste as a colloid composed of spheres, where each sphere represents gel substance together with its associated gel pores. The interstices between the spheres represent capillary pores outside the C–S–H gel. The C–S–H gel is composed of particles that have a layered structure, made up of two to three layers, arranged randomly, and are bonded together by surface forces, as in clay, with occasional strong ionic–covalent bonds linking adjacent particles (see Figure 1.1a). Powers

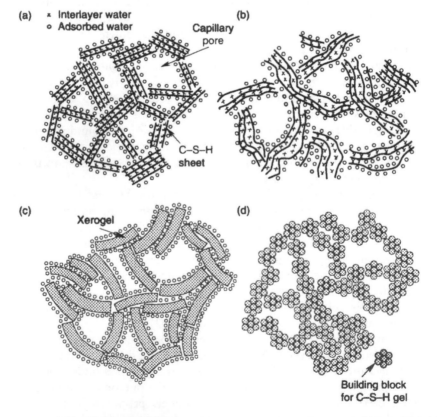

Figure 1.1 Schematic representation of the most commonly mentioned models for C–S–H gel. (a) Model proposed by T. Powers and T. Brownyard, 1948 [1.10]; (b) Model proposed by R. Feldman and P. Sereda, 1970 [1.18]; (c) Munich model proposed by F. Wittmann, 1977 [1.22]; and (d) Model proposed by H. Jennings and P. Tennis, 1994 [1.23].

suggested that, since penetrating water cannot break the bonds holding cement paste together, inter-particle chemical bonds exist, which cannot be severed by the spreading pressure of water. The hysteresis loops observed in the isotherms of water and nitrogen sorption (see also Chapter 4) are commonly associated with materials containing slit-shaped pores, which provided support for their proposed layered structure of C–S–H gel [1.10,1.11].

From the surface area values of cement paste determined using water vapor, Powers estimated that the C–S–H layers are separated by about 1.8 nm. Water vapor can penetrate all the spaces between the particles to provide a measure of their surface area, whereas nitrogen can only penetrate larger spaces and does not measure the whole surface area. The water between the layers is held until strong drying occurs, and then it is lost irreversibly. S. Brunauer and colleagues, 1962–1972, made significant contributions to the experimental study of the pore system in cement pastes referring to the model proposed by Powers and Brownyard using water and nitrogen adsorption measurements [1.12,1.13,1.14,1.15,1.16].

Based on their studies, Powers and Brownyard classified water held in cement pastes into three categories:

- *Water of constitution* involves water of crystallization and otherwise chemically bound, non-evaporable water. This category includes: Hydroxide groups that are strongly bound by the metallic ions, e.g. $Ca(OH)_2$; water that is bound by covalent bonds and retains its identity to a large degree, e.g. $MgCl_2 \cdot 6H_2O$; and water bound by hydrogen bonds by one or both its hydrogen atoms, e.g. $CuSO_4 \cdot 5H_2O$.
- *Gel water* includes adsorbed water and water bound by physical surface forces. This category includes: Zeolitic water that is regarded as being packed between the layers of the crystal hydrate that comprises the zeolites and several basic salts and hydroxides of bivalent metals; lattice water, which is water of crystallization that cannot be associated chemically with the principal constituents of the crystal lattice; adsorbed water that is held by surface forces (van der Waals).
- *Capillary water* is free water inside the pores, not bound by surface forces.

H. Taylor, 1950, based on X-ray diffraction studies, concluded that there are two distinct semi-crystalline phases of C–S–H that have a layered structure [1.17]. S. Brunauer, 1962, based on water-sorption studies, proposed that C–S–H gel consists of two or three layers [1.12]. Upon drying, water is removed from interlayer spaces and the layers collapse. Upon wetting, water cannot reenter the interlayer spaces, and this explains the irreversible shrinkage.

2. R. Feldman and P. Sereda, 1970, proposed a model similar to the Powers–Brownyard model with some important differences [1.18]. They considered that the sheets composing the C–S–H gel do not have an ordered layered structure, but are rather an irregular array of single layers, two to four molecules thick, which may come together randomly to create interlayer space, as it is observed in clay minerals (see Figure 1.1b). Based on gas sorption isotherm results, they concluded that water enters vacated interlayer spaces of

C–S–H at low relative humidities and is structurally incorporated (absorbed) in the C–S–H structure. The interlayer water can reversibly enter and leave the interlayer spaces, a feature that challenged the model by Powers and Brownyard and questioned the interpretation of adsorption data experiments. The Feldman–Sereda model can deduce specific characteristics of the microstructure by finding a "hidden" pore structure that is consistent with a variety of observations, even though the pore structure cannot be observed directly. Bonding between layers is considered to be through solid–solid contacts, which are visualized as bonds intermediate in character between weak van der Waals and strong ionic–covalent bonds. The solid–solid contacts between the layers form during drying of the cement paste and are disrupted by wetting. Interlayer bonding is a special kind of chemical bonding and cannot be considered as resulting from interactions between free surfaces of the layers. Thus, they concluded that nitrogen adsorption measures the surface area of pores within the C–S–H gel more closely than water adsorption, since adsorption of water molecules within the interlayer region of the C–S–H gel distorts the measurements made with water vapor.

P. Sereda *et al.*, 1980, subdivided the water contained in cement pastes into three states, in a way similar to that proposed by Powers described earlier, although the distinction between these states is not absolute [1.19]:

- *Chemically bound water*, also known as non-evaporable water, is chemically bound with cement particles and becomes part of the cement gel. The amount of chemically bound water can be determined from the ignition loss of specimens at 1000°C.
- *Physically-bound water*, also termed chemisorbed water, occupies the gel pores. The water separating the layers is called interlayer hydrate water.
- *Free water*, also known as pore water, stays inside the capillary pores. It is realistic to consider pore water as water that can be removed from a porous solid without significant changes in the solid's dimensions and cannot be described as structural water to the unit cell.

Based on adsorption studies on hydrated C_3S paste after the removal of $Ca(OH)_2$, M. Daimon *et al.*, 1977, modified the Feldman–Sereda model and proposed the existence of two kinds of pores: wide inter-gel particle pores, which can be seen in the C–S–H gel by high resolution Scanning Electron Microscopy (SEM), and small intra-gel pores existing within the gel particles, which cannot be observed by SEM [1.20]. The intra-gel pores are further classified into intercrystallite pores, which may enlarge in the presence of water, and intracrystallite pores corresponding to the interlayer space mentioned by Feldman and Sereda. They defined bound water as water that at normal temperatures (10–30°C) and relative humidities (10–90% RH) does not form part of the pore solution. According to them, bound water includes water present in the crystalline phases, water present in the gel (both in the layers and in the interlayer spaces), and adsorbed or monolayer water, of which most will be on the surfaces of the micropores.

3. The Munich model developed by F. Wittmann, 1977, was proposed based on sorption measurements and conceived the structure of C–S–H gel as a three dimensional network of amorphous colloid gel particles forming a xerogel (see Figure 1.1c) [1.21,1.22]. The C–S–H gel particles are separated by strongly adsorbed thin films of water that adopt a repulsive swelling pressure to counteract the attractive van der Waals interactions between the solid surfaces. Van der Waals forces contribute to the binding energy between gel particles, but strong ionic–covalent bonds still predominate. The Munich model suggests that the disjoining pressure of the water in a capillary pore could cause the C–S–H gel to swell or shrink as a function of relative humidity in the pressure ranges associated with capillary condensation. The Munich model has been used to explain the behavior of hardened cement paste under different moisture conditions.

4. H. Jennings and P. Tennis, 1994, 2000, based on nitrogen adsorption observations presented a model for simplified representation of cement paste's microstructure within the size range of about 1–100 nm [1.23,1.24,1.25]. They suggest that the basic building blocks are spherical units that cluster together (flocculate) to make globules with a radius of 2.5 nm or less (see Figure 1.1d). The Jennings–Tennis model proposes that the basic building blocks that form into globules have two possible packing arrangements: high density and low density. The interglobular spaces correlate to gel porosity and the intraglobular spaces correlate to interlayer porosity that can be reversibly re-entered by nitrogen. As the water–cement ratio increases, the C–S–H particles become accessible to nitrogen. Each type of C–S–H contains a specific amount of total gel porosity: none of the pores in high density C–S–H is accessible to nitrogen, while only some of the pores in low density C–S–H are accessible to nitrogen. The surface area of the high-density type is not detected by several techniques, and this is the main cause of so many different values in the literature.

1.2.2 Capillary pores

The initially water-filled spaces exist in the hardened paste as interconnected channels or, if the cement paste is dense enough, as cavities interconnected only by gel pores, and are called capillary pores or capillary cavities. The capillary porosity of the cement paste depends on the original water–cement ratio and on the amount of cement that has become hydrated. Capillary pores change with time by the precipitation of hydrates, mainly C–S–H gel, in the originally water-filled space. These precipitation processes and the nature of the C–S–H gel leads to an extremely disordered physical structure, the surface of which are the boundaries of the gel and capillary pores.

Capillary pores have highly irregular shape and their size could, in theory, range from very small to large values, i.e. from 2 nm to 10 μm. Capillary pores are divided into two types: large and small capillary pores. One reason for categorizing capillary pores in two types is related to the influence of mineral admixtures on the cement paste microstructure: mineral admixtures result in the formation of a large amount of pores in the range between 2 and 50 nm [1.1].

In well-cured cement paste having a water–cement ratio of 0.50, the volume of capillary pores would be about 20% of the total volume of the paste [1.11]. The fluid contained in capillary pores is not pure water, but an ionic solution that is in equilibrium with the hydrated paste. The water in capillary pores is almost completely evaporable at relative humidities below about 40%. At relative humidities above 40%, capillary condensation occurs, that is, the capillaries become filled with water by condensation from the vapor phase [1.26].

Capillary pores are assumed to have a major effect on transport processes, but only a minor effect on hydration rates (as opposed to gel pores) [1.1]. In addition, damage due to freezing happens due to destruction of the capillary pores. Water can freeze in capillary pores, but cannot freeze in gel pores, which are too small to permit nucleation of ice crystals.

D. Winslow and S. Diamond, 1970, observed that pores having sizes between 10 and 100 nm constitute a pore volume that is substantial at all ages; this is the pore volume generally considered to lie between gel pores and capillary pores [1.27]. From this evidence, they suggested that most of the pores in cement paste are neither capillary pores nor gel pores, but consist of spaces left between particulate hydration products.

1.2.3 *Hollow-shell pores*

Traditionally, the intrinsic pores of cement hydration have been considered to be gel and capillary pores; however, hollow-shell pores can be considered to be a third type of intrinsic pore of cement hydration. The occurrence of hollow-shell hydration grains as a significant feature of the hydration process at the cement paste aggregate interface was first demonstrated by D. Hadley in 1972, and was later confirmed to occur in bulk cement pastes by B. Barnes *et al.*, 1978 [1.28]. For various reasons the occurrence of these pores has been widely ignored, even though it has been repeatedly confirmed in SEM studies (e.g. [1.29,1.30,1.31]). Recently, in a review paper, S. Diamond, 2000, drew attention to the hollow-shell pores and their importance in pore structure interpretations [1.32].

Hollow-shell pores are closed, very distinct pores and generally have the shape of the relicts of cement particles. After a rim of hydration products forms on the cement grains early in the hydration process, the cement gel mostly grows outwards, into the capillary pore space. Hydrates formed within the peripheries of the original cement grains are termed inner products, whereas hydrates formed outside, in the capillary pore space are termed outer products. However, there is no clear distinction between inner and outer products using the electron microscope. When stable hydrates do not form within the original cement grain boundaries, a void space (i.e. hollow shell) may develop within the boundary of the original cement grain as the cement grain recedes on continued hydration. These pores are often called hollow-shell pores or hollow shells or Hadley grains [1.30].

Hollow-shell pores are in the size range of about 1–15 μm [1.30]. In mature systems, in particular, hollow-shell pores are embedded in cement gel and appear to be connected to the continuous capillary pore system by much smaller gel pores. At low water–binder ratios, they can be larger than the capillary pores by more than two orders of magnitude. Because of their strong "ink-bottle" nature they may not easily be detected by many indirect methods for pore structure characterization, e.g. by mercury intrusion porosimetry.

Hollow-shell pores seem to largely remain saturated with pore fluid despite self-desiccation effects. The smaller capillary pores desiccate before the larger hollow-shell pores, apparently because the hollow-shell pores are not drained until their small entryway pores, the gel pores, are drained. Hollow-shell pores have been estimated to constitute approximately 20% of the total porosity in cement pastes containing 10% silica fume, and about 10% or less in cement pastes without silica fume [1.31].

1.2.4 Air voids

Air voids in hardened cement paste can be entrained or entrapped. *Entrapped air voids* occur inadvertently during mixing and placing of concrete. The size of entrapped air voids ranges up to several millimeters. Entrapped air voids are typically isolated from other entrapped air voids, and have little influence on the permeability of concrete. While the small entrapped air voids in concrete are generally observed to be nearly spherical in shape, the large entrapped air voids are often seen to be nonspherical and irregularly formed. *Entrained air voids* are introduced intentionally during production of concrete by using an air-entraining chemical admixture. Entrained air voids are discrete, individual bubbles of spherical shape, usually in an amount of about 2–6% by volume of concrete. They are uniformly distributed throughout the cement paste, are not interconnected with each other, and, therefore, do not affect the permeability of concrete.

ASTM terminology relating to concrete and concrete aggregates (ASTM C 125) sets a size criterion of 1 mm and defines as entrapped the air voids having size above this nominal limit, and as entrained the air voids with diameter between 10 μm and 1 mm. Even though protection to concrete against freezing and thawing is mainly provided by the spherical entrained air voids, large entrapped air voids are not altogether without benefit to frost resistance. Air voids protect cement paste from frost damage in two ways: (a) they limit the hydraulic pressure in the cement paste pores during the initial stages of freezing; (b) they limit or prevent the growth of microscopic bodies of ice in cement paste when the temperature is below the normal freezing point [1.33]. Each air void may be considered to be surrounded with a thin "shell" of paste, within which conditions producing expansion cannot develop.

1.2.5 Pore size ranges

In general, it is convenient to categorize pores into several size ranges. A feature of special interest is the width of the pores, e.g. the diameter of a cylindrical pore or the distance between the sides of a slit-shaped pore. The terminology mentioned in the previous sections (gel pores, capillary pores, hollow-shell pores, and air voids) is used in concrete science. A more general classification of pores in solids according to their average width was originally proposed by M. Dubinin, 1960, and is now officially adopted by the International Union of Pure and Applied Chemistry (IUPAC) [1.34,1.35]. According to IUPAC, the following sizes of diameters characterize pores:

- Micropores have sizes less than 2 nm
- Mesopores have sizes ranging from 2 to 50 nm
- Macropores have sizes larger than 50 nm.

The borderlines between the different classes are not strict and depend on the shape of the pores. Dubinin also suggested the terms *ultramicropores*, with pore size less than about 0.6 nm and *supermicropores*, with pore size ranging from 0.6 to 1.6 nm [1.34,1.36]. The ultramicropores are sometimes referred in the literature as submicropores.

Attention should be drawn to the fact that the size ranges for micro-, meso-, and macropores are sometimes used differently in concrete science. For example, J. Young *et al.*, 1998, define pores with size smaller than 2.5 nm as micropores, pores with size between 2.5 and 100 nm as mesopores, and pores with size greater than 100 nm as macropores [1.37]. The suggestion of defining macropores as pores with sizes greater than 100 nm had also been suggested by P. Mehta and D. Manmohan, 1980 [1.38]. Even though the difference in the size definitions between concrete science and applied physics/materials science is not significant, attention should be drawn on how the pores are defined when referring to experimental results from different fields.

The mechanisms for transport processes in the different categories of pores are quite different. For example, in mesopores, electrostatic interactions between the pore walls and the pore liquid extend over a significant fraction of the pores' cross-sectional area; for this reason, electrostatic effects may hinder transport processes through pores having sizes between 2 and 50 nm. Electrostatic effects do not happen in macropores [1.1].

A comparison of the different pore size ranges according to the IUPAC convention and to concrete scientists' terminology is presented in Table 1.1 [1.39]. It should be mentioned here that in the technical literature the terms pore radius, pore diameter and pore width are used for defining the size of a pore. Since pores in cement paste do not have a specific shape that would make any of the previously mentioned terms reasonable, the term "pore size" is mostly used in the following chapters, unless the term radius or diameter is used in particular cases.

From Table 1.1, it is obvious that the classification according to P. Mehta, 1986, does not classify the pores having sizes between 3 and 10 nm, between 50 nm and 3 μm, and between 5 and 50 μm. The classification used by S. Mindess *et al.*, 2002, does not classify the pores having sizes between 10 and 100 μm. Other researchers have proposed similar classifications [1.41,1.42]. From the size ranges and terms used, it is apparent that there is not yet a distinct terminology used for the various sizes of pores in cement pastes. In addition, the size for division between capillary and gel pores is to a large extent arbitrary, because the spectrum of pore sizes in cement paste is continuous.

Sometimes the general terms for pores and the terms used in concrete science are mixed in technical papers. Throughout this book, the terms gel pores

Table 1.1 Classification of pore sizes according to the general classification by the IUPAC and to concrete science terminology [1.39]

According to IUPAC [1.35]		According to P. Mehta, 1986 [1.40]		According to S. Mindess et al., 2002 [1.4]				
Name	Diameter	Pore type	Size range	Group	Name	Diameter	Role of water	Paste properties affected
Micropores	Up to 2 nm	Interparticle space between C–S–H sheets	1 nm to 3 nm	Gel pores	Micropores "inter layer"	Up to 0.5 nm	Structural water involved in bonding	Shrinkage, creep at all RH
				Gel pores	Micropores	0.5 nm to 2.5 nm	Strongly adsorbed water; no menisci form	Shrinkage, creep at all RH
Mesopores	2 nm to 50 nm	Capillary pores (low w/c)	10 nm to 50 nm	Gel pores / Capillary pores	Small (gel) capillaries	2.5 nm to 10 nm	Strong surface tension forces generated	Shrinkage between 50% and 80% RH
				Capillary pores	Medium capillaries	10 nm to 50 nm	Moderate surface tension forces generated	Strength, permeability, shrinkage at high RH, >80%
Macropores	>50 nm	Capillary pores (high w/c)	3 µm to 5 µm	Capillary pores / Hollow-shell pores	Large capillaries	50 nm to 10 µm	Behaves as bulk water	Strength, permeability
		Entrained voids	50 µm to 1 mm		Entrained air	0.1 mm to 1 mm		Strength

Notes
These are the most commonly mentioned classifications with reference to hydrated cement pastes. The hollow-shell pores were not included in the classification by S. Mindess et al.

and capillary pores used in concrete science are going to be used, with the classification proposed by S. Mindess *et al.* For micro-, meso-, and macropores, the classification according to IUPAC is going to be followed, when theoretical considerations are discussed that have been developed in Applied Chemistry.

1.3   Methods for characterizing pore structure

A wide variety of techniques has been used for the characterization of the pore structure of cement pastes. Total pore volume is probably the easiest pore parameter to obtain, since there are varieties of techniques that provide values for the pore volume that agree fairly well with one another. Each of the various techniques available is most suitable for a specific size range. Some methods have only access to open pores (e.g. penetration and adsorption methods), while others have access to both open and closed pores (e.g. methods using radiation). Sorption methods are widely used to characterize pores with an average pore size of up to about 30 nm. Pores with size between 2.5 nm and 100 μm can be measured by mercury intrusion porosimetry. The mercury intrusion porosimetry assumes that the pores intruded are cylindrical, a condition that is never fulfilled for cement paste; as a consequence, a characteristic pore size distribution is measured rather than the real pore size distribution. The results from mercury intrusion porosimetry, however, can be used for comparison of similar systems and to study the effect of different factors on the cement paste's pore structure. Although mercury intrusion porosimetry has several limitations, the pore structure parameters of cement paste determined by this technique have been related to the factors that control permeation of fluids and diffusion of ions [1.43]. The small-angle scattering and the nuclear magnetic resonance (NMR) techniques are used for specimens that have not been dried. Large pores, e.g. spherical entrained air voids, can be analyzed using optical methods. Computerized image analyzing techniques have become popular during the last 20 years. Table 1.2 lists the various methods used for characterization of pore structure in cement-based materials that are discussed in this book and the range of pore sizes where each technique is applicable [1.39].

The types of techniques used to experimentally measure the pore volume and/or determine the pore size distribution in porous materials can be categorized into direct and indirect methods [1.44,1.45].

In *indirect* methods, an external stimulus is applied to the material and the material's response is measured using a suitable detector. The pore structure parameters are determined indirectly from properties such as adsorptive capacity, density, etc. The indirect methods most commonly used for pore structure characterization include the following techniques:

- Mercury intrusion porosimetry
- Gas adsorption
- Displacement methods
- Thermoporometry
- NMR
- Small-angle scattering.

Table 1.2 Methods used to characterize the pore structure in cement-based materials, including the range of pore sizes where they are applicable [1.39]

Size range	1 nm	10 nm	100 nm	1 µm	10 µm	100 µm	1 mm
Name of pores according to IUPAC	Micropores	Mesopores			Macropores		
Name of pores according to concrete science terminology	Gel pores		Capillary pores			Air voids	
Method							
Mercury intrusion porosimetry							
Gas adsorption							
Water absorption							
Helium pycnometry							
Thermoporometry							
Nuclear magnetic resonance							
Small-angle scattering							
Optical microscopy							
Electron microscopy							

Direct methods produce a direct physical image of the microstructure examined, revealing the size and shape of the phases, and extrapolating the relationship of the phases in space by using mathematical analysis techniques. The most commonly used direct observation techniques include:

- Optical microscopy
- Scanning electron microscopy
- Image analysis, which is actually an advanced form of optical and electron microscopy.

1.4   Definition of pore structure parameters

Characterization of a porous material provides information on the material's physical characteristics, such as total pore volume, surface area of pores, pore size distribution, pore shape, and pore connectivity. In order to reliably describe the properties and performance of hardened cement pastes, more than one pore structure parameter needs to be determined. Even a detailed description of one parameter alone is unlikely to be sufficient to predict the material's properties and performance. For example, as it was mentioned earlier, total porosity is required in order to determine strength, but total porosity, pore size distribution, and pore connectivity are needed in order to determine the permeability of a material. The pore structure parameters determined using various techniques are described in the following sections separately for general pores and for air voids.

1.4.1   General pores

The parameters most commonly used to characterize the pore structure of a cement-based material are the following:

- Porosity
- Hydraulic radius
- Specific surface area
- Threshold diameter
- Pore size distribution.

These parameters can be determined by the methods described in the following chapters. Any additional parameters obtained from a specific method used, is defined in the chapter describing the method. It should be kept in mind that no experimental method provides yet the absolute value of parameters such as total porosity, surface area of pores, or pore size. Each method gives a characteristic value that depends on the principles of the technique used and on the nature of the specimen tested. Therefore, one should not look for a perfect agreement between the values of the parameters determined by different methods; such an agreement would not necessarily prove the validity of the derived quantities.

Other parameters often mentioned are the continuity and tortuosity of pores, which indicate the interconnection of pores, and the distortion of the pore network, respectively. These parameters mostly deal with aspects of transportation of fluids through a porous medium and are outside the scope of this book.

1.4.1.1 Porosity

Two classes of porosity can be defined: total (or absolute) porosity, and effective porosity.

Total porosity is the fractional volume of pores with respect to bulk volume of the material, as it is expressed by Equation (1.1). The total porosity is a dimensionless quantity, usually expressed as a percentage value, and it is obtained regardless of pore connections. Total porosity includes both open and closed pores. Since some techniques can only determine open pores, the method used to obtain total porosity must be stated when reporting results:

$$\varepsilon = \frac{V_p}{V_b} \times 100 \tag{1.1}$$

where ε is the total porosity (%), V_p is the total pore volume in the bulk material (mm^3), V_b is the bulk volume of the material (mm^3).

Effective porosity is the fraction of pores with respect to the bulk volume of the material constituted only by open and interconnecting pores. The effective porosity is always less than or equal to the total porosity.

Closed pores influence macroscopic properties such as bulk density, mechanical strength, and thermal conductivity, but are inactive in processes such as liquid flow and adsorption of gases. Interconnecting pores provide a continuous channel of communication with the external surface of the material; therefore, the effective porosity can be an indication of permeability.

If we ignore the pore volume in the aggregates, well-compacted concrete has a total porosity about 20–25% at a young age, and about 10–15% at a mature age [1.46]. It is not possible for ordinary concrete to have a total porosity less than about 10%; therefore, concrete will always have pores that can provide passageways for liquids to move through concrete. The total pore volume of hardened cement paste depends on the initial water–cement ratio, the degree of compaction, the degree of hydration of cement, and the presence of mineral admixtures. As the pore size reduces and approaches molecular dimensions, it becomes difficult to determine the total pore volume using certain techniques. This is another reason to clearly specify the technique used for determining the pore volume.

1.4.1.2 Hydraulic radius

The concept of "pore size" cannot, in general, be given any precise definition and various measures are used to define it. The hydraulic radius is a useful geometric parameter for defining the size of the pores. The hydraulic radius can be defined for a single pore or for the entire pore system.

The hydraulic radius of a *single pore* is defined as the cross-sectional area of the pore divided by the perimeter of the pore space, as it is given by Equation (1.2). For definite pore shapes, the hydraulic radius is a fixed multiple of the pore size. For example, for a cylindrical pore, the hydraulic radius is equal to half the radius of the cylinder. For parallel plates, the hydraulic radius is half the distance separating the plates:

$$r_h = \frac{A_p}{P_p} \tag{1.2}$$

where r_h is the hydraulic radius (mm), A_p is the cross-sectional area of a pore (mm^2), P_p is the perimeter of the pore cross section (mm).

Usually, we do not have area and perimeter data available from point to point, so we can only define an average (or mean) hydraulic radius for the *entire pore system*. The average hydraulic radius is obtained by dividing the total pore volume by the total surface area of the pores, as it is expressed by Equation (1.3):

$$r_h = \frac{V_p}{A_s} \tag{1.3}$$

where r_h is the hydraulic radius (m), V_p is the total pore volume in the bulk material (m^3), A_s is the total surface area of the pores (m^2).

Even though, in general, the hydraulic radius contains only a limited amount of information about the pore structure, it is often found to be a useful characteristic in many practical problems. The hydraulic radius is a convenient parameter for pores of irregular or unknown shapes. Two cement pastes having the same hydraulic radius may have widely differing pore size distributions, but when the two cement pastes are made from the same Portland cement, identity of hydraulic radii should imply at least a rough similarity in pore size distribution.

For hardened Portland cement pastes having water–cement ratios of 0.40, 0.60, 0.80 in the initial mixture, the hydraulic radius has been determined by R. Feldman, 1989, to be 3.94 nm, 6.42 nm, and 10.70 nm respectively [1.47]. These values of hydraulic radius were determined using nitrogen adsorption and helium pycnometry (see Chapters 4 and 5), and represent the cement paste pore system excluding the gel pores. T. Powers, 1960, estimated the hydraulic radius of gel pores to be of the order of 0.7–1.5 nm, which implies an average distance between gel pore walls of 1.5–3.0 nm, depending on the shape of the pores [1.8]. Powers favors a value of 1.8 nm as the best estimate of the average distance between the gel pore walls.

1.4.1.3   Specific surface area

Another common pore structure parameter is the surface area of the boundary between the pore space and the solid, in simplified terms the surface area of the pore walls. The surface area of the pore walls is properly expressed on some specific basis, such as per unit of material mass, unit of material volume, pore volume, or bulk volume.

Specific surface area (or simply specific surface) of a porous material is usually defined as the surface area of the pores per unit mass or per unit bulk volume of the porous material and is given by Equation (1.4). It is obvious that for two materials with the same total pore volume, the material with fine pores has a much greater specific surface than the material with large pores. The specific surface is an important parameter with regard to the permeability of a porous material:

$$S = \frac{A_s}{V_b} \quad \text{and alternatively} \quad S = \frac{A_s}{m} \tag{1.4}$$

where S is the specific surface (m^{-1} or m^2/g), A_s is the total surface area of the pores (m^2), V_b is the bulk volume of the material (m^3), m is the mass of the material (g).

Specific surface has dimensions m^{-1} when it is expressed as the ratio of area to volume, or m^2/g when it is expressed as the ratio of area to mass. When the specific surface is expressed as the ratio of surface area of pores to the total pore volume, it is the reciprocal of the hydraulic radius, which was described in Section 1.4.1.2. The specific surface of pores is also analogous, in some ways, to the fineness of unhydrated cement, which is expressed as the estimated total cement surface area per unit mass of cement. The unhydrated Portland cement has a Blaine surface area of about $0.3–0.4\,m^2/g$.

The specific surface measured depends on the experimental method used, on the size of the specimen tested, but also on the assumptions inherent to the shape of the pores. The specific surface of hydrated cement paste has been determined by water vapor adsorption to be about $250\,m^2/g$ irrespective of the water–cement ratio of the paste. When nitrogen adsorption is used, the specific surface measured is determined to be $120\,m^2/g$ [1.19,1.27]. The specific surface determined using the small-angle X-ray scattering technique has been reported to be as high as $600\,m^2/g$ [1.48]. The specific surface is even higher than this value, when it is computed for the C–S–H gel only.

1.4.1.4 Threshold diameter

D. Winslow and S. Diamond, 1970, have defined the term threshold diameter of pores to correspond approximately to the minimum diameter of channels that are essentially continuous through the paste at a given age [1.27]. The threshold diameter was introduced to take into account the possibility that large pores can be present in the interior of the cement paste but have access to the exterior only through channels of small size of the continuous pore system. More information about the threshold diameter is given in Section 3.2.

1.4.1.5 Pore size distribution

When the radius of a cylindrical pore changes from r to $r - dr$, the corresponding decremental change in the pore volume V is given by Equation (1.5):

$$dV = -(2\pi r_p)l_p dr \quad \text{and in general} \quad dV = -n(2\pi r_p)l_p dr \tag{1.5}$$

where dV is the change in the pore volume (mm³), r_p is the radius of the pore (mm), l_p is the length of the cylindrical pore (mm), n is the number of pores having radius r_p and length l_p, dr is the change in the pore radius (mm).

Pore size distribution presents the fraction of the pore volume in which the pores lie within a stated size range. The pore size distribution as a function of the pore radius is represented by the derivative expressed in Equation (1.6):

$$D_V(r) = -\frac{dV}{dr} \tag{1.6}$$

where $D_V(r)$ is the pore size distribution function, dV is the change in the pore volume (mm³), dr is the change in the pore radius (mm).

If there are various volumes of pores of various sizes, the pore size distribution can be represented in two ways. A plot of the volume of voids that are smaller (or larger) than a given size vs. the pore size is the *cumulative (or integral) pore size distribution*. The slope of this curve, plotted against pore size, is the *differential pore size distribution*. Both curves are used in concrete science and are discussed in detail in Chapter 3.

Attempts have been made to devise mathematical functions to represent the pore size distributions of cement paste that are found experimentally. The pore size distribution may be represented either as a continuous function or as a histogram, it may be broad or narrow, and show one peak or more according to the complexity of the pore structure concerned. The mathematical treatment is necessarily based on the assumption that the number of pores in the sample is large enough for statistical considerations to be applicable. The two distributions that are considered for pores in concrete are the Gaussian (normal) distribution and the log-normal distribution.

The Gaussian distribution of pore sizes is given by Equation (1.7):

$$P(r) = \frac{1}{s\sqrt{2\pi}}\exp\left[-\frac{(r_i-\bar{r})^2}{2s^2}\right] \quad \text{where} \quad s = \sqrt{\frac{\sum(r_i-\bar{r})^2 n_i}{n}} \tag{1.7}$$

where $P(r)$ is the probability density, r_i is the radius of a given pore i, $\bar{r}$ is the arithmetic mean of the radius of all pores, s is the standard deviation, n is the total number of pores, n_i is the number of pores with radius r_i.

A plot of $P(r)$ vs. r gives a curve of the well-known bell-shaped form. The sharpness of the peak is determined by the value of the standard deviation, the peak becoming narrower as the value of the standard deviation decreases (see Figure 1.2). In Figure 1.2, curve A with the sharper peak corresponds to a more uniform size distribution (smaller standard deviation) than curve B.

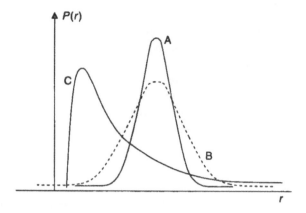

Figure 1.2 (a) Gaussian (normal) pore size distribution. Curve A represents a more uniform pore size distribution than curve B. (b) Log-normal pore size distribution (curve C). $P(r)$ is the probability function.

The Gaussian distribution is rare for pores in hardened cement paste, and the skewed curve, which is a log-normal distribution, is much more common (curve C in Figure 1.2). S. Diamond and W. Dolch, 1972, described the pore size distributions in cement pastes with a single log-normal distribution that was continuous over the range from gel to capillary pores without evidence of bi-modality [1.49]. In a similar way, the diameters of air voids in entrained concrete also have a log-normal distribution. To obtain the expression for the log-normal distribution, it is only necessary to substitute the pore size and the standard deviation in Equation (1.7) with the logarithms of these quantities, and obtain Equation (1.8):

$$P(r) = \frac{1}{\ln s_g \sqrt{2\pi}} \exp\left[-\frac{(\ln r_i - \ln r_g)^2}{2 \ln^2 s_g} \right] \tag{1.8}$$

where $P(r)$ is the probability density, r_g is the geometrical mean of the radius of all pores, s_g is the standard deviation of $\ln(r)$, r_i is the radius of a given pore.

1.4.1.6 *Other parameters*

Other parameters of interest in pore structure analysis are the shape factor that relates to the pore shape, parameters that relate to irregular-shaped particles, and the fractal dimension. These parameters are briefly discussed in the following paragraphs.

Shape factor The most commonly assumed shape in the absence of specific knowledge of pore shape is cylindrical. If an irregular cross section is assumed, the problem is treated by the inclusion of shape factor in the appropriate expressions. The shape factor is not a parameter determined during an experiment, but a parameter inherent in the assumed model.

Originally, the shape factor of a pore was defined by H. Rootare and A. Nyce, 1971, as the total pore volume divided by the product of the area and average radius of the pore [1.50]. The researchers provided a table with values of the shape factor for different pore geometries. More recently, G. Mason and N. Morrow, 1991, defined the shape factor of a pore as the hydraulic radius of a pore made dimensionless by division through the pore perimeter [1.51]. In a different way, the shape factor is expressed as the pore cross-sectional area divided by the square of the perimeter of the pore, as it is defined by Equation (1.9). It is obvious that a single value of the pore shape factor may correspond to different pore shapes. Other researchers have proposed different formulas for the shape factor [1.52]:

$$G = \frac{r_h}{P_p} \quad \text{or} \quad G = \frac{A_p}{P_p^2} \tag{1.9}$$

where G is the shape factor (unitless), A_p is the cross-sectional area of a pore (mm^2), P_p is the perimeter of the pore cross section (mm), r_h is the hydraulic radius (mm).

Two other related parameters are the *roundness*, R, and the *circularity*, C, expressed by Equations (1.10) and (1.11) respectively [1.53]:

$$R = \frac{P_p^2}{4\pi A_p} \tag{1.10}$$

$$C = \frac{4\pi A_p}{P_p^2} \tag{1.11}$$

where P_p and A_p are the perimeter of the pore cross section and the cross-sectional area, respectively, as they were defined in Equation (1.9).

Circularity provides a convenient way to describe the roundness of a pore: circularity values range between a minimum value of 0 and a maximum value of 1, with 1 corresponding to a circular cross section, and the value decreases as the irregularity of the cross section increases. For this reason, circularity is frequently referred to as the shape factor or form factor during image analysis [1.54]. An example of values for the shape factor and circularity (form factor), as they are defined by Equations (1.9) and (1.11), is given in Table 1.3 for different cross sections.

It should be noted here that there are variations in the way the shape factor is used by researchers. For example, the shape factor as defined by Equation (1.9) has been used in fluid flow through porous media [1.55]. D. Lange *et al.*, 1994, have defined and used in their work the shape factor of a particle in the way the roundness is defined in Equation (1.10) [1.56]. Even though there is no standardized universal definition of this parameter, attention must be drawn on how the parameter is defined and used, since interpretation of results by some methods make use of the term.

R. Jenkins and M. Rao, 1984, introduced the *aspect ratio* for elliptical pores, which is the ratio of the major and minor axes of the ellipse, to analyze mercury intrusion porosimetry results [1.57]. The case of slit-shaped pores (parallel plates) can then be approximated with an ellipse that has a very high aspect ratio.

Table 1.3 Examples of values of the shape factor and circularity (which is also mentioned sometimes as form factor in image analysis) for various shapes of cross section of pores

Shape of pore cross section	Perimeter, P_p	Area, A_p	Shape factor $G = \dfrac{A_p}{P_p^2}$	Circularity (form factor) $C = \dfrac{4\pi A_p}{P_p^2}$
Circle of radius r	$2\pi r$	πr^2	0.080	1
Hexagon of side a	$6a$	$\dfrac{3\sqrt{3}a^2}{2}$	0.072	0.907
Ellipse (2:1)	$2\pi\sqrt{\dfrac{(a^2+b^2)}{2}}$	πab	0.064	0.800
(4:1)			0.037	0.471
(8:1)			0.020	0.246
(16:1)			0.010	0.125
Square of side a	$4a$	a^2	0.062	0.785
Triangle of side a	$3a$	$\dfrac{\sqrt{3}a^2}{4}$	0.048	0.604

Note
In ellipse, the ratios in parentheses indicate the aspect ratio (the ratio of the semiaxes $b:a$), for which the values of shape factor and circularity have been calculated.

Parameters for irregularly shaped particles The shape factor mentioned in Section 1.4.1.6 is used for direct and indirect techniques, and is mentioned in mercury intrusion porosimetry and image analysis. In image analysis, in addition to the shape factor, several other parameters are defined and used to characterize irregular pores, such as capillary pores, or particles in a cross section. A section through a cement paste specimen in order to get a polished surface or a thin section, will intersect the irregularly shaped capillary pores in random orientation, and result in sections of irregular shapes. Therefore, for microscopical analysis and in order to quantify the measurements by the techniques described in Section 8.5, it is necessary to provide a satisfactory definition of the term "pore size." Various techniques are used to measure the size of irregularly shaped particles when viewed through a microscope and are used to classify the two-dimensional particle images in terms of an equivalent spherical particle. Some of the most commonly used parameters in concrete science are shown in Figure 1.3.

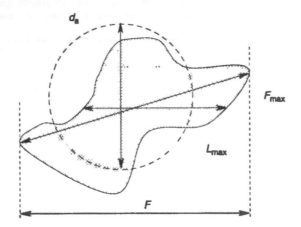

Figure 1.3 Parameters used to define the size of irregularly shaped cross sections of particles or pores: the Feret's diameter, F; the maximum Feret's diameter, F_{max}; the projected area diameter, d_a; and the maximum horizontal intercept, L_{max}. The circle drawn to define the projected area diameter, has the same area as the irregular particle, which is defined by the shadowed area.

- The *Feret's diameter*, F, is the maximum length of the particle measured in a fixed direction, and is the easiest parameter to measure manually.
- The *maximum Feret's diameter*, F_{max}, is the longest dimension for a particle, measured at no set direction.
- The *projected area diameter*, d_a, is the diameter of a circle with the same area as the two-dimensional image of the particle. The projected area diameter gives the best estimate of the true cross-sectional area of the particle.
- The *maximum horizontal intercept*, L_{max}, is the length of the longest line that can be drawn through the particle in a fixed direction.

Other parameters mentioned in the technical literature, but rarely used in concrete science are the Martin's diameter, which is the length of a line that bisects the area of the particle image, and the perimeter diameter, which is the diameter of a circle having the same circumference as the perimeter of the particle [1.54].

The chosen parameter to be measured depends on the parameter of interest that needs to be determined during analysis and the equipment available. However, it is very important when reporting experimental results that the parameter used and the approach followed be clearly stated.

Fractal dimension   Fractal geometry was described initially by B. Mandelbrot, 1977, as a geometric language born out of the need to characterize irregular, fragmented structure [1.58]. Although cement paste microstructure is an example of a disordered system, research has been carried out only in recent years to characterize the fractal nature of cement paste.

Fractal geometry offers important advantages over traditional stereology (described in Chapter 8). Stereology has its foundation in Euclidean geometry, using lines, circles, ellipses, and polyhedra to model shapes. In conventional Euclidean geometry, forms can be broadly grouped into points, lines, surfaces, and volumes. Each class has a characteristic number of dimensions needed to describe it; e.g. lines have one dimension and volumes have three. Euclidean forms are also usually considered to be smooth and well-defined; e.g. a circle has a perimeter that is smooth (everywhere differentiable) and has a definite length. Fractal concepts divide irregular forms into the same broad categories as Euclidean geometry, but they permit the forms to have non-integer numbers of dimensions, called fractal dimensions. The term fractal was selected because these irregular forms have a *fractional* number of dimensions. The amount by which a fractal form's dimension exceeds its smooth, Euclidean counterpart, is a quantitative measure of its irregularity. In this way, an irregular line with a fractal dimension of 1.5 is more irregular than a line with dimension 1.2, and both lines are more irregular than a smooth Euclidean line that has dimension 1. Not all irregular shapes are necessarily fractals.

The governing characteristic of a fractal boundary is its *self-similarity* and not its roughness. When upon appropriate magnification, a smaller region of an object appears to be either exactly or approximately similar to a larger region viewed at a lesser magnification, then the object is a fractal, and this property is called the self-similarity of a fractal object. In nature (and in cement paste microstructure) irregularity and roughness are the norm, rather than the exception, and fractal geometry can describe these two parameters by the fractal dimension. An important characteristic of fractals when applied to two phase materials, e.g. solid-pore systems, is the relationship between the fractal dimensions of the solid phase and the pore structure. The pore structure is essentially the inverse of the solid and therefore they share the same boundaries and the same fractal dimension.

A simplified example used to explain the concept of fractal dimension is shown in Figure 1.4 using a line. The fractal dimension is defined in its simple form by Equation (1.12) [1.58]:

$$L = r^{(1-D)} \tag{1.12}$$

where L is the total length of the line, r is the length of the short segment generating the line, D is the fractal dimension of the line.

Assuming that the length of the line in Figure 1.4a is $L = 1$, the fractal dimension of the line is $D = 1$. In Figure 1.4b the line is made irregular and the total length of the lines connecting the original distance is $L = \frac{4}{3}$, the length of the short segment is $r = \frac{1}{3}$, and the fractal dimension is calculated using Equation (1.12) to be 1.26. The fractal dimension is $D = 1.26$ also for the lines shown in Figure 1.4c, where 16 segments of length $\frac{1}{9}$ each are used and the total length of the lines is $L = 1.78$. The process of segmenting each line can be repeated iteratively. The fractal dimension does not depend upon the number of iterations; it depends upon the fraction by which the length scale is divided and the number of segments after each iteration.

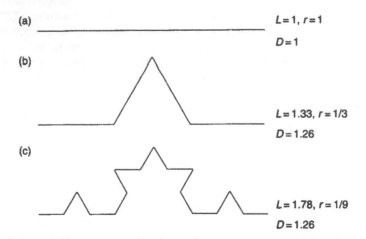

Figure 1.4 Example of a line with non-integer (fractal) dimensions, known as the Koch curve. (a) The initial line has a length of 1 and a fractal dimension $D = 1$. (b) The middle third of the line is replaced by two lines, each having a length $r = \frac{1}{3}$. The length of the curve is $L = 1.33$ and the fractal dimension is calculated using Equation (1.12) to be $D = 1.26$. (c) Each segment of the curve in (b) is replaced in the same way, increasing the line of the length to $L = 1.78$. The fractal dimension is calculated to be again $D = 1.26$.

A number of researchers have shown, using a variety of pore structure characterization techniques, that hydrated Portland cement paste has a fractal microstructure on the length scale of 1 nm to 1 µm [1.25,1.59,1.60,1.61,1.62, 1.63,1.64]. The C–S–H gel has been reported to have a fractal dimension between 2 and 3, with the value differing with the method used for characterization; this is due to the fact that different techniques have different sensitivities to different scales.

Two terms often used in cement paste characterization are *mass (or volume) fractal*, and *surface fractal*. In a surface fractal, the surface or the interface with another material is fractal. In a mass fractal, both the solid itself and its surface or interface with another material is fractal. Mass fractals are frequently formed by aggregations of small subunits. The distinguishing characteristic of a mass fractal is that its mass or volume may be enclosed in a sphere and varies in proportion to a power of the radius of that sphere; the power is the fractal dimension of the mass fractal. Porous materials that are also mass fractals tend to have fractal dimensions between 2 and 3.

1.4.1.7   Factors affecting the parameters measured

Pore structure parameters measured depend not only on the technique used for pore structure characterization, but also on inherent features of the microstructure of the specimen examined, or on the specimen preparation technique. Some of the microstructural variables include type of cement used, water–cement ratio, presence of mineral and/or chemical admixtures, curing conditions, age, and degree of hydration. The specimen preparation variables include crushing procedure, drying techniques (temperature and solvent exchange), and storage after drying. From these variables, the effect of the drying procedures (pretreatment) is going to be discussed systematically in the following chapters when describing the different pore structure characterization techniques. The most commonly used pretreatment techniques are reviewed in Chapter 2.

1.4.2 Air voids

The parameters used to describe air voids in concrete, and more generally in cement-based materials, are the following:

- Total air content
- Specific surface
- Spacing factor.

The air void parameters are determined in practice using mainly microscopic techniques that are described in Chapter 8. Total air content and specific surface are defined in a way similar to the way pore volume and specific surface are defined for general pores. A brief reference to them is given in the following sections.

1.4.2.1 Total air content

Total air content is the proportion of the total volume of air voids to the bulk volume of concrete including all the constituents of concrete and the total volume of air, as expressed by Equation (1.13). Given that air voids contribute to the frost resistance of the hardened cement paste only, it is also useful to express the air content as a percentage of the cement paste volume, i.e. volume of cementitious materials, water, and air:

$$A_a = \frac{V_\alpha}{V_{con}} \times 100 \ (\%) \tag{1.13}$$

where A_a is the total air content (%), V_α is the total volume of air voids (mm³), V_{con} is the bulk volume of concrete (mm³).

The amount of air content required for concrete durability to freezing and thawing varies according to the severity of exposure and the maximum size of aggregates used in concrete. The air content required can range from 1.5% to 7.5%.

1.4.2.2 Specific surface

The specific surface of the air void system is a statistical parameter and is a measure of the air void size distribution. Size and distribution of air voids are variables that are as significant for the durability of concrete to freezing and thawing as the total air content. The specific surface is defined as the total surface area of the air voids divided by their total volume. The specific surface is expressed as area per unit volume resulting in units of mm^{-1}. ASTM C 457 mentions that for concrete to be durable to freezing and thawing effects, the specific surface of air voids in concrete should be between $23.6\,mm^{-1}$ and $43.3\,mm^{-1}$.

High values of specific surface imply a fine air void system. However, specific surface cannot provide any information about the actual number of air voids having a specific size. Therefore, the specific surface cannot uniquely define the size distribution, since the same value of specific surface could represent broadly different size distributions. For air void systems that have similar total air contents, the specific surface can be a useful indicator of the air void distribution.

1.4.2.3   Spacing factor

The spacing factor has been considered to be the best indicator of the adequacy of an air void system to provide durability of concrete to freezing and thawing. Total air content, perhaps the most widely used parameter, often fails to be by itself a reliable indicator of concrete durability to freezing and thawing, because it does not reflect the quality of the air void system. The most widely used air void spacing equation is the Powers spacing factor. The spacing factor is believed to be the farthest distance that water would have to travel through paste in order to get to the nearest air void. In this way, the spacing factor can be roughly determined as half of the greatest distance between any two adjacent air voids. However, K. Snyder, 1998, pointed out that there are misconceptions about what the Powers spacing factor represents and that the spacing factor actually calculates the fraction of paste within some distance of an air void [1.65].

T. Powers, 1954, proposed a simple model to determine the spacing factor, by determining the relative proportion of the hardened cement paste within the beneficial zone of influence of one or more air voids [1.33]. Powers assumed that all air voids have the same size and are arranged in a simple cubic lattice, where each void is equidistant from its nearest neighbor, and the imaginary lines connecting the voids are mutually perpendicular (see Figure 1.5).

Powers proposed that the spacing factor be calculated from air content, paste content, and specific surface of air voids by one of the two formulas in Equation (1.14) [1.66]. Equation (1.14) is simply the volume of air-free paste divided by the boundary area of the air voids, giving in some way the "hydraulic radius" of the paste-filled spaces among the air voids. Therefore, when calculating the spacing factor of an air void system, the air content must be expressed as a fraction of the air-free paste content.

$$\bar{L} = \frac{p_{cem}}{\alpha A_a} \quad \text{if} \ \frac{p_{cem}}{A_a} < 4.342 \tag{1.14a}$$

$$\bar{L} = \frac{3}{\alpha}\left[1.4\left(\frac{p_{cem}}{A_a} + 1\right)^{1/3} - 1\right] \quad \text{if} \ \frac{p_{cem}}{A_a} \geq 4.342 \tag{1.14b}$$

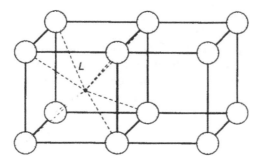

Figure 1.5 Schematic layout of air voids according to Powers' model. The spherical air voids are considered to be of the same size and at equal distances from each other. The spacing factor, *L*, can be visualized as the distance of each void to the center of a cube.

where $\bar{L}$ is the spacing factor (mm), p_{cem} is the cement paste content (%), A_a is the total air content (%), α is the specific surface of air voids (mm^{-1}).

In average air-entrained concretes, the ratio p_{cem}/A is generally greater than 4.342 and Equation (1.14b) is used in most cases. For example, for an air-entrained concrete with paste content of 26%, air content 5%, and specific surface of 23.6 mm^{-1}, the spacing factor is determined to be 0.20 mm.

The spacing factor proposed by Powers is the basis for the ASTM Standard C 457, which is used to characterize the air void system in concrete. T. Powers, 1965, indicated that every point in the cement paste must be no more than about 0.75 mm away from an air void [1.66]. The spacing factor required by ASTM C 457 to ensure the production of concrete durable to freezing and thawing is between 0.10 mm and 0.20 mm, which means that the average distance between two voids in the cement paste should not exceed about 0.30 mm. Considering that the size of air voids is 10 μm–1 mm, the spacing of the air voids is of the same order as their size. The spacing factor proposed by Powers is applicable only to concretes that have similar air void size distributions. In addition, it is determined based on the assumption that all air voids have the same size, which does not represent the reality. Therefore, the validity of the equation used to determine the spacing factor has been questioned, because it is possible that two concretes with the same calculated spacing factor will not show the same resistance to frost attack.

Several researchers have proposed equations to determine the spacing between air voids or the air void size distribution. R. Philleo, 1983, introduced a parameter, which he termed the *protected paste volume* concept and is most known as the Philleo factor [1.67]. The Philleo factor indicates the distance around an air void up to which hardened cement paste is protected from the effects of freezing and thawing. This distance is selected in a way that only a small portion of the cement paste (usually 10%) lies further than that distance from the perimeter of the nearest air void. The Philleo factor gives a more realistic concept of the internal geometry of air-entrained paste because it requires only the total volume of air voids in the paste and the number of air voids per unit volume of paste. E. Attiogbe, 1993 proposed an equation that estimates half the minimum spacing among neighboring air voids in concrete [1.68]. R. Pleau and M. Pigeon, 1996, have proposed an equation for the air void spacing distribution [1.69,1.70]. They considered the distances from a random point in the cement paste to the nearest air void and the distribution of the radius of the air voids. Further details on the equations proposed by these authors are given in their original papers.

References

1.1   Brown P.W., Shi D. "Porosity/permeability relationships," in *Materials Science of Concrete II*, J. Skalny, S. Mindess (eds), The American Ceramic Society, Westerville, OH, 1991, pp. 83–109.

1.2   Hearn N., Hooton R.D., Mills R.H. "Pore structure and permeability," in *Significance of Tests and Properties of Concrete and Concrete Making Materials*, 4th Edition, P. Klieger, J.F. Lamond (eds), ASTM STP 169C, American Society for Testing and Materials, West Conshohocken, PA, 1994, pp. 240–262.

1.3   Beaudoin J.J., Feldman R.F., Tumidajski P.J. "Pore structure of hardened Portland cement pastes and its influence on properties," *Advanced Cement Based Materials*, Vol. 1, No. 5, 1994, pp. 224–236.

1.4   Mindess S., Young J.F., Darwin D. *Concrete*, 2nd Edition, Prentice Hall, Englewood Cliffs, NJ, 2002.

1.5   Taylor H.F.W. *Cement Chemistry*, Thomas Telford Services Ltd, London, 1997.

1.6   Diamond S. "Cement paste microstructure – an overview at several levels," in *Proceedings of Conference on Hydraulic Cement Pastes: Their Structure and Properties*, Sheffield, Cement and Concrete Association, April 1976, pp. 2–30.

1.7   Jennings H.M., Dalgleish B.J., Pratt P.L. "Morphological development of hydrating tricalcium silicate as examined by electron microscopy techniques," *Journal of the American Ceramic Society*, Vol. 64, 1981, pp. 567–572.

1.8   Powers T.C. "Physical properties of cement paste," in *Proceedings of the Fourth International Symposium on the Chemistry of Cement*, Washington, DC, 1960, pp. 577–613.

1.9   Powers T. "Structure and physical properties of hardened Portland cement paste," *Research Bulletin 94*, Portland Cement Association, Chicago, IL, March 1958, Reprint from *Journal of the American Ceramic Society*, Vol. 41, 1958, pp. 1–6.

1.10   Powers T.C., Brownyard T.L. "Studies of the physical properties of hardened Portland cement paste," *Research Bulletin 22*, Portland Cement Association, Chicago, IL, 1948, Reprint from *Journal of the American Concrete Institute*, October 1946–April 1947, in *Proceedings Vol. 43*, 1947.

1.11   Powers T.C. "The physical structure and engineering properties of concrete," *Portland Cement Association R&D Bulletin*, Vol. 90, 1958, pp. 1–28.

1.12   Brunauer S. "Tobermorite gel: the heart of concrete," *American Scientist*, Vol. 50, 1962, pp. 210–229.

1.13   Mikhail R.Sh., Copeland L.E., Brunauer S. "Pore structure and specific surface area of hardened Portland cement pastes as determined by nitrogen adsorption," *Canadian Journal of Chemistry*, Vol. 42, 1964, pp. 436–438.

1.14   Hagymassy J., Odler I., Yudenfreund M., Skalny J., Brunauer S. "Pore structure analysis by water vapor adsorption III. Analysis of hydrated calcium silicates and Portland cements," *Journal of Colloid and Interface Science*, Vol. 38, No. 1, 1972, pp. 20–34.

1.15   Bodor E.E., Skalny J., Brunauer S., Hagymassy J., Yudenfreund M. "Pore structure of hydrated calcium silicates and Portland cement determined by nitrogen adsorption," *Journal of Colloid and Interface Science*, Vol. 34, No. 4, 1970, pp. 560–568.

1.16 Odler I., Hagymassy J. Jr, Bodor E.E., Yudenfreund M., Brunauer S. "Hardened Portland cement pastes of low porosity IV. Surface area and pore structure," *Cement and Concrete Research*, Vol. 2, No. 5, 1972, pp. 577–589.

1.17 Taylor H.F.W. "Hydrated calcium silicates. Part I. Compound formation at ordinary temperatures," *Journal of the Chemical Society*, 1950, Part IV, pp. 3682–3690.

1.18 Feldman R.F., Sereda P.J. "A new model for hydrated Portland cement and its practical implications," *Engineering Journal of Canada*, Vol. 53, No. 8–9, 1970, pp. 53–59.

1.19 Sereda P.J., Feldman R.F., Ramachandran V.S. "Structure formation and development in hardened cement pastes," Sub-Theme VI-1, in *Proceedings of the Seventh International Congress on Chemistry of Cement*, 1980, pp. VI-1/3–VI-1/44.

1.20 Daimon M., Abo-El-Enein S.A., Hosaka G., Goto S., Kondo R. "Pore structure of calcium silicate hydrate in hydrated tricalcium silicate," *Journal of the American Ceramic Society*, Vol. 60, No. 3–4, 1977, pp. 110–114.

1.21 Wittmann F. "Interaction of hardened cement paste and water," *Journal of the American Ceramic Society*, Vol. 56, No. 8, 1973, pp. 409–415.

1.22 Wittmann F. "Grundlagen eines Modells zur Beschreibung charakteristischer Eigenschaften des Betons," Deutscher Ausschuss für Stahlbeton, Heft 290, Berlin 1977, pp. 45–100.

1.23 Jennings H.M., Tennis P.D. "Model for the developing microstructure in Portland-cement pastes," *Journal of the American Ceramic Society*, Vol. 77, No. 12, 1994, pp. 3161–3172; and correction, Vol. 78, No. 9, 1995, p. 2575.

1.24 Jennings H.M. "A model for the microstructure of calcium silicate hydrate in cement paste," *Cement and Concrete Research*, Vol. 30, No. 1, 2000, pp. 101–116.

1.25 Tennis P.D., Jennings H.M. "A model for two types of calcium silicate hydrate in the microstructure of Portland cement pastes," *Cement and Concrete Research*, Vol. 30, No. 6, 2000, pp. 855–863.

1.26 Verbeck G. "Pore structure," in *Significance of Tests and Properties of Concrete and Concrete Aggregates*, STP 169A, American Society for Testing and Materials, West Conshohocken, PA, 1966, pp. 211–219.

1.27 Winslow D.N., Diamond S. "A mercury porosimetry study of the evolution of porosity in Portland cement," *Journal of Materials*, Vol. 5, No. 3, 1970, pp. 564–585.

1.28 Barnes B.D., Diamond S., Dolch W.L. "Hollow shell hydration of cement particles in bulk cement paste," *Cement and Concrete Research*, Vol. 8, No. 3, 1978, pp. 263–272.

1.29 Kjellsen K.O., Jennings H.M., Lagerblad B. "Evidence of hollow shells in the microstructure of cement paste," *Cement and Concrete Research*, Vol. 26, No. 4, 1996, pp. 593–599.

1.30 Kjellsen K.O., Helsing Atlassi E. "Pore structure of cement silica fume systems: presence of hollow-shell pores," *Cement and Concrete Research*, Vol. 29, 1999, pp. 133–142.

1.31 Kjellsen K.O., Lagerblad B., Jennings H.M. "Hollow-shell formation – an important mode in the hydration of Portland cement," *Journal of Materials Science*, Vol. 32, 1997, pp. 2921–2927.

1.32 Hadley D.W., Dolch W.L., Diamond S. "On the occurrence of hollow-shell hydration grains in hydrated cement paste," *Cement and Concrete Research*, Vol. 30, 2000, pp. 1–6.

1.33 Powers T.C. "Void spacing as a basis for producing air-entrained concrete," *Research Bulletin 49*, Portland Cement Association, Chicago, IL, July 1954. Reprint from *Journal of the American Concrete Institute*, May 1954, in *Proceedings Vol. 50*, 1954, pp. 741–760.

1.34 Dubinin M.M. "The potential theory of adsorption of gases and vapors for adsorbents with energetically nonuniform surfaces," *Chemical Reviews*, Vol. 60, No. 2, 1960, pp. 235–241.

1.35 *Manual of Symbols and Terminology for Physicochemical Quantities and Units – Appendix II Definitions, Terminology and Symbols in Colloid and Surface Chemistry, Part I*, International Union of Pure and Applied Chemistry, Division of

Physical Chemistry, Washington, DC, prepared for publication by D.H. Everett, L.K. Koopal, 1971.

1.36   Dubinin M.M. "Microporous structures of carbonaceous adsorbents," in *Characterisation of Porous Solids, Proceedings of International Symposium,* Neuchatel, Switzerland, 9–12 July 1978, S.J. Gregg, K.S.W. Sing, H.F. Stoeckli (eds), Society of Chemical Industry, London, 1979, pp. 1–11.

1.37   Young J.F., Mindess S., Bentur A., Gray R.J. *The Science and Technology of Civil Engineering Materials,* Prentice Hall, Engelwood Cliffs, NJ, 1998.

1.38   Mehta P.K., Manmohan D. "Pore size distribution and permeability of hardened cement pastes," in *Proceedings of the Seventh International Congress on the Chemistry of Cement,* Paris, 1980, Vol. III, pp. VII-1–VII-5.

1.39   Aligizaki K.K. "Determination of pore structure parameters in hardened cementitious materials," MS Thesis, The Pennsylvania State University, 1995, p. 265.

1.40   Mehta P.K. *Concrete Structure Properties and Materials,* Prentice Hall, Engelwood Cliffs, NJ, 1986.

1.41   Setzer M.J. "Interaction of water with hardened cement paste," in *Ceramic Transactions Vol. 16: Advances in Cementitious Materials,* S. Mindess (ed.), The American Ceramic Society, Westerville, OH, 1991, pp. 415–439.

1.42   Oberholster R.E. "Pore structure, permeability and diffusivity of hardened cement paste and concrete in relation to durability: status and prospects," in *Proceedings of the Eigth International Congress on the Chemistry of Cement,* Vol. I, 1986, pp. 323–335.

1.43   Roy D.M. "Relationships between permeability, porosity, diffusion and microstructure of cement pastes, mortar, and concrete at different temperatures," in *Symposium Proceedings Vol. 137: Pore Structure and Permeability of Cementitious Materials,* L.R. Roberts, J.P. Skalny (eds), Materials Research Society, Warrendale, PA, 1989, pp. 179–189.

1.44   Haynes J.M. "Determination of pore properties of constructional and other materials – general introduction and classification of methods," *Matériaux et Constructions,* Vol. 6, No. 33, 1973, pp. 169–174.

1.45   Pratt P.L. "Physical methods for the identification of microstructures," *Materials and Structures,* Vol. 21, 1988, pp. 106–117.

1.46   Diamond S. "Methodologies of PSD measurements in HCP: postulates, peculiarities and problems," in *Symposium Proceedings Vol. 137: Pore Structure and Permeability of Cementitious Materials,* L.R. Roberts, J.P. Skalny (eds), Materials Research Society, Warrendale, PA, 1989, pp. 83–89.

1.47   Feldman R.F. "The porosity and pore structure of hydrated Portland cement paste," in *Symposium Proceedings Vol. 137: Pore Structure and Permeability of Cementitious Materials,* L.R. Roberts, J.P. Skalny (eds), Materials Research Society, Warrendale, PA, 1989, pp. 59–73.

1.48   Winslow D.N., Diamond S. "Specific surface of hardened Portland cement paste as determined by small-angle X-ray scattering," *Journal of the American Ceramic Society,* Vol. 57, No. 5, 1974, pp. 193–197.

1.49   Diamond S., Dolch W.L. "Generalized log-normal distribution of pore sizes in hydrated cement paste," *Journal of Colloid and Interface Science,* Vol. 38, No. 1, 1972, pp. 234–244.

1.50   Rootare H.M., Nyce A.C. "Use of porosimetry in the measurement of pore size distribution in porous materials," *International Journal of Powder Metallurgy,* Vol. 7, No. 1, 1971, pp. 3–11.

1.51   Mason G., Morrow N.R. "Capillary behavior of a perfectly wetting liquid in irregular triangular tubes," *Journal of Colloid and Interface Science,* Vol. 141, No. 1, 1991, pp. 262–274.

1.52   Tuller M., Or D., Dudley L.M. "Adsorption and capillary condensation in porous media: liquid retention and interfacial configuration in angular pores," *Water Resources Research,* Vol. 35, No. 7, 1999, pp. 1949–1964.

1.53 *Practical Guide to Image Analysis*, ASM International, Materials Park, OH, 2000.

1.54 Russ J.C. *The Image Processing Handbook*, CRC Press, Boca Raton, FL, 4th Edition, 2002.

1.55 Patzek T.W., Silin D.B. "Shape factor and hydraulic conductance in noncircular capillaries I. One-phase creeping flow," *Journal of Colloid and Interface Science*, Vol. 236, 2001, pp. 295–304.

1.56 Lange D.A., Jennings H.M., Shah S.P. "Image analysis techniques for characterization of pore structure of cement-based materials," *Cement and Concrete Research*, Vol. 24, No. 5, 1994, pp. 841–853.

1.57 Jenkins R.G., Rao M.B. "The effect of elliptical pores on mercury porosimetry results," *Powder Technology*, Vol. 38, 1984, pp. 177–180.

1.58 Mandelbrot B. *Fractals: Form, Chance, and Dimension*, W.H. Freeman and Co., New York, 1977.

1.59 Allen A.J., Oberthur R.C., Pearson D., Schofield P., Wilding C.R. "Development of the fine porosity and gel structure of hydrating cement systems," *Philosophical Magazine B*, Vol. 56, No. 3, 1987, pp. 263–288.

1.60 Blinc R., Lahajnar G., Zumer S., Pintar M.M. "NMR-study of the time evolution of the fractal geometry of cement gels," *Physical Review B*, Vol. 38, No. 4, 1988, pp. 2873–2875.

1.61 Winslow D., Bukowski J.M., Young J.F. "The fractal arrangement of hydrated cement paste," *Cement and Concrete Research*, Vol. 25, No. 1, 1995, pp. 147–156.

1.62 Wang Y.T., Diamond S. "A fractal study of the fracture surfaces of cement pastes and mortars using a stereoscopic SEM method," *Cement and Concrete Research*, Vol. 31, No. 10, 2001, pp. 1385–1392.

1.63 Winslow D. "The fractal nature of the surface of cement paste," *Cement and Concrete Research*, Vol. 15, No. 5, 1985, pp. 817–824.

1.64 Heinemann A., Hermann H., Haussler F. "SANS analysis of fractal microstructures in hydrating cement paste," *Physica B*, Vol. 276, 2000, pp. 892–893.

1.65 Snyder K.A. "A numerical test of air void spacing equations," *Advanced Cement Based Materials*, Vol. 8, No. 1, 1998, pp. 28–44.

1.66 Powers T. "Topics in concrete technology 4. Characteristics of air void systems," *Journal of the PCA Research and Development Laboratories*, Vol. 7, No. 1, 1965, pp. 23–41.

1.67 Philleo R.E. "A method for analyzing void distribution in air-entrained concrete," *Cement, Concrete and Aggregates*, Vol. 5, No. 2, 1983, pp. 128–130.

1.68 Attiogbe E.K. "Mean spacing of air voids in hardened concrete," *ACI Materials Journal*, Vol. 90, No. 2, 1993, pp. 174–181.

1.69 Pleau R., Pigeon M. "The use of the flow length concept to assess the efficiency of air entrainment with regards to frost durability: Part I – Description of the test method," *Cement, Concrete and Aggregates*, Vol. 18, No. 1, 1996, pp. 19–29.

1.70 Pleau R., Pigeon M., Laurencot J.L., Gagné R. "The use of the flow length concept to assess the efficiency of air entrainment with regards to frost durability: Part II – Experimental results," *Cement, Concrete and Aggregates*, Vol. 18, No. 1, 1996, pp. 30–41.

2 Specimen pretreatment

The specimens used in the techniques for pore structure determination described in the following chapters, must first be crushed or sawed, prepared and stored properly, and then used for testing. Specimen preparation and pretreatment is very important because total porosity values and pore size distributions determined are affected by the pretreatment technique used. The general procedures and principles of pretreatment that have been developed and are commonly used by researchers are described in this chapter. However, several researchers might develop a combination of pretreatment techniques, if they find it more appropriate for their needs. The preparation techniques described include procedures for water removal before microstructural analysis in mature cement pastes, and preparation techniques for optical observations of microstructure and air void analysis using microscopic techniques.

2.1 Water removal

Most of the available techniques used in pore structure studies require that the hardened cement paste be thoroughly treated to remove water and evacuated prior to testing. Water removal can change significantly the pore structure of a water-saturated cementitious material and there is much concern with regard to the collapse of the structure or with dehydration of poorly crystalline hydration products through drying.

From the types of water mentioned in Chapter 1, non-evaporable water is lost when the paste is heated (ignited) to 1000°C. Hydrate water from the C–S–H gel is held inside the gel pores, including interlayer spaces, and can be removed by heating, desiccation, evacuation, or a combination of these techniques. In addition to the water that is part of the hydration products, the pores of the sample may contain evaporable water that must be removed. The walls of even empty pores are generally covered with an adsorbed layer of water vapor. Pore water removal will generate temporary capillary pressure that may be sufficient to cause shrinkage of the sample. Any such shrinkage will result in a decrease in the volume of pores and will alter the fine pore structure of the cement paste and consequently affect the results. For samples that are highly prone to shrinkage, there are several techniques available that will minimize permanent volume changes. The ideal outgassing procedure would remove water and leave the specimen's solid matrix unchanged during the process, which may be extremely difficult or impossible to accomplish. Therefore, frequently, some specimen alteration is accepted in order to obtain any measurement at all.

Water removal is done by two general ways: using a drying technique, or using a solvent replacement technique.

2.1.1   Drying techniques

Standard drying techniques used to remove water from hydrated cement pastes are the following: oven-drying, vacuum-drying, P-drying, D-drying, desiccant drying, and freeze-drying. These drying techniques are briefly described in the following sections. Critical point drying is also mentioned here.

2.1.1.1   Oven-drying

Concrete scientists find oven-drying at 105°C to be one of the most convenient drying techniques and to be satisfactory for most purposes that require cement paste drying. Oven-drying is much quicker than most drying techniques, and pore water removal is complete after 24 hours of heating. The specimen is first weighed and placed into an oven that is set at a temperature of 105°C and at atmospheric pressure (101 kPa or 760 mmHg). A flow of dry, CO_2-free air is circulated through the oven. The mass of the specimen can be monitored periodically to follow the drying process. When the specimen reaches a constant mass, then there is no more pore water removed, and the drying is complete. Constant mass is generally considered to be achieved when in subsequent weighings the difference in mass measured is less than 0.1% of the mass of water originally held in the cement paste. The specimen can then be put into a desiccator for storage.

Although oven-drying is an effective way to remove evaporable water from a specimen, it has the drawback that microcracking may be induced into the specimen as a result of differential thermal expansion and contraction of the constituent components (aggregates and hardened cement paste) during the drying cycle. During oven-drying the surface tensions of the receding water menisci produce stresses that cause microstructural changes. Drying at temperatures above 105°C is believed to remove structural water from the cement paste and results in an irreversible decomposition of the cement paste itself [2.1]. It is also believed that some hydration products might dehydrate at temperatures well below 105°C. R. Detwiler *et al.*, 2001, have suggested that oven-drying should be avoided above temperatures of 35°C, because of its destructive effect on the cement paste microstructure and its tendency to induce cracking to the cement paste [2.2]. As a compromise to oven-drying and avoidance of microcracking and decomposition of hydration products, the researchers propose the use of vacuum oven-drying at lower temperatures, e.g. at about 50°C. In this case, specimens may need longer than the normal time of 24 hours to attain constant mass, but the specimens do not experience the damages of oven-drying that happen at higher temperatures. Therefore, moisture removal is relatively effective and microcracking occurs to a lesser extent.

2.1.1.2   *Vacuum-drying*

During vacuum-drying the specimen is placed in a sealed container that is attached to a vacuum-generating mechanical pump. As vacuum is generated, pore water turns into vapor and is removed from the cement paste specimen. The technique is relatively slow, making it most suitable for specimens older than about 28 days, which have fairly low contents of evaporable water at usual humidities. Vacuum-drying is also believed to induce some microcracking in the cement paste specimen.

2.1.1.3   *P-drying*

P-drying (or Perchlorate-drying) is carried out at ambient temperature by placing the specimen to be dried over magnesium perchlorate hydrates (dihydrate and tetrahydrate) $[Mg(ClO_4)_2 \cdot 2H_2O{-}Mg(ClO_4)_2 \cdot 4H_2O]$. This arrangement yields a partial pressure of 1.1 Pa (8 μmHg) at 25°C for the H_2O vapor over the magnesium perchlorate hydrates, leading to evaporation of the pore water from the cement paste specimen [2.1,2.3,2.4]. The vapor from the pore solution can be removed by the application of vacuum, and subsequently the specimen is placed in a desiccator, and is sealed for storage. It is generally considered that P-dried cement pastes contain some residual pore water. It has been calculated that at a water vapor pressure of 8 μm, the C–S–H gel contains 2.8 molecules of water per molecule [2.4].

2.1.1.4   *D-drying*

D-drying (or Dry-ice drying) is the most widely adopted technique and is the most rigorous vacuum-drying condition carried out at ambient temperature. The method was first mentioned by T. Powers, 1948 [2.5] and was proposed as a drying technique by L. Copeland and J. Hayes, 1953 [2.3]. The D-drying apparatus consists of a sealed glass container that is attached to a mechanical vacuum pump through a cold trap. The cold trap is cooled by a mixture of solid CO_2 and alcohol at a temperature −79°C (see also Figure 2.1). The partial

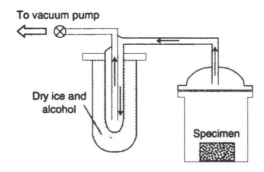

Figure 2.1 Schematic setup for a D-drying apparatus. The arrows show the direction of the flow of the vapor removed from the specimen as it is dried.

pressure of water vapor over dry ice is 0.07 Pa (0.5 μmHg). Under these conditions, removal of pore water solution is very slow and drying for several days is required. The vacuum applied to facilitate the drying process should keep the pressure in the system below 4.2 Pa (30 μmHg). The time needed until constant mass of the specimen is reached is approximately 14 days. D-drying is a common drying technique, because it is generally assumed to remove all of the physically adsorbed water in the cement paste pores, like the oven-drying technique at 105°C, but without damaging the microstructure of the cement paste. It has been calculated that at a water vapor pressure of 0.07 Pa (0.5 μmHg), the C–S–H gel contains just a little more than 2 molecules of water [2.4]. J. Thomas *et al.*, 1998, have found using Small-Angle Neutron Scattering that D-drying removes not only water in the gel pores but also some of the water bound into the interlayer spacings [2.6].

2.1.1.5 Direct freeze-drying

Direct freeze-drying, which is sometimes mentioned also as vacuum freeze-drying, is a very fast drying technique and causes little damage to the microstructure of the cement paste. A specimen is immersed directly in liquid nitrogen (−196°C) for about 5 minutes, or, alternatively, it can be immersed in a mixture of solid CO_2 and methanol (−80°C). For small specimens, freezing of pore water is almost instant, the ice crystals form very rapidly and have a much smaller size than when they are formed at 0°C. In this way, they do not exert pressure on the surrounding hardened cement paste, avoiding the formation of excessive microcracking in the cement paste. After freezing, the specimens are transferred to a sealed container and vacuum is applied, and then the specimens are placed in a desiccator until testing. Alternatively, after vacuuming the specimens can be placed in a drying oven at 105°C for one hour and then stored in a desiccator until testing. Results have shown that after 24 hours of vacuum freeze-drying, water is removed effectively from a water-saturated hardened cement paste [2.7].

According to the *capillary stress theory* increasing internal hydrostatic stresses develop in a capillary during drying as the meniscus radius becomes smaller [2.5]. Under vacuum the water molecules sublime directly from solid ice crystals to gas without passing through the liquid state and the danger of capillary stresses is eliminated. In addition, any solutes present in the pore solution are deposited on the pore walls without crystallizing [2.2,2.8].

It has been stated by J. Gillott, 1976, that direct immersion in liquid nitrogen appears to cause rapid freezing, but the rate of cooling is less than required because of the formation of an insulating envelope of gas surrounding the specimen [2.9]. Even when ice crystals are formed in the specimen, there is a considerable possibility that they will devitrify and recrystallize during sublimation. It has been reported that this occurs at about −130°C, which is close to the recrystallization temperature, a temperature of about half of the melting point, in degrees Kelvin, at which many materials recrystallize. However, the rate of sublimation is extremely slow, owing to the low vapor pressure of ice below this temperature. In order to avoid this pitfall, Gillot recommended that the specimen be immersed first in propane, isopentane, or Freon 22*, and then supercooled in liquid nitrogen. These compounds have a low freezing point, so they are liquids at very low temperatures. In addition, they have a relatively high boiling point, so they are not vaporized as readily by heat from the specimen as is nitrogen; therefore, cooling is more efficient. Pretreatment with Freon before freeze-drying has been used in cement pastes by M. Moukwa and P. Aitcin, 1988 [2.10] (see also Section 2.1.3).

It has been shown that freeze-drying causes less damage to the structure than oven-drying [2.11]. However, freeze-drying alone may not completely dry the specimens, since the efficiency of sublimation lies in the openness of the network through which water must move during drying [2.10,2.12]. For dense cement pastes, sublimation must be prolonged to be efficient, and the specimen's mass could be monitored regularly to determine when constant mass has been achieved. It is also believed to be unlike that bound water from poorly crystal-lized hydrated phases is removed, because the freeze-drying method does not allow for a large supply of energy to free the hydroxyl from the hydrated phases.

2.1.1.6 Indirect freeze-drying

Instead of being immersed directly in liquid nitrogen, as it was described in the previous paragraph, the specimen is placed in a small glass vial, which is submerged into liquid nitrogen for about 15 minutes. In this way, there is no direct contact of the specimen with the liquid nitrogen. The specimen is then removed, vacuumed for 24 hours, and is further treated as it was described for direct freeze-drying [2.8].

2.1.1.7 Desiccant drying

Other drying techniques used by researchers to remove water from cement pastes include desiccant drying using silica gel or anhydrous $CaCl_2$ [2.13]. The desiccator containing the desiccant is evacuated separately at the beginning of the drying period and then sealed during the drying process so that the total internal pressure is due to water vapor from the specimens. The specimen mass losses are measured at regular intervals during the drying period. The drying is continued until weighing on successive days indicates that the specimen has reached a steady state. Steady state can be defined in different ways; e.g. it has been defined as the condition when the rate of mass loss is reduced to less than 1 mg of water per 1 g of paste per day at ~20°C [2.13].

2.1.1.8 Critical point drying

Critical point drying has been mentioned as an alternate drying technique and involves raising the temperature and pressure of the liquid phase to values above those of the critical point of the liquid [2.14,2.15,2.9]. The phase

---

* Freon, the commercial name for chlorofluorocarbons (CFC) has been banned for use due to its adding to the depletion of the earth's ozone shield.

diagram of water is presented in Figure 2.2, which shows the relationship between pressure, temperature, and state. Each state of water (and matter in general) can only exist within certain ranges of pressure and temperature. Two important points are the "triple point," marked as T, and the critical point marked as C in Figure 2.2. At the triple point, the three phases of solid, liquid, and gas exist simultaneously and are at equilibrium. The critical point is characterized by a fixed temperature, pressure, and density, at which the phys-ical properties of a liquid and its vapor become the same and the distinction between the liquid and gas phase disappears. At the critical point, the interface between liquid and vapor vanishes, so surface tension forces cease to exist. The critical point of water is at a temperature of 374°C and at a pressure of 2.25×10^7 Pa and occurs at lower values of temperature and pressure for other fluids. The critical pressure and temperature for some substances are reported in Table 2.1 [2.20].

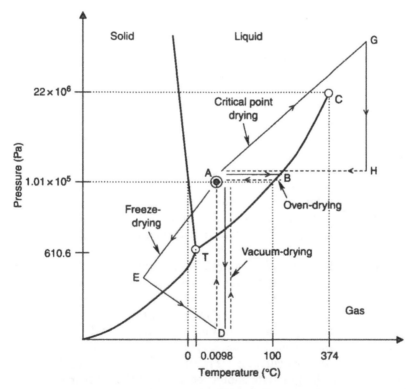

Figure 2.2 The phase diagram of water showing the triple point T and the critical point C. The curve TC is the area at which transition needs to be avoided during drying. The transitions between the phases are shown using the diagram for the processes of freeze-drying, oven-drying, vacuum-drying, and critical point drying. The initial state of water is taken at point A at atmospheric pressure and room temperature (approximately 23°C). The solid lines correspond to a condition in which the specimen contains water and the doted lines to a condition in which the specimen is dry. The arrows indicate the paths followed during the drying process.

Table 2.1 Critical temperature and pressure for some of the liquids used for critical drying in materials [2.20]

Solvent	Molecular formula	Critical temperature T_c (K)	Critical pressure P_c (kPa)
Acetone	C_3H_6O	508.1	4.70
Benzene	C_6H_6	562.0	4.90
Carbon dioxide	CO_2	304.1	7.38
Ethanol	C_2H_6O	514.0	6.14
Hexane	C_6H_{14}	507.6	3.02
Isopropanol	C_3H_8O	508.3	4.76
Methanol	CH_4O	512.5	8.08
Nitrogen	N_2	126.2	3.39
Nitrogen tetroxide	N_2O_4	431.0	10.10
Oxygen	O_2	154.6	5.04
Toluene	C_7H_8	591.8	4.11
Water	H_2O	647.1	22.06

Several researchers have used the method for a variety of materials, in particular biological tissues and materials with a delicate layered structure, which need to be prepared for Scanning Electron Microscopy (SEM) observations by removing the water without collapsing the structure of the material. The main body of the apparatus is a pressure vessel that can withstand high pressures including a water jacket for heating and cooling. Drying can be done by evaporating the pore water itself under supercritical conditions; however, it is done most commonly by replacing pore water by another liquid and then removing the new liquid subsequently under supercritical conditions. Usually carbon dioxide is used because of its low critical point. However, to the writer's knowledge, this technique is not used in cement-based materials, since critical point drying is associated, even though not exclusively, with the use of carbon dioxide, which generates carbonation problems with cement pastes.

Using the phase diagram in Figure 2.2, the effect of the different drying techniques can be illustrated with reference to the water removal. The line TC in Figure 2.2 between the critical and triple point is the state at which liquid water exists with vapor above it. The interfacial tension from the boundary between the liquid and gas phase has the damaging effect during preparation (drying) of the specimen, because it results in collapse of the delicate microstructure of the cement paste. Therefore, during drying, the transition from the liquid to the gas state should be avoided (i.e. crossing the line TC should be avoided). For simplicity, it is assumed that a cement paste specimen is at room temperature and atmospheric pressure (point A in Figure 2.2) before the drying technique is applied.

During oven-drying, the temperature is raised while the pressure does not change from atmospheric pressure. The transition is from point A to point B, and this path reaches the boundary of the liquid–gas interface leading to collapse of the cement paste microstructure. The path also indicates why oven-drying recommended at a temperature of approximately 60°C, before crossing the TC curve, is effectively drying the specimen without causing damage to the microstructure.

Vacuum-drying without temperature increase leads to point D, and after the water is converted to water vapor and is removed by the vacuum process, the specimen returns to atmospheric pressure back at point A.

P-drying has a similar effect as vacuum-drying.

During freeze-drying, the temperature is reduced to the temperature of the liquid nitrogen and the pressure is also reduced, and the state of water is at point E. Placing the specimen under vacuum results in sublimation converting ice into vapor and the temperature is slowly increased to room temperature (point D in Figure 2.2). Then by removing the vacuum, the specimen returns dry to its initial conditions of room temperature and atmospheric pressure (point A). D-drying has a similar effect as freeze-drying.

Critical point drying is done by increasing the temperature (and consequently the pressure in the sealed vessel) until the critical point C is exceeded and the liquid is at the supercritical state (point G in Figure 2.2), in which the liquid and gas boundary disappears. While keeping the temperature high, the pressure can be lowered to atmospheric pressure (point H) as vapor gas is released from the vessel. Then by reducing the temperature, the specimen is left dry at room temperature and atmospheric pressure.

2.1.2 Solvent replacement

The solvent replacement procedure, also mentioned as solvent exchange procedure, is used as an alternate drying technique prior to porosity measurements because it is considered to be gentle to the cement paste microstructure [2.16,2.17]. The solvent replacement technique is also used to determine the total porosity of cement pastes and in that aspect, it is discussed in more detail in Chapter 5.

Solvents are organic liquids miscible with water but with a higher vapor pressure than water. Organic solvents have physical properties that present numerous advantages for drying before pore structure characterization. For instance, they have a much lower surface tension than water (see Table 2.2) and this property minimizes pore structure damage upon drying. The specific gravity of most organic solvents is also significantly lower than that of water; therefore, organic solvents tend to naturally replace the pore solution in a hydrated cement paste by a simple counter-diffusion process [2.18,2.19]. The solvents most commonly used for drying cement pastes are the alcohols methanol [CH_3OH], ethanol [CH_3CH_2OH], and isopropanol [$CH_3CH(OH)CH_3$]. Other solvents used to a lesser extent are acetone, benzene, toluene, xylene, and pentane.

During solvent replacement, the specimen is first weighed and then immersed in an organic liquid. The solution-to-specimen volume ratio is important, since the point of saturation of water in most organic solvents is quite low; this ratio is usually taken as 100:1. In addition, in order to facilitate the diffusion process, the solvent should be renewed regularly (e.g. every hour during the first 24 hours of the experiment and less frequently

Table 2.2 Physical properties of some common liquids that have been used in the past for pretreatment or intrusion in cement-based materials [2.20]

Substance	Molecular formula	Density (g/cm^3) at $20°C$	Vapor pressure (kPa) at $25°C$	Surface tension γ (mN/m) at	
				25°C	50°C
Acetone	C_3H_6O	0.7899	30.8	23.46	20.66
Benzene	C_6H_6	0.8765	12.7	28.22	25.00
Cyclohexane	C_6H_{12}	0.7785	13.0	24.65	21.68
Ethanol	C_2H_6O	0.7893	7.87	21.97	19.89
Hexane	C_6H_{14}	0.6594	20.2	17.89	15.33
Isopropanol	C_3H_8O	0.7855	6.02	20.93	18.96
Mercury	Hg	13.5458	2.26×10^{-4}	485.48	480.36
Methanol	CH_4O	0.7914	16.9	22.07	20.14
Pentane	C_5H_{12}	0.6262	68.3	15.49	—
Toluene	C_7H_8	0.8669	3.79	27.93	24.96
Water	H_2O	0.9982	3.17	71.99	67.91

afterwards) [2.11,2.21]. Once the specimen is immersed in the solvent, the solvent immediately penetrates into the pores of the material and replaces the cement paste pore solution. During the experiment, the penetration of the solvent within the specimen is monitored by the change in mass of the specimen determined at regular intervals. When there is no further mass change, the specimen is removed, placed into a desiccator, and evacuated for 2 days to remove the solvent. Of particular importance is the time needed for exchange of pore water with the alcohol solution, which depends on the density and the dimensions of the specimens tested. For example, it has been reported that for cylindrical specimens with 5.5 mm diameter and 10 mm length the time needed can vary from 1 to 2 days for pastes with w/c = 1.0 and 14 to 21 days for pastes with w/c = 0.30 [2.22]. These results refer to 2–3-year-old cement pastes made of white cement that were water-saturated at immersion into ethanol, and complete exchange was defined as the state at which 95% of the maximum possible exchange has been reached.

The replacement of pore water by solvent is believed to reduce microcracking, because when the solvent is removed, surface tension forces are much smaller than when the original pore water is removed. Because of the lower surface tension of the solvent compared to water, one would expect that the stresses induced in the cement paste pore walls are smaller and less damage occurs as the pore liquid evaporates; several investigators have experimentally confirmed this theory [2.11,2.21,2.23]. The solvents may partially dehydrate the calcium silicate hydrates and ettringite; however, it is believed that the outward morphology of the crystals is not seriously affected [2.2].

The suitability of organic solvents for use with cement-based materials has been tested in the past by many researchers in order to determine the ease of penetration of the solvents inside the pores, and the degree of their physical or chemical interaction with the cement paste. The results of some of these studies are discussed in the following sections.

2.1.2.1 Ease of penetration

The ease of penetration of a solvent in the cement pastes depends not only on the density of the specimen tested, but also upon the drying history, and the size of the solvent molecule. The starting condition of the cement paste to be tested (e.g. oven-dried, D-dried or water-saturated), the sequence of exchange, i.e. the drying history, and the type of the solvent, have an effect on the volume stability of the cement pastes. L. Parrott, 1981, 1983, studied the exchange of methanol with pore water in C_3S pastes [2.17,2.21]. He found that the rate of exchange was dependent upon the *drying history* of the specimen. Oven-dried Portland cement pastes had increased solvent diffusivity compared to pastes that had not been predried. Successive weighings showed that solvent replacement treatments not involving pre-application of heat did not result in true mass loss equilibrium; in contrast, the heated specimens reached mass loss equilibrium in 48 hours. The rate of methanol exchange with pore water and the irrecoverable drying shrinkage results suggested that previous drying at intermediate relative humidities caused a partial collapse in the smaller pores and a corresponding increase in volume of the larger pores [2.21]. Studies by G. Litvan, 1976, have shown that a combination of methanol–pentane replacement is the most effective method of preserving the microstructure of C–S–H in its saturated state [2.24]. In Litvan's proposed method, pore water is replaced by methanol to a great extent, which is then replaced by pentane. Pentane can be successively evaporated through a cold trap using a vacuum pump.

Different liquids have different *molecule sizes* and are thus able to penetrate only pores greater than their diameters. R. Mikhail and S. Selim, 1966, in sorption tests on pre-dried samples, reported a difference in solvent accessibility to fine pores [2.25]. Propanol has a larger molecule than methanol: the molecular area of propanol is calculated to be $0.272\,nm^2$ and that of methanol $0.181\,nm^2$. Interpretation of solvent exchange data is complicated by the generation of vacated pore space during the exchange process. Methanol appears able to largely refill the vacated pore space, while propanol only achieves it at early ages. For this reason, the amount of water replaced by methanol may be slightly greater than the volume of evaporable water contained in the sample. In addition, methanol molecules may be able to penetrate the layered structure of C–S–H gel and replace some of the structural water [2.16].

D. Hughes and N. Crossley, 1994, tested the penetration of methanol and isopropanol in cement pastes [2.16,2.26]. Their experiments indicate that, while the penetration of methanol in saturated pastes containing silica fume appears to be a standard counter-diffusion process, the immersion of similar specimens in ethanol and isopropanol yields unexpected results. This difficulty could be either due to difficulty in penetration in the cement matrix by the larger isopropanol molecules, or due to interactions of isopropanol with the cement paste.

Studies by L. Parrott, 1983, 1984, have shown that methanol provides the fastest exchange with pore water compared to other solvents [2.18,2.17]. The length of exchange time does not have an effect upon the total porosity of C_3S, although methanol adsorption experiments indicated a change in pore size distribution. Pore structure alterations caused by removing water from the pores could be minimized by relatively short exchange periods so that about half of the pore water was exchanged.

2.1.2.2   Physical and chemical interactions

Several observations on diffusion rates and chemical analysis of solvent exchanged pastes has led researchers to believe that some solvents may react with the hydration products of cement during solvent replacement, which questions the suitability of solvent replacement as a drying technique and its effect on microstructural determinations. Straight-chain aliphatic alcohols react with $Ca(OH)_2$. Methanol is highly polar and can react with calcium, sodium, and potassium to form methoxides [2.27].

At first R. Day, 1981, and later on other researchers checked the possible *chemical interactions* of alcohols with cement hydration products [2.23,2.28,2.24]. Day showed that specimens of hardened cement paste or of calcium hydroxide that had been soaked in methanol had different mass loss characteristics compared to unsoaked samples during thermal decomposition. He suggested that these differences might be explained by a reaction between methanol and calcium hydroxide, forming calcium methoxide $Ca(OCH_3)_2$ as a product. J. Beaudoin, 1987, checked the interaction of methanol with calcium hydroxide at 22°C [2.29]. Using several characterization techniques for chemical composition (e.g. X-ray diffraction, infrared spectroscopy, thermogravimetric analysis), he observed similarities between the reaction products and calcium methoxide. He concluded that calcium hydroxide reacts with methanol and either calcium methoxide, or a methylated complex, or a carbonate-like product is formed. H. Taylor and A. Turner, 1987, checked the interaction of acetone and other organic liquids with cement paste, and found that acetone could react with the solid to form mesityl oxide, phorone, and isophorone [2.30]. More recently, length change experiments reported by J. Beaudoin *et al.*, 1998, suggest that other solvents, such as benzene and isopropanol, could also react chemically with calcium hydroxide [2.31]. R. Feldman, 1987, reported no signs of chemical interaction between isopropanol and cement paste and concluded that isopropanol is an acceptable fluid for diffusion coefficient measurements of hydrated Portland cement [2.19]. He also reported that water replacement using isopropanol followed by immediate evacuation and heating at 100°C for 20 hours causes the least damage to the cement paste [2.32].

Several authors have challenged the conclusions of various similar studies with respect to interaction of alcohols with cement paste [2.33,2.17,2.34]. L. Parrott, 1983, could not identify any signs of chemical interaction of cement pastes with organic liquids and specifically methanol, if methanol was removed from the pore systems of the samples by vacuum-drying before thermal analysis [2.17]. M. Thomas, 1989, examined the interaction between three organic solvents (methanol, propanol, and ethanol) with calcium hydroxide, and with cement paste using thermogravimetric analysis, X-ray diffraction, infrared spectroscopy, and mercury intrusion porosimetry [2.34]. X-ray diffraction and infrared spectroscopy techniques produced no evidence of a reaction product due to the interaction of either methanol and Ca(OH)$_2$ or methanol and cement paste. He reported that although there was evidence of a reaction between the solvents (particularly methanol) and cement paste during thermal analysis, there was no indication of a reaction at room temperature (22°C). The reactions that occur during thermal analysis are uncertain, but it is possible that the sorbed solvent may react chemically with calcium hydroxide during heating in the thermobalance to produce the corresponding alkoxide. The pore size distributions of vacuum-dried cement samples were unaffected by methanol soaking, and Thomas concluded that solvent exchange techniques are suitable for the preparation of samples for pore structure determinations.

Several investigations have also demonstrated the *physical interaction* of alcohols with cement pastes. Most solvents tend to be strongly adsorbed on the pore walls and cannot be entirely removed from the specimen by conventional drying techniques like D-drying, vacuum- or oven-drying [2.34,2.35]. H. Taylor and A. Turner, 1987, found that solvents are strongly sorbed by tricalcium silicate pastes and cannot be completely removed by vacuum-drying or oven-drying at temperatures that do not profoundly alter the cement paste [2.30]. R. Mikhail and S. Selim, 1966, have found that drying temperatures exceeding 100°C are needed to remove the sorbed solvent molecules from the pore walls of hydrated cement paste [2.25]. R. Feldman, 1987, found by length change measurements that thin cement paste specimens immersed in methanol result in large expansions, while specimens immersed in isopropanol shrink [2.19]. He associated the volume increase with the penetration of the relatively small methanol molecules into the layered silicate structure.

It is obvious that the presence of residual solvent molecules on the surface of the pore walls complicates the interpretation of pore structure results. Even though it is believed that solvent exchange does not affect pore structure characterization if the solvent is removed from the cement paste completely, solvent exchange techniques should be used with caution for the preparation of specimens for pore structure determinations. Additional research is needed to fully understand the nature of such interactions in order to select a suitable solvent for which the interaction with the solid is minimal and that can easily penetrate the pore structure of dense cement paste systems.

2.1.3 Comparison of different water removal techniques

The temperature and pressure conditions under which each drying technique is used in cement pastes mentioned earlier and carried out are summarized for comparison in Table 2.3.

An example of the effect of different water removal techniques, oven-drying, direct freeze-drying, and vacuum-drying, on total porosity of cement pastes as they were measured by mercury intrusion porosimetry is shown in Figure 2.3 from experiments carried out by L. Gallé, 2001 [2.36]. The water removal method used for pretreatment prior to pore structure measurements can affect the results obtained. Several researchers have applied different drying techniques to test the effectiveness in water removal and/or to identify damage caused to the

Table 2.3 Comparison of the drying methods commonly used to pretreat cement-based materials before testing

Method	Temperature (°C)	Pressure (Pa)	Condition
P-drying	23	1.1	Vacuum over perchlorates
D-drying	23 (−78)	0.07	Vacuum over dry ice
Oven-drying	105	1.01×10^5	Heating in oven
Vacuum-drying	23	0.07	Vacuum (sometimes with heating)
Direct freeze-drying	−80	2.67	Submersion in liquid nitrogen
Indirect freeze-drying	−80	2.67	Submersion of vial containing the specimen in liquid nitrogen
Solvent replacement	23	1.01×10^5	Soaking in solvent for several days

Notes
Ambient temperature is at 23(±1)°C and atmospheric pressure is at 1.01×10^5 Pa. The vacuum applied can vary.

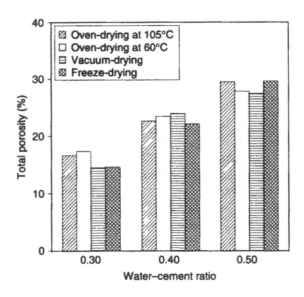

Figure 2.3 Effect of pretreatment on porosity determined by mercury intrusion porosimetry on cement pastes made from ordinary Portland cement for different water–cement ratios, after being cured for 4 months in $Ca(OH)_2$ solution.

Source: Drawn from experimental results of C. Gallé, 2001 [2.36].

pore structure [2.13,2.12]. M. Moukwa and P. Aitcin, 1988, studied the effect of oven-drying, freeze-drying, and freeze-drying followed by 24 hour oven-drying at 105°C on cement pastes for 14 days [2.10]. Their results showed that oven-drying may affect the cement paste pore structure by opening pores in the range of 0.02–0.1 µm. In the freeze-drying procedure, the samples were immersed in freon at −196°C, prior to immersion in liquid nitrogen as suggested by J. Gillott, 1976 [2.9]. Based on their observations, they concluded that the solvent replacement technique preserves the pores in the finest pore size region better than the freeze-drying technique. The solvent replacement technique and the freeze-drying technique caused less damage to the microstructure compared to oven-drying. In addition, the freeze-drying technique is less time-consuming than the solvent replacement technique. They recommended choosing the drying technique based on the property that needs to be tested. For example, the solvent replacement technique is suitable for drying when the examined properties are related to the fine pores (e.g. shrinkage). For the evaluation of properties such as strength and permeability, which are related mostly to larger pores, the less time-consuming freeze-drying technique should be sufficient.

A. Kumar and D. Roy, 1986, suggested the "loss factor" as an estimate of the effectiveness of the freeze-drying method in removing pore water [2.12]. After freeze-drying the mass of the specimen is recorded. Then the specimen is oven-dried and the mass of the specimen after oven-drying is also recorded. The mass loss due to oven-drying is expressed by Equation (2.1) as a percentage of the mass after oven-drying, and is called the loss factor. The smaller the loss factor, the more efficient the freeze-drying is. Even though the researchers proposed the loss factor for freeze-drying, the parameter could be used to test the effectiveness of other water-removal techniques:

$$LF = \frac{m_{f\text{-}d} - m_{OD}}{m_{OD}} \times 100 \tag{2.1}$$

where LF is the loss factor (%), $m_{f\text{-}d}$ is the mass of specimen after freeze-drying (g), m_{OD} is the mass of specimen after oven-drying (g).

It is obvious that the selection of the appropriate technique used for water removal depends on a number of factors, such as the time available, the property of interest of the material examined, the equipment available, etc. It is up to the needs and experience of the researcher to use the technique most suitable for the purpose of the research. Additional information on the effect of pretreatment techniques on the pore structure parameters as determined by various techniques is given in the chapters describing the techniques used.

2.2 Preparation for microscopy

Specimen preparation for analysis using microscopical techniques is very important because it facilitates examination and interpretation of microstructural features. Improper preparation methods may obscure features and even create artifacts that may be misinterpreted. Analysis using Optical Microscopy (OM) or SEM requires a highly polished surface for optimum imaging. Rough-textured surfaces, such as those produced using only saw-cutting, diminish the image quality by reducing contrast and by loss of feature definition. Additionally, the lack of a well-polished specimen makes quantitative estimates arduous, because the surface is no longer planar (see also Chapter 8).

The specimens used for microscopical observation can have polished or fracture surfaces or be thin sections. Polished and fracture surfaces are used for reflected light microscopy, and thin sections are used for transmitted light microscopy. ASTM Standard C 856 outlines the practical procedures for preparing samples, making petrographic observations, and interpreting the observations made.

2.2.1   Polished surface

A two-dimensional image can provide information that may be used to quantify the microstructure and pore structure of a cement-based material. A polished surface is a plane surface that contains a proportionate amount of each of the ingredients in the composite material. The quality of the polish will affect the image: a low quality polish can produce pitting and scarring that are interpreted as porous regions introducing in this way error into the analysis. A polished surface is used mostly for OM in order to carry out air void analysis (see Chapter 8). The cutting, grinding, and polishing steps are common to all concrete specimen preparations in order to expose a fresh surface for microscopic observation. The observation of finer pores usually involves the intrusion and hardening of an epoxy resin.

To minimize structural damage in the section plane, and to prevent the pores from being clogged with detritus, it is advisable before grinding to fill all open pores with a suitable epoxy resin, introduced by vacuum impregnation and polymerized in-situ. Contrast during microscopical analysis may be enhanced by adding a colored, or fluorescent, dye to the resin, or by using a fusible metal or alloy. These steps are discussed briefly in the following sections.

2.2.1.1   Cutting, grinding, and polishing

The specimen is cut to a suitable size using a diamond saw lubricated with water, propylene glycol, or isopropyl alcohol in order to prevent the specimen from drying while cutting. Then the specimen surface to be observed follows several grinding steps in order to remove the marks of the saw blade and any damage to the surface that might have been introduced during the cutting process. The details of grinding and polishing methods have been well described in the technical literature [2.37]. For cement-based materials all grinding and polishing must be carried out using non-aqueous fluids to avoid leaching of $Ca(OH)_2$ from the cement paste matrix, and by experienced technicians.

Grinding is an important process in order to obtain a flat surface so that information is obtained about the composition of the specimen and its phases. Abrasive papers, usually of silicon carbide of 320, 400, and 600 grits* are

---

* Grit size is the nominal size of abrasive particles corresponding to the number of openings per inch in a screen through which the particles can just pass. Usually more than 600 grits (particle size of 16 μm) is categorized as ultra-fine, 240–360 grits (53.5–28.8 μm) as very fine, 180–220 grits (78.0–66.0 μm) as fine, 100–150 grits (141–93 μm) as medium, and 50–80 grits (351–192 μm) as coarse [2.38].

suitable for rapid removal of material by grinding [2.2]. After the 600 grit grind, the surface is even enough for subsequent grinding with finer papers or with diamond pastes. Grinding striations on the specimen surface indicate that grit has completely removed a layer of material. To ensure that the entire surface has been ground equally, one can alternate grinding directions by 90°. Each grinding step, while producing damage itself, must remove the damage from the previous step. The depth of damage and the material removal rate decreases as the abrasive size increases [2.37]. Between steps, the specimens should be cleaned in an ultrasonic bath of isopropyl alcohol for about 30 seconds to ensure that all contaminants are removed before moving to a finer grit.

Polishing removes the damage imparted by the sawing and grinding operations. Polishing is the final step (or steps) used to produce a deforation-free surface, which is flat, scratch-free, and mirror-like in appearance. The polishing stage involves use of a sequence of diamond polishing pastes with successively finer particle sizes ranging from 6 to 0.25 μm, and a lap wheel covered with a low-relief polishing cloth [2.2]. When continued polishing ceases to improve the surface, the specimen should be thoroughly cleaned by ultrasonic cleaning before moving to the next smaller grit size. Subsequent polishing with finer diamond pastes removes fine scratches and improves definition of constituent elements on the concrete surface.

The quality of the polish will affect the image. A low quality polish can produce pitting and scarring that is visible as dark regions in the image. Such flaws might be interpreted as porous regions introducing error into the analysis.

There is a distinction between *finely ground* and *polished* surfaces [2.37]. Usually a surface produced, when the smallest grinding medium used is about 5 μm, is referred to as finely ground. When the final grinding medium is submicron in size, the surface is referred to as polished.

2.2.1.2 *Impregnation by epoxy resin*

The process of impregnating the pores of cement-based materials with low-viscosity epoxy resin is a common specimen preparation method for microscopic evaluation of total porosity and pore size distribution. Epoxy-resin impregnation is usually applied after the cutting and before the grinding and polishing processes. Epoxy impregnation of the pore system serves two purposes: (a) it fills the voids and upon cutting supports the microstructure restraining it against shrinkage cracking, and (b) when containing a dye, it enhances contrast between the pores and the cement paste (see Figure 2.4). For cement pastes having high permeability, an epoxy of low viscosity is necessary, while for the less permeable cement pastes and concretes, an ultra low-viscosity epoxy aids in rapid infiltration into the pore structure [2.39]. Cement pastes, mortars, and concretes may be impregnated in two ways: by dry vacuum impregnation and by solvent replacement. Dry vacuum impregnation is used when the specimen has been oven-dried before being impregnated; this happens when cracking due to drying shrinkage is not of concern, or when rapid preparation is needed. Solvent replacement is used to prepare a polished

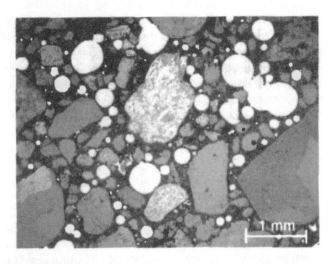

Figure 2.4 Photograph of a thin section of air-entrained concrete using polarized
transmitted light. The concrete has been impregnated with epoxy containing
fluorescent dye (white areas) in order to enhance the contrast of the
air voids with the concrete matrix. Photograph obtained at magnification
25×.

Source: Picture courtesy of Dr Joe Larbi, TNO Building and Construction Research, Delft,
The Netherlands.

section when the material has not been oven-dried before impregnation, and
therefore has not undergone any drying shrinkage.

Dry vacuum impregnation   Dry vacuum impregnation involves taking a
sawn section or block of concrete and oven-drying the specimen at low temperature, usually at 30–50°C to remove as much moisture as possible with
minimal damage to the microstructure. Removal of water is necessary as it can
interfere with intrusion and hardening of the epoxy resin. Filling the pores by
epoxy resin is generally achieved using vacuum impregnation. The dried specimen is immersed in epoxy solution while under vacuum, then brought to atmospheric pressure while still immersed in the epoxy resin. Upon release of the
vacuum, the epoxy is forced into the pore system. The epoxy resin is cured at low
temperature (65°C) and then the specimen is ready for grinding and polishing.
Dry vacuum impregnation is the most widely adopted method for preparing
cement paste, mortar, and concrete specimens for microscopic investigation.
    Epoxies containing fluorescent or colored dyes are used to enhance optical
contrast between the pores and the cement paste matrix, and to emphasize certain
features such as cracks and voids under the optical microscope. The fluorescent
dye is added to the epoxy at a quantity of 0.0275% by mass of epoxy. The technique can be applied to impregnate thick sections, as well as standard thin sections. For thin sections, the entire thickness of the specimen must be impregnated
with epoxy, as it is described in Section 2.2.2. The strong contrast between fluo-
rescing epoxy and nonfluorescing mineral grains makes porosity highly visible.

Fluorescent liquid replacement The procedure of epoxy impregnation containing fluorescent dye after alcohol replacement has been described by L. Struble and P. Stutzman, 1989 [2.39]. Specimens can be immersed in ethanol at a slightly elevated temperature, e.g. 60°C, until replacement of pore solution by ethanol is complete. Carrying out the replacement of pore water by ethanol at 60°C rather than at room temperature (22°C) reduces the time required to replace the pore solution by ethanol, which in a typical concrete specimen can range from 2 to 8 days. After the solvent replacement, the specimen is placed in vacuum for several hours to remove the ethanol. While still under vacuum, a low-viscosity epoxy, also laced with fluorescent dye, is poured over the specimen. The surface of the specimen is then covered with a thin coat of fluorescent epoxy. After the specimen is submerged in epoxy, the vacuum is broken and atmospheric pressure forces the epoxy into the pores. The epoxy solution is replaced every 60 hours with a fresh mixture. It is important to keep the epoxy at low temperatures to facilitate diffusion of epoxy by preventing the gradual polymerization and consequent increase in viscosity that occurs at room temperature. Finally, specimens are immersed in fresh epoxy solution and hardened by heating at 60°C for approximately 24 hours (this is considered to be an optimum temperature for low dehydration of hydrated cement paste and high curing of epoxy) [2.39]. In the case that the specimen is to be used as a plane section, the preparation is completed by a slight grinding to remove excess epoxy from the specimen surface.

H. Gran, 1995, proposed the use of a technique called Fluorescent Liquid Replacement (FLR) that makes it possible to do plane section and thin section analysis on concretes of very low water–cement ratios [2.40]. The technique was adapted from a similar procedure followed for impregnating biological samples. The specimens do not need to be oven-dried at 30–50°C before the impregnation process; instead, the specimen is immersed in a solution of alcohol (ethanol) and fluorescent dye for 12 hours to 4 days depending on the thickness and porosity of the specimen. The concentration of fluorescent dye in the solution is 1% by mass of solvent. The ethanol replaces some of the pore water in the cement paste as it was described in Section 2.1.2, which reduces the tendency of cracking in concrete during the subsequent drying procedure and serves to carry fluorescent dye into the pore solution of the specimen. Then the alcohol is replaced by an ultra-low viscosity (ULV) epoxy solution and finally the epoxy is cured. The ULV epoxy solution is miscible with the alcohol, so there is no need for a transitional solvent such as polypropylene glycol.

A problem with this technique is that large air voids may not be filled with fluorescent material during impregnation and thus these voids will be difficult to detect with the microscope. The problem can be easily remedied by dripping fluorescent epoxy over the surface and removing the excess material before final grinding and polishing [2.40]. An important point and source of error is the solubility of dye in water. Differences in the solubility of dye in water and alcohol may lead to changes in the relative amount of dye within the cement paste both in time and in space. This means that theoretically it should be possible to distinguish the alcohol front by the color intensity of the dye in the cement paste. In his studies, Gran found that the dye he used had a very low-exchange rate compared to the exchange rate of water and ethanol. However, this is a point that needs attention when applying the method.

Another very important aspect in thin section and plane section analysis is the penetration capability of the fluorescent epoxy. Compared to the traditional epoxy impregnation, the FLR technique has been shown to provide a significant increase in impregnation capability in concrete. It also greatly reduces the risk of introducing cracks and damage to the pore structure during impregnation, compared to the dry impregnation. Further research by E. Hansen and H. Gran, 2002, studied the fate of the fluorescent dye during the exchange process. Of particular importance is whether the dye is inert or if it may react with the inner/outer surface of the cement paste [2.41]. Their studies showed that both the dye and the ethanol are chemically inert, i.e. they do not react chemically with the cement paste. The researchers found that the exchange rate of water and ethanol are independent of the presence of dye, provided that the volume of the surrounding solution is large compared to the pore volume of the cement paste.

2.2.2   Thin sections

Thin sections have been used by geologists and petrographers to study rocks and minerals for more than 150 years. Thin sections were first used in concrete in Denmark in the 1950s to investigate alkali–silica reaction. One of the major advantages of thin sections is that they provide an opportunity to identify the individual mineral constituents, their morphology, and their amounts. Thin sections generally allow more detailed examinations to be made, particularly in conjunction with polarized light and/or fluorescent epoxy impregnation. Thin sections are also much more stable compared to polished specimens, and therefore, are more suitable for long-term storage.

Thin sections of cement paste, mortar, and concrete may be prepared in various ways. Concrete thin sections can be particularly problematic in their manufacture, because of the frail and brittle nature of the concrete microstructure when reduced to fine dimensions. The exact size of thin sections varies, but e.g. for concrete with 20 mm nominal size of coarse aggregate, the section should have a surface area of at least 65 mm × 45 mm. This means that a thin section is large compared to sand, cement paste pores, air voids, and microcracks in concrete, but small compared to coarse aggregate and large cracks. The thickness required for analysis is approximately 20–30 μm for standard thin sections of concrete or cementitious materials. The thickness chosen for the final section will depend on the problem being investigated and the features of interest. A concrete thin section must be prepared with greater care than what is required for most rock thin sections.

For thin section study, the typical range of magnification used by petrographers is 2.5× to 600× and is determined by the objective lens and the ocular of the microscope (see also Chapter 8). For a general examination of a thin section, usually a magnification of 25× is used, while a magnification of 65× is used for a more detailed examination. Thin sections are the most versatile type of microscope specimen, since they can be used for both transmitted and reflected light microscopy.

A thin section consists of a thin slice of concrete usually impregnated with fluorescent epoxy, glued to an object glass and protected by a cover glass. Fluorescent thin sections are prepared by first impregnating the specimen with a dyed epoxy resin, which fills the pores, and then grinding and polishing a slice of this specimen to its final thickness. Under the petrographic microscope, porosity is identified by its coloration. In general, the procedure followed for the making of a thin section is as follows (see also Figure 2.5). The specimen is sectioned with a fine blade cut-off saw to a suitable size and shape of approximately 10 mm thickness and is dried to remove as much moisture as possible with minimal damage to the microstructure. The specimen is then

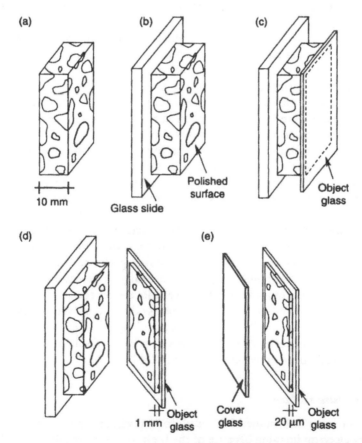

Figure 2.5 Steps followed to construct a thin section from a block of concrete. (a) A concrete specimen is used that has been impregnated with fluorescent epoxy and is finely polished. (b) The concrete sample is glued to a glass slide. (c) An object glass is glued on the finely polished surface of concrete. (d) The specimen is cut approximately 1 mm above the object glass. (e) The concrete slice on the object glass is ground down to a thickness of 20 μm, and finally it is protected by placing a cover glass on it, forming a "sandwich."

impregnated with a cold-setting resin under vacuum, which helps to support microstructure during later grinding and polishing stages. As it was already mentioned in the previous section, it is helpful if the resin is colored or contains a fluorescent dye, usually yellow, which makes porosity and microcracks visible under the microscope. After hardening of the epoxy, one face of the specimen is re-sectioned to remove excess resin, is glued on a window glass to facilitate handling, and is ground and polished to prepare for bonding to an object glass. The concrete specimen is then glued to the object glass, and is subsequently cut approximately 1 mm above the object glass. The concrete slice bonded to the object glass is polished again with successively finer particle sizes down to a thickness of 20 μm. Finally, a thin cover glass is glued to the thin section, forming a sandwich. At a thickness of about 20 μm, the concrete is translucent and semi-transparent. It should be noted though that thin sections prepared for SEM analysis that requires conductive coating, cannot have a cover glass.

D. Bager and E. Sellevold, 1979, proposed ethanol replacement treatment instead of drying during thin section preparation to reduce cracking during drying [2.42]. However, the ethanol treatment does not prevent visible cracks from developing during drying. Cement paste disks soaked in ethanol for 24 hours and then dried in room conditions did show visible cracks, even though the cracks were less extensive than those developed in control specimens dried from a wet state without ethanol replacement.

For thin sections, epoxy and fluorescent dye penetration through the whole thickness of the specimen is required; this is normally only achieved through application of a vacuum to assist epoxy impregnation. The problems with insufficient impregnation and possible introduction of microcracks during preparation can be solved by using the fluorescent liquid replacement technique described previously, instead of the vacuum impregnation technique with a fluorescent epoxy [2.40]. A simple method has also been proposed for fluorescent dye application and involves staining the surface of an epoxy-impregnated thin section with a dye. A film of dye is spread onto the surface of a clean, polished thin section. The dye adheres to epoxy-filled pores and after an appropriate amount of time, the excess dye is removed by washing the thin section with alcohol. The dye is adsorbed by most epoxies and, unlike previous fluorescent techniques, this method may be applied to thin sections that have already been epoxy-impregnated [2.43].

2.2.3   Fracture surface

Fracture surfaces are sometimes used for specimens to be examined by secondary electron imaging because of the high degree of resolution and depth of field obtained with this imaging technique. Polished surfaces provide a two-dimensional view of the microstructure but fracture surfaces expose the microstructural features unaltered. Fracture surfaces also have the advantage that they do not require further treatment other than drying the specimen and coating it with a conductive layer. Thus, whatever is seen on the fracture surface is not altered by other treatments. It should however be kept in mind that fracture surfaces do not represent the material as a whole, but rather its weakest surface: a fracture surface will not contain a proportional amount of each of the ingredients in the concrete, as the fracture will most likely occur through the cement paste matrix and very little aggregate will be fractured. Therefore, even though it is useful for qualitative observations of the pore structure, it is considered of rather limited use for quantitative analysis.

2.2.4 Intrusion alloys

Electron microscopy observations of porosity, and in particular backscattered electron imaging, can give a two-dimensional indication of the physical distribution of pores larger than 0.5 μm, but the pore contrast may be difficult to separate unambiguously from the contrast of cement phases. In an attempt to address this problem, a technique has been developed that employs the principles of mercury intrusion porosimetry (described in Chapter 3) but is capable of utilizing modern microscopy techniques to eliminate some of the errors associated with mercury intrusion porosimetry. This technique was introduced by F. Dullien and G. Dhawan, 1974, to characterize the porosity of sandstones [2.44]. The method includes first penetrating the pore network with a fusible metal using a procedure similar to the mercury intrusion. Then the specimen is cooled so that the metal solidifies and is sectioned for microscopic observations. The principles of quantitative stereology are used to relate the two-dimensional information to the size distribution of the three-dimensional pore network (see also Chapter 8). The most commonly used fusible alloys for cement-based materials are the alloys commonly referred to as Wood's metal and Newton's metal, and the metal gallium. The composition of these two alloys is shown in Table 2.4. It should be noted that the compositions can vary slightly depending on the manufacturer.

Under the name *fusible metal* or fusible alloy, is understood a metal or mixture of metals which becomes liquid at temperatures at or below the boiling point of water. Fusible alloys contain cadmium, which like bismuth has the valuable property of lowering the melting point of many alloys.

Newton's metal has a melting point of 95°C. Wood's metal has a slightly lower melting point, between 66°C and 72°C [2.47]. Wood's metal belongs in the family of cerro alloys, which are low melting fusible alloys consisting of varying combinations of bismuth, lead, tin, cadmium, and indium. Some of the

Table 2.4 Composition of two fusible alloys used in concrete science, Newton's metal [2.46] and Wood's metal [2.45]

Alloy	Composition (% by mass)			
	Bismuth	Lead	Tin	Cadmium
Newton's metal	50.0	31.2	18.8	0
Wood's metal	50.0	26.7	13.3	10.0

common grades of cerro alloys available commercially are cerrolow, cerrosafe, cerroflow, cerromatrix, cerroseal, cerrobase, etc. Wood's metal has a density of $9.7 g/cm^3$ and is insoluble in water. The contact angle for Wood's metal in contact with mortar has been reported to be 130°, a value commonly used for mercury [2.48].

Intrusion of the fusible metal in cement paste pores occurs in a similar way as with mercury intrusion (see Chapter 3) after maintaining the system above the alloy's melting temperature to ensure it is liquid. The cement paste specimen needs to be oven-dried first. The specimen remains under pressure and at high temperature for 15 minutes, in order to ensure that the metal has sufficient time to intrude. Then the heat is turned off and the vessel is allowed to cool to room temperature with the applied pressure still in the chamber. After solidification, the specimen is removed and, using appropriate procedures, it is sectioned, polished, and used as described earlier in this chapter using diamond pastes. In the photomicrographs, the pores containing the Wood's metal appear as white features in a black matrix, which is the cement paste. A Wood's metal porosimeter and the entire procedure of Wood's metal porosimetry have been described in detail in the technical literature [2.49].

At first A. Rahman, 1984, and later other researchers, used Wood's metal intrusion and subsequent microscopical observation (also mentioned as Wood's metal intrusion porosimetry – WMIP) on cement-based materials [2.48,2.50,2.51]. K. Willis *et al.*, 1998, used the technique to compare porosity results obtained by image analysis to mercury intrusion porosimetry [2.48]. The technique can also be used to enable a qualitative understanding of the mechanisms and processes occurring during the intrusion process. I. Richardson *et al.*, 1989, preferred the use of Newton's metal, instead of Wood's metal, for observations in an electron microscope, because Newton's metal does not contain cadmium [2.52]. Cadmium is volatile and under the SEM vacuum can generate voids on the surface of the specimen, leading to errors.

Using Wood's metal intrusion it is also possible to obtain a "pore cast," which is a three-dimensional pore network replica [2.49]. A pore cast is obtained after the metal solidifies inside the porous material by dissolving the solid matrix with the appropriate solution and obtaining the metal solid phase.

K. Tanaka and K. Kurumisawa, 2002, used gallium as an alternative intrusion liquid because of its property of being solid at room temperature (its melting point is 29.8°C) [2.53]. Gallium can be used as an alternative intrusion liquid because it is non-wetting, and has a high contact angle and surface tension. The procedure followed is similar to the procedure followed for the intrusion of Wood's metal. Using an Electron Probe Microanalyzer, the researchers could map the distribution of solid gallium in the cement paste matrix and observe the shapes of pores of cement pastes. The gallium intrusion method is capable of indicating the shapes and locations of pores with sizes greater than $1 \mu m$. For pores smaller than $1 \mu m$, the method can roughly provide the pores' presence and distribution due to the limitation in resolution of the Electron Probe Microanalyzer. Further information on this research is presented in Chapter 8.

References

2.1 Mindess S., Young J.F., Darwin D. *Concrete*, Prentice Hall, Engelwood Cliffs, NJ, 2nd Edition, 2003.

2.2 Detwiler R.J., Powers L.J., Jakobsen U.H., Ahmed W.U., Scrivener K.L., Kjellsen K.O. "Preparing specimens for microscopy," *Concrete International*, Vol. 23, No. 11, 2001, pp. 51–58.

2.3 Copeland L.E., Hayes J.C. "Determination of non-evaporable water in hardened Portland-cement paste," *ASTM Bulletin*, Vol. 194, December 1953, pp. 70–74.

2.4 Brunauer S. "Tobermorite gel: the heart of concrete," *American Scientist*, Vol. 50, 1962, pp. 210–229.

2.5 Powers T.C., Brownyard T.L. "Studies of the physical properties of hardened Portland cement paste," *Research Bulletin 22*, Portland Cement Association, Chicago, IL, March 1948, Reprint from *Journal of the American Concrete Institute*, October 1946–April 1947, in *Proceedings*, Vol. 43, 1947.

2.6 Thomas J.J., Jennings H.M., Allen A.J. "Determination of the neutron scattering contrast of hydrated Portland cement paste using H_2O/D_2O exchange," *Advanced Cement Based Materials*, Vol. 7, Nos 3–4, 1998, pp. 119–122.

2.7 Taylor H.F.W. *Cement Chemistry*, 2nd Edition, Thomas Telford Publishing, London, 1997.

2.8 Konecny L., Naqvi S.J. "The effect of different drying techniques on pore size distribution of blended cement mortars," *Cement and Concrete Research*, Vol. 23, No. 5, 1993, pp. 1223–1228.

2.9 Gillott J.E. "Importance of specimen preparation in microscopy," Soil Specimen Preparation for Laboratory Testing, ASTM STP 599, American Society for Testing and Materials, West Conshohocken, PA, 1976, pp. 289–307.

2.10 Moukwa M., Aitcin P.C. "The effect of drying on cement pastes," *Cement and Concrete Research*, Vol. 18, No. 5, 1988, pp. 745–752.

2.11 Day R.L., Marsh B.K. "Measurement of porosity in blended cement pastes," *Cement and Concrete Research*, Vol. 18, No. 1, 1988, pp. 63–73.

2.12 Kumar A., Roy D.M. "The effect of desiccation on the porosity and pore structure of freeze dried hardened Portland cement and slag-blended pastes," *Cement and Concrete Research*, Vol. 16, No. 1, 1986, pp. 71–78.

2.13 Zhang L., Glasser F.P. "Critical examination of drying damage to cement pastes," *Advances in Cement Research*, Vol. 12, No. 2, 2000, pp. 79–88.

2.14 Anderson T.F. "The use of critical point phenomena in preparing specimens for the electron microscope," *Journal of Applied Physics*, Vol. 21, No. 7, 1950, p. 724.

2.15 Anderson T.F. "Techniques for the preservation of 3-dimensional structure in preparing specimens for the electron microscope," *Transactions of the New York Academy of Sciences*, Vol. 13, No. 4, 1951, pp. 130–134.

2.16 Hughes D.C., Crossley N.L. "Pore structure characterization of GGBS/OPC Grouts using solvent techniques," *Cement and Concrete Research*, Vol. 24, No. 7, 1994, pp. 1255–1266.

2.17 Parrott L.J. "Thermogravimetric and sorption studies of methanol exchange in an alite paste," *Cement and Concrete Research*, Vol. 13, No. 1, 1983, pp. 18–22.

2.18 Parrott L.J. "An examination of two methods for studying diffusion kinetics in hydrated cements," *Materials and Structures*, Vol. 17, 1981, pp. 131–137.

2.19 Feldman R.F. "Diffusion measurements in cement paste by water replacement using propan-2-ol," *Cement and Concrete Research*, Vol. 17, No. 4, 1987, pp. 602–612.

2.20 Lide D.R. *CRC Handbook of Chemistry and Physics*, 84th Edition, CRC Press, Boca Raton, FL, 2003–2004.

2.21 Parrott L.J. "Effect of drying history upon the exchange of pore water with methanol and upon subsequent methanol sorption behaviour in hydrated alite paste," *Cement and Concrete Research*, Vol. 11, No. 5–6, 1981, pp. 651–658.

2.22   Gran H.C., Hansen E.W. "Exchange rates of ethanol with water in water-saturated cement pastes probed by NMR," *Advanced Cement Based Materials*, Vol. 8, 1998, pp. 108–117.

2.23   Day R.L. "Reactions between methanol and Portland cement paste," *Cement and Concrete Research*, Vol. 11, No. 3, 1981, pp. 341–349.

2.24   Litvan G.G. "Variability of the nitrogen surface area of hydrated cement paste," *Cement and Concrete Research*, Vol. 6, No. 1, 1976, pp. 139–143.

2.25   Mikhail R.Sh., Selim S.A. "Adsorption of organic vapors in relation to the pore structure of hardened Portland cement pastes," *Special Report 90: Symposium on Structure of Portland Cement Paste and Concrete*, Highway Research Board, National Research Council, Washington, DC, 1966, pp. 123–134.

2.26   Hughes, D.C. "The use of solvent exchange to monitor diffusion characteristics of cement pastes containing silica fume," *Cement and Concrete Research*, Vol. 18, No. 2, 1988, pp. 321–324.

2.27   Monick J.A. "Alcohols: their chemistry, properties and manufacture," Reinhold Book, New York, 1968.

2.28   Pratt P.L. "Physical methods for the identification of microstructures," *Materials and Structures*, Vol. 21, 1988, pp. 106–117.

2.29   Beaudoin J.J. "Validity of using methanol for studying the microstructure of cement paste," *Materials and Structures*, Vol. 20, No. 115, 1987, pp. 27–31.

2.30   Taylor H.F.W., Turner A.B. "Reactions of tricalcium silicate paste with organic liquids," *Cement and Concrete Research*, Vol. 17, No. 4, 1987, pp. 613–623.

2.31   Beaudoin J.J., Gu P., Marchand J., Myers R.E., Liu Z. "Solvent replacement studies of hydrated Portland cement systems – the role of calcium hydroxide," *Advanced Cement Based Materials*, Vol. 8, 1998, pp. 56–65.

2.32   Feldman R.F., Beaudoin J.J. "Pretreatment of hardened cement pastes for mercury intrusion measurements," *Cement and Concrete Research*, Vol. 21, Nos 2–3, 1991, pp. 297–308.

2.33   Chandra S., Flodin P. "A discussion on the paper, 'Reactions between methanol and Portland cement paste' by R.L. Day (Vol. 11, 1981, pp. 341–349)," *Cement and Concrete Research*, Vol. 12, No. 2, pp. 261–262.

2.34   Thomas M.D.A. "The suitability of solvent exchange techniques for studying the pore structure of hardened cement paste," *Advances in Cement Research*, Vol. 2, No. 5, 1989, pp. 29–34.

2.35   Sereda P.J., Feldman R.F., Ramachandran V.S. "Structure formation and development in hardened cement pastes," Sub-Theme VI-1 in *Proceedings of Seventh International Congress on Chemistry of Cement*, 1980, pp. VI-1/3–VI-1/44.

2.36   Gallé C. "Effect of drying on cement-based materials pore structure as identified by mercury intrusion porosimetry: a comparative study between oven-, vacuum-, and freeze-drying," *Cement and Concrete Research*, Vol. 31, No. 10, 2001, pp. 1467–1477.

2.37   St John D.A., Poole A.B., Sims I. *Concrete Petrography: A Handbook of Investigative Techniques*, Butterworth-Heinemann, London, 1998.

2.38   [Abrasives info, Coated Abrasive Fabricators Association, Coated Division of the Unified Abrasive Manufacturers' Association.]

2.39   Struble L., Stutzman P.E. "Epoxy impregnation of hardened cement for microstructural characterization," *Journal of Materials Science Letters*, Vol. 8, No. 6, 1989, pp. 632–634.

2.40   Gran H.C. "Fluorescent liquid replacement technique. A means of crack detection and water:binder ratio determination in high strength concretes," *Cement and Concrete Research*, Vol. 25, No. 5, 1995, pp. 1063–1074.

2.41   Hansen E.W., Gran H.C. "FLR technique exchange kinetics of ethanol/fluorescent dye with water in water-saturated cement paste examined by ^{1}H- and ^{2}H-NMR," *Cement and Concrete Research*, Vol. 32, No. 5, 2002, pp. 795–801.

2.42 Bager D.H., Sellevold E.J. "How to prepare polished cement product surfaces for optical microscopy without introducing visible cracks," *Cement and Concrete Research*, Vol. 9, No. 5, 1979, pp. 653–654.

2.43 Cather M.E., Morrow N.R., Klich I. "Characterization of porosity and pore quality in sedimentary rocks," in *Proceedings of the IUPAC Symposium, on Studies in Surface Science and Catalysis, Vol. 62: Characterization of Porous Solids II*, Alicante, Spain, 6–9 May 1990, F. Rodriguez-Reinoso, J. Rouquerol, K.S. Sing, K.K. Unger (eds), Elsevier Science Publishers, Amsterdam, The Netherlands, 1991, pp. 727–736.

2.44 Dullien F.A.L., Dhawan G.K. "Characterization of pore structure by a combination of quantitative photomicrography and mercury porosimetry," *Journal of Colloid and Interface Science*, Vol. 47, No. 2, 1974, pp. 337–349.

2.45 J.T. Baker Inc. Material Safety Data Sheet No. W3500, "Wood's alloy, sticks," 2003.

2.46 Kamal M., El-Bediwi A. B. "Structure, mechanical metallurgy and electrical transport properties of rapidly solidified $Pb_{50}Sn_{50-x}Bi_x$ Alloys," *Journal of Materials Science: Materials in Electronics*, Vol. 11, No. 6, 2000, pp. 519–523.

2.47 Hopkins A.A. (ed.), *The Scientific American Cyclopedia of Formulas*, Scientific American Publishing Co., Inc., New York, 1910.

2.48 Willis K.L., Abell A.B., Lange D.A. "Image-based characterization of cement pore structure using Wood's metal intrusion," *Cement and Concrete Research*, Vol. 28, No. 12, 1998, pp. 1695–1705.

2.49 Dullien F.A.L. "Wood's metal porosimetry and its relation to mercury porosimetry," *Powder Technology*, Vol. 29, No. 1, 1981, pp. 109–116.

2.50 Rahman A.A. "Characterization of the porosity of hydrated cement pastes," *Proceedings of the British Ceramic Society*, Vol. 35, 1984, pp. 249–266.

2.51 Scrivener K.L., Nemati K.M. "The percolation of pore space in the cement paste/aggregate interfacial zone of concrete," *Cement and Concrete Research*, Vol. 26, No. 1, 1996, pp. 35–40.

2.52 Richardson I.G., Groves G.W., Rodger S.A. "The porosity and pore structure of hydrated cement pastes as revealed by electron microscopy techniques," in *Symposium Proceedings Vol. 137: Pore Structure and Permeability of Cementitious Materials*, L.R. Roberts, J.P. Skalny (eds), Materials Research Society, Warrendale, PA, 1989, pp. 313–318.

2.53 Tanaka K., Kurumisawa K. "Development of technique for observing pores in hardened cement paste," *Cement and Concrete Research*, Vol. 32, No. 9, 2002, pp. 1435–1441.

3 Mercury intrusion porosimetry

Mercury intrusion porosimetry has proven over several decades to be a useful technique in characterizing many porous materials, whose pore sizes range over several orders of magnitude. In 1921, E. Washburn set the theoretical foundations for the technique, and proposed an equation (mentioned in Section 3.1.3) that holds true for any non-wetting liquid in contact with a porous material [3.1]. In 1945, H. Ritter and L. Drake built a mercury intrusion porosimeter and tested a variety of porous materials by penetration of mercury into pores as small as 10 nm in radius initially, and 2 nm few years later [3.2,3.3]. L. Joyner *et al.*, 1951, calculated surface areas of assumed cylindrical pores from mercury intrusion porosimetry data [3.4]. At around 1960, hydraulic pressurized instruments became commercially available, and since then the technique has been developed and improved to the extent that it is now possible to determine a wide variety of pore structure parameters in porous materials. Such parameters include total pore volume, pore size distribution, density of solids and powders, and specific surface area of pores. In this way, mercury intrusion porosimetry has become a convenient and fast technique for pore structure characterization. The rise in popularity of mercury intrusion porosimetry for pore structure analysis has been due to the fact that the technique is applicable to a broad range of pore sizes more than any other method, since it is comparatively easy to apply a broad range of pressures. Practical advantages of mercury as an intrusion liquid include low vapor pressure, relative inertness in terms of chemical reactivity with many materials, and normally non-wetting properties for most surfaces.

Mercury Intrusion Porosimetry (MIP) has become the most widely used method for determining the pore size distribution of hardened cement pastes, because equipment is readily available, and the method is relatively easy to perform and quick: an experiment operated by an automated system can take as little as 1 hour to complete. In addition, MIP is the only available technique that is supposed to cover nearly the whole range of sizes that must be analyzed in cement pastes. The technique was introduced to concrete science by L. Edel'man *et al.*, 1961, who used MIP to determine the pore size distribution in cement pastes, and obtained cumulative pore volume curves and differential curves for different water–cement ratios [3.5]. Since then, the technique has been used extensively by researchers in order to analyze pore structure of cement pastes, study the effects of several parameters in pore structure development, and identify the pore structure parameters that relate to permeability of cement-based materials.

3.1 Theory and testing procedure

Mercury has a high surface tension and therefore non-wetting properties towards many solids. Wetting of a solid surface by a liquid occurs when the attraction between the liquid and solid molecules is greater than the attraction of liquid molecules with each other. Water has a contact angle between 20° and 30° and wets most solids. Mercury, on the contrary, has a contact angle greater than 90° with most materials, and, therefore, is a non-wetting liquid. For this reason, a drop of water on a glass surface will obtain a flattened shape, while a drop of mercury on the same surface will form a ball. An example of the contact angle between water and mercury on a flat surface and in a capillary tube is shown in Figure 3.1. The surface tension of a liquid opposes entrance into the pores of a material if the liquid has a contact angle with the material greater than 90°. This means that if a porous specimen is immersed in mercury at atmospheric pressure, mercury will not enter into the pores of the specimen.

Mercury wets certain materials such as lead, zinc, cadmium, silver, and copper. For materials that are wetted by mercury, or form amalgams, MIP may still be used, if the solid surface is coated with a polymer [3.6].

In principle, every non-wetting liquid could be used for intrusion into a porous material, but few alternatives have been investigated. The use of Wood's metal for intrusion in combination with micrography seems to be a promising technique to determine pore structure by microscopy techniques and is further discussed in Section 3.4.7 later, and in Chapter 8.

3.1.1 Instrument description

There are several different instruments commercially available for performing MIP experiments, which are commonly called *porosimeters*. Despite the variety of modifications and perfections, the concept of the first instrument developed by Ritter and Drake has been preserved. All commercial instruments have

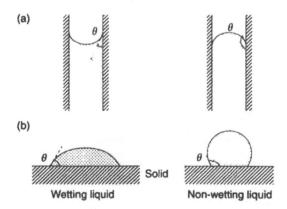

Figure 3.1 Contact angle, θ, for a wetting liquid e.g. water and a non-wetting liquid e.g. mercury (a) inside a capillary, and (b) on a flat surface. The contact angle is defined so as to include the liquid.

means of evacuating the specimen holder and the specimen it contains, and then, while under vacuum, flooding them with mercury, and apply high pressures. All gases should be removed from the system prior to a test, so that the mercury is free to penetrate into the pores without additional impediment; this means that a vacuum pump and vacuum indicator are also required. The penetration technique requires a means for generating a pressure and a means for determining how much mercury is forced into the pores or void spaces of the material being tested under a given pressure. In principle, a mercury intrusion porosimeter consists of the following parts (see also Figure 3.2) [3.7,3.8]:

- The specimen holder, which is variously called a specimen cell, a dilatometer, or most frequently, a *penetrometer*. The penetrometer is a glass capillary tube of known volume with a sealable bulbous body at one end of a long stem, which is used to hold the specimen during testing. The stem attached to the cell also holds the mercury used for filling the pores during increase of pressure while running the test.
- A high pressure vessel made of heavy-walled steel which contains the specimen cell.
- A vacuum apparatus used to remove air from the cell and the specimen, and for transferring mercury into the specimen cell upon release of vacuum.
- A pressure generator.
- A hydraulic fluid, usually oil, needed to transmit the pressure from the pressure generator to the penetrometer.
- The equipment used for monitoring the progress of the penetration of mercury into the specimen tested, by measuring the applied pressure at any stage of the experiment, either using a conventional pressure gauge or an electronic pressure transducer.

The material to be analyzed is first dried by one of the techniques described in Section 2.1, and evacuated to remove adsorbed gases and vapors. The specimen typically can range in volume from a few cm³ up to 15 cm³ depending

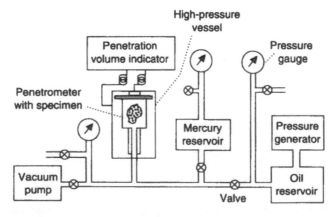

Figure 3.2 Schematic out-of-scale layout of the MIP system.

upon the capacity of the penetrometer, and should nearly fill the penetrometer bulb. For cement paste specimens, usually few broken fragments from the material are placed in the penetrometer. It is advantageous to use large specimens in order to reduce the effect of local heterogeneities of the material and to obtain results that are more representative. The effect of the specimen size on the results is further discussed in Section 3.4.2 later.

3.1.2 Testing procedure

The dried specimen to be tested is first weighed and then placed in the penetrometer. The unit is assembled and weighed prior to testing to provide known parameters for use in later calculations. The penetrometer assembly with the porous specimen is placed into the mercury-filling device, where it is evacuated. The evacuation of the specimen in the penetrometer has two functions: removing adsorbed species from the specimen, and removing air out of the penetrometer. The removal of adsorbed species is important in order to clean the surface of the solid, and decisive in obtaining the correct contact angle when mercury comes in contact with the specimen. Inadequate pumping of air from the penetrometer will leave some air in the porous material and thus affect the intrusion of mercury into the pores [3.9]. The cell is evacuated until the pressure is lower than 13.33 Pa (approximately 1.3×10^{-4} atm), and then it is pressurized.

The pressurization test is completed in two parts: the low pressure part and the high pressure part. Increases in pressure are achieved by pneumatic means in the low pressure part up to a pressure of 1 or 2 atm, and by hydraulic means in the high pressure part up to the maximum capacity of the instrument.

3.1.2.1 Low pressure

After evacuation, air is introduced into the system to raise the pressure progressively from vacuum to atmospheric pressure manually or automatically. The penetrometer bulb is filled with mercury via the stem, until the specimen is completely covered with mercury. The bulk volume of the specimen can be determined by the displaced volume of mercury by the specimen in the specimen cell, which has a known volume.

The pressure is transmitted from the far end of the capillary tube to the mercury surrounding the specimen in the specimen chamber. The pressure is increased to 1 atm and even above atmospheric pressure up to 3 or 4 atm in some instruments. The first reading of pressure and the volume of mercury penetrating into the specimen is usually taken at a pressure of 3–4 kPa (3×10^{-2} – 4×10^{-2} atm) although lower pressure readings are possible [3.8,3.10]. The minimum controllable pressure that will cause flooding of the specimen with mercury is of considerable importance; this is because, in addition to forcing mercury into the penetrometer and surrounding the specimen, the low pressure also causes the intrusion of mercury into sufficiently large pores. Thus, the value of the low "filling" pressure places an upper limit on the size of pores that can be subsequently included in the pore size distribution being determined [3.11]. Then the penetrometer is removed from the filling device, weighed again to determine the amount of mercury held, and is placed back in the porosimeter for the high pressure part.

3.1.2.2 High pressure

After initial filling, the penetrometer is separated from the supply of mercury, and the pressure on the mercury in the penetrometer is raised. The chamber around the specimen cell is filled with hydraulic fluid (oil) and then, using a hydraulic pump and a pressure intensifier, the pressure can be further increased to values up to 405 MPa (4000 atm) depending on the apparatus. Prior to intrusion, mercury resides in the capillary stem of the penetrometer. As intrusion progresses into the specimen, the thread of mercury in the stem shortens as some of the mercury enters into the specimen's pores. After each increment, the pressure is monitored and may be maintained by applying additional pressure until the system comes into equilibrium. The volume of mercury intruded into the specimen is determined for each pressure increment. The maximum achievable pressure is of great importance, because it deter-mines the smallest size of pores to be included in the measured distribution. In instruments where the maximum pressure may be as high as about 4000 atm, mercury will intrude pores having a diameter as small as 2 nm. During depres-surization, the lowering of the mercury level is automatically recorded as a function of the lowering pressure. A full analysis, which involves 50 or more separate points, can be completed in as little as 30 minutes for a specimen having mesopores (pores with sizes in the range 2–50 nm). However, it may take several hours for a specimen with small pores due to the time required for pressure equilibrium to be reached at each step. For cement paste, the time required depends upon the total porosity of the specimen and may last several hours for dense pastes.

The pressure on mercury can be increased in two ways: in incremental mode or in continuous mode. Earlier equipment operated in incremental mode, while modern porosimeters operate mainly in continuous mode.

In *incremental* mode, the pressure is increased in steps with a pause after each step so that the system is allowed to come to equilibrium, which takes between 30 seconds and 1 minute [3.11]. After each step, the intruded volume of mercury is recorded. The incremental mode offers a better assurance that true equilibrium is reached for each data point than the continuous mode, as long as the equilibration time interval is chosen sufficiently long. In order to avoid temperature effects due to compressive heating, equilibration times of 5 minutes or more may have to be selected at high pressures in the incremen-tal mode [3.10]. The incremental mode has several drawbacks, such as inability to hold the pressure constant at any given step, long analysis times, collection of relatively limited data points in order to construct pore size distribution curves, and the possibility of skipping important information in pressure ranges corresponding to the filling of narrow or multimodal pore sizes [3.12].

In *continuous* (or scanning) mode, the pressure is increased continuously at a predetermined rate and mercury intrusion is recorded continuously. In the continuous mode, the system does not come to equilibrium. The rate of pressure build-up must not be too high, because a certain time is needed to allow the transport of mercury through the porous network of the material. The continuous mode is suitable for low porosity specimens and for quality control purposes. In general, pore size distributions determined in the continuous mode will differ from the distributions obtained in the incremental mode [3.8]. The continuous mode offers the possibility to run an experiment in a very short time, in as low as 15 minutes, depending on the specimen; however, this requires a careful consideration of a variety of correction factors. A large number of data points can be recorded and even small differences between specimens can be observed.

Above pressures of about 20 MPa (200 atm), there is a slight compressibility of mercury, and therefore, the measured pore volume of the material is larger than the actual pore volume. It is obvious that the larger the volume of the specimen and the smaller the pore volume of the specimen, compared to the amount of mercury in the penetrometer, the smaller will be the error due to the compressibility of mercury. At very high pressures, reaching 400 MPa, some additional problems may appear: the temperature of the pressured system may rise appreciably, resulting into change of the volume of mercury intruded into the specimen. A way to determine high-pressure correction due to compressibility of mercury is by a *"blank" intrusion experiment*, which is performed using a penetrometer filled only with mercury. The blank test should be made in the same manner as the actual intrusion tests so that the system will experience as much as possible the same pressure and temperature changes as the actual test. The results of the blank test will either be an apparent intrusion, which means that compressibilities are dominant, an apparent extrusion, which indicates that heating is dominant, or no apparent change in volume, in which case, no correction for the instrument is required and only the specimen's compressibility needs to be considered. Several researchers have used the blank intrusion technique in cement paste testing, in order to correct for mercury compression and heating [3.13,3.14,3.15].

3.1.3   Calculation of pore size

The pressures that force mercury into the specimen pores can be converted to equivalent pore sizes using the equation derived by E. Washburn in 1921 from Young and Laplace's work on interfacial tension for the rise or depression of liquids in capillaries [3.1]. Washburn assumed that the cross section of a pore is circular and that the surface tension of mercury acts along the circle of contact equal to the perimeter of the circle. The surface tension acts to drive the non-wetting mercury out of the pore. The force developed due to interfacial tensions acts normal to the plane of the circle of contact (see Figure 3.3), and can be expressed as the product of the surface tension of mercury and the circumference of the pore, by Equation (3.1). The factor $\cos\theta$ in Equation (3.1) is introduced because the force tending to drive the mercury out of the pore in

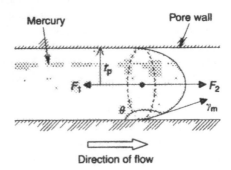

Figure 3.3 Equilibrium of forces acting inside a capillary, forcing (F_1) and opposing (F_2) mercury intrusion into the capillary. The radius of the capillary, r_p, the surface tension, γ_m, and the contact angle, θ, are also marked.

order to decrease the area being wetted acts through the contact angle θ. The negative sign enters in Equation (3.1) because for $\theta > 90°$, the term $-\cos\theta$ is intrinsically positive:

$$F_1 = -(2\pi r_p)\gamma_m \cos\theta \tag{3.1}$$

where F_1 is the force driving mercury out of the pore (N), r_p is the pore radius (m), γ_m is the surface tension between mercury and the pore wall (N/m), θ is the contact angle between mercury and the pore wall (degrees).

The opposing force due to the applied pressure that drives mercury into the pore, acts over the area of the circle of contact, and can be expressed as the product of the applied pressure P and the cross-sectional area of the pore, by Equation (3.2):

$$F_2 = (\pi r_p^2)P \tag{3.2}$$

where F_2 is the force driving mercury into the pore (N), r_p is the pore radius (m), P is the pressure applied on mercury to intrude the pore (N/m²).

At equilibrium, the opposing forces are equal, which results in Equation (3.3):

$$2\pi r_p \gamma_m \cos\theta = -P\pi r_p^2 \quad \text{or} \quad r_p = \frac{-2\gamma_m \cos\theta}{P} \tag{3.3}$$

where r_p is the pore radius (m), γ_m is the surface tension between mercury and the pore wall (N/m), θ is the contact angle between mercury and the pore wall (degrees), P is the pressure applied on mercury to intrude the pore (N/m²).

Use of Equation (3.3) presumes that the surface tension of mercury, γ_m, and the contact angle, θ, between mercury and the pore walls are independent of the applied pressure [3.16]. When the contact angle is greater than 90°, as is in the case of mercury in contact with cement paste, $\cos\theta$ is negative and pressure greater than ambient must be applied to the mercury in the reservoir containing the specimen to force mercury into the pore.

Equation (3.3) can also be derived in a different way by equating the work required to increase the area wetted by mercury to the work required to force mercury into the pore. Work equations were used by H. Rootare and C. Prenzlow, 1967, and are mentioned in Section 3.1.5 for determining the surface area [3.17].

When using Equation (3.3) to determine pore size from pressure values, attention needs to be drawn to certain parameters, such as the shape of the pore, the surface tension, and the contact angle.

The cylindrical *shape of pores* is a common assumption for pore geometry. Most researchers assume a circular cross section for the pores so that their experimental results about the pore structure become directly comparable to results obtained by other researchers and/or other materials. However, if one assumes pores of a cross section of different shape, then the shape factor (or a related parameter) mentioned in Section 1.3.1.6 must be introduced in the analysis. Several researchers have drawn attention to the shape of pores in MIP analysis and have used correction factors incorporating the shape factor to interpret better the MIP results from cement pastes [3.18,3.19,3.20]. Since the shape factor or any factor for correction of the pore shape can be defined in a variety of ways, it is essential that the way of determining the correction factor be clearly defined so that comparison with previous results is meaningful.

In cement pastes, the slit-shaped pores can be analyzed using the approach of elliptical cross section. The effect of assuming pores with elliptical cross section rather than cylindrical pores tends to shift the pore size distribution curves towards smaller pore dimensions but does not affect the total surface area, which is independent of the pore geometry. For elliptical cylinders with minor and major axes r_1 and r_2 respectively, Equation (3.3) is modified and expressed by Equation (3.4). Slit pores can be approximated with elliptical pores, in which the dimension of the major axis, r_2, is significantly greater than the dimension of the minor axis, r_1, and Equation (3.3) is modified to Equation (3.5):

$$P = -\gamma_m \cos\theta \left(\frac{1}{r_1} + \frac{1}{r_2} \right) \tag{3.4}$$

$$P = -\gamma_m \frac{\cos\theta}{r_1} \quad \text{or} \quad P = \frac{-2\gamma_m \cos\theta}{d} \tag{3.5}$$

where P is the pressure applied on mercury to intrude the pore (N/m²), γ_m is the surface tension between mercury and the pore wall (N/m²), r_1 is the minor axis of the elliptical cross section of the pore (m), r_2 is the major axis of the elliptical cross section of the pore (m), d is the width of the slit pore (m), θ is the contact angle between mercury and the pore wall (degrees).

Surface tension does not have a constant value, but is variable depending on a number of factors. The value for the surface tension of mercury equal to $\gamma_m = 0.480\,\text{N/m}$ was initially adopted by H. Ritter and L. Drake, 1945 [3.2]. Temperature has a minor effect on surface tension: e.g. at 25°C the surface tension is 0.485 N/m, while at 50°C it is calculated to be 0.479 N/m (*). Using Equation (3.3) and with all other parameters remaining the same, a difference in surface tension from 0.484 to 0.472 N/m results in a difference to the calculated pore size of about 2.5%. The choice of the surface tension value within the range of commonly accepted values has a much smaller effect on the determination of pore radius than the choice of the value for the contact angle. Slight variation in the surface tension of mercury is not of great concern with respect to errors in the determination of the pore size distribution at 25°C. A value of 0.485 N/m for surface tension is commonly accepted by most researchers.

The *contact angle* between mercury, or any other liquid, and a solid surface depends on whether the mercury is advancing or receding from the solid surface, and on the physical and chemical state of the surface itself. The contact angle between mercury and the pore walls of hardened cement paste is affected by several factors, such as the structure of the solid surface, the specimen's age, the drying method used for pretreatment, and the mercury purity [3.15]. Values of contact angle commonly used are between 130° and 140°. More information about the factors affecting the contact angle and its effect on measurements is given in Section 3.4.4.1 later.

3.1.4 Pore size distribution

Differentiation of the Washburn equation (Equation (3.3)), assuming that the surface tension and contact angle are constant, gives Equation (3.7):

$$P \, dr + r_p \, dP = 0 \tag{3.7}$$

By combining Equation (1.6) mentioned in Chapter 1 and Equation (3.7), we obtain Equation (3.8):

$$r_p \, dP = P \frac{dV}{D_V(r)} \quad \text{or} \quad D_V(r) = \frac{P}{r_p}\left(\frac{dV}{dP}\right) \tag{3.8}$$

where $D_V(r)$ is the pore size distribution function, P is the pressure applied on mercury to intrude the pore (N/m²), r_p is the pore radius (m), dV is the change in the pore volume (m³), dP is the change in the applied pressure (N/m²).

---

* The surface tensions of mercury at two different temperatures, T_1 and T_2, can be calculated from Equation (3.6).

$$\gamma_2 = \gamma_1 + (T_2 - T_1)\frac{d\gamma}{dT} \tag{3.6}$$

where γ_2 is the surface tension of mercury at temperature T_2 (N/m²), γ_1 is the surface tension of mercury at temperature T_1 (N/m²), $d\gamma/dT$ is the temperature coefficient for surface tension (N/m/K).

The surface tension of mercury is $\gamma_1 = 0.485$ N/m at a temperature $T_1 = 25$ °C, and the value of temperature coefficient is $d\gamma/dT = -0.22 \cdot 10^{-3}$ N/m/K, which is independent of temperature (see also Table 2.2) [3.21].

From Equation (3.8) it is obvious that, as the pressure applied to mercury is increased, the amount of mercury forced into the pores increases, and consequently, the corresponding pore volume intruded by mercury increases. From a plot of differential volume dV vs. pore size dr one can determine the pore size distribution. Pore size distributions are frequently biased in favor of small pore sizes due to the hysteresis effect caused by ink-bottle pores, which is discussed in more detail in Section 3.3.1.1 [3.8]. This is because the MIP does not give a true pore size distribution, but rather indicates the accessible pore volume as a function of pore size.

3.1.5 *Specific surface area*

The specific surface area can be calculated from the equation derived by H. Rootare and C. Prenzlow, 1967, directly from standard porosimeter curves of applied pressure vs. intruded volume [3.17]. For their derivations they used work equations, assuming constant surface tension and contact angle throughout the experiment. Even though the derivation of the equation by Rootare and Prenzlow for specific surface areas does not contain any assumption about pore geometry, a cylindrical pore shape is assumed for simplification of the calculations shown in the following paragraph. In addition, it is assumed that the movement of the mercury meniscus is reversible, an assumption that is not valid for interconnected or irregularly shaped pores [3.22]. Irreversible work is spent during penetration of mercury, therefore, predictions based on the following Equation (3.11) derived here tend to be too high.

The work required to increase the surface area of a cylindrical pore wetted by mercury is the product of surface area and surface tension and is given by Equation (3.9). The work required to force mercury into the cylindrical pore is the product of force and distance and is given by Equation (3.10):

$$dW_1 = -dS\,\gamma_m \cos\theta \tag{3.9}$$

$$dW_2 = P\,dV \tag{3.10}$$

where dW_1 is the work required to increase the area wetted by mercury (Nm), dS is the surface area of the pore wetted by mercury (m^2), γ_m is the surface tension between mercury and the pore wall (N/m), dW_2 is the work required to force mercury into the pore (Nm), P is the pressure applied on mercury to intrude the pore (N/m^2).

By equating the work expressions from Equations (3.9) and (3.10) with Equation (3.3), we obtain Equation (3.11):

$$dS\gamma_m \cos\theta = P\,dV \quad \text{or} \quad dS = \frac{P\,dV}{\gamma_m \cos\theta} \tag{3.11}$$

where dS is the surface area of the pore wetted by mercury (m^2), γ_m is the surface tension between mercury and the pore wall (N/m), θ is the contact angle between mercury and the pore wall (degrees), P is the pressure applied on mercury to intrude the pore (N/m^2), dV is the change in the pore volume (m^3).

The surface area of pores that are filled with mercury is given by integrating all surface areas over the range of the pore radii intruded by mercury. Therefore, by integrating Equation (3.11), Equation (3.12) is derived.

$$S = \frac{1}{\gamma_m \cos \theta} \int_0^V P \, dV \tag{3.12}$$

where S is the pore surface area (m^2), γ_m is the surface tension between mercury and the pore wall (N/m), P is the pressure applied on mercury to intrude the pore (N/m^2), dV is the change in pore volume (m^3), θ is the contact angle between mercury and the pore wall (degrees).

3.2 Plots obtained

Mercury porosimetry data is obtained by recording the volume of mercury that penetrates the porous specimen as a function of pressure. A pressurization curve is the curve of intruded volume of mercury V vs. applied pressure P and is sometimes called a porogram (see Figure 3.4a). However, a porogram cannot be used to obtain information about the pore structure parameters. The readings of intruded volume need to be corrected and can be normalized in a variety of ways, such as dividing the intruded volume by the specimen mass (resulting in units of m^3/g) or by the specimen bulk volume. In order to determine the material's porosity or density, no further interpretation is needed. However, accurate determination of the total porosity and the true density of the solid is limited by the maximum applied pressure.

The specimen's true density at any pressure is obtained by dividing the mass of the dry specimen by the specimen's volume, determined when all pores have been filled with mercury. The specimen volume can be determined from the difference between the volume of mercury filling the empty specimen cell and the volume of mercury filling the specimen cell when the specimen is present. The volume of mercury filling the empty cell is usually given by the manufacturer or can be obtained by a "blank" test as it was noted previously. The volume of mercury filling the cell when the specimen is present is determined by weighing the cell including both the specimen and mercury and back-calculating to any desired pressure condition [3.13].

Pore structure parameters of a material can be determined from various plots that can be obtained using MIP. Such plots are the following:

- cumulative intrusion volume vs. pressure (or vs. diameter);
- percent intrusion volume vs. pressure (or vs. diameter);
- cumulative porosity curve;
- incremental intrusion volume vs. diameter;
- differential intrusion volume vs. diameter;
- logarithmic differential volume vs. diameter;
- incremental surface area vs. diameter;
- differential surface area vs. pressure (or vs. diameter);
- logarithmic differential pore area vs. diameter.

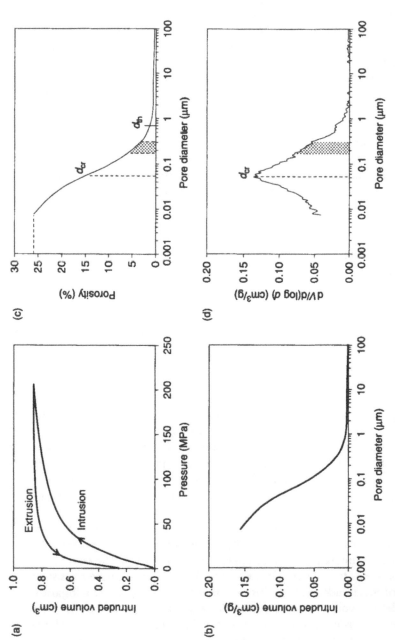

Figure 3.4 Plots used most frequently in MIP to report experimental results. (a) Pressurization curve, also mentioned as porogram; (b) cumulative intruded volume curve; (c) cumulative intrusion curve giving the cumulative pore size distribution; (d) differential pore size distribution. Curves in b, c, and d are based on intrusion data only. The total porosity is marked at 26% in the cumulative intrusion curve. The critical diameter, d_{cr}, and the threshold diameter, d_{th}, are also noted in the cumulative intrusion curve. The shadowed area in the differential pore size distribution represents the total volume within the specific range of pores.

On the same graphs showing the intrusion data, the plots obtained using the extrusion data can also be presented. Values of pore radius corresponding to the specific value of pressure P at any time during the experiment are calculated using Equation (3.3) or in modern automated equipment are read directly from the experimental plot of data. Sometimes the pore diameter (or equivalent pore diameter of a cylindrical pore) is used during plotting instead of the pore radius.

The volumes of the intruded pores are accumulated starting with pores having the largest diameter measured up to pores with the smallest diameter, in the order in which the pores are intruded by mercury. Pore size should not be understood as the cavity size, but rather the larger entrance or throat opening towards the pore intruded. Pore size distribution information is most often obtained from a cumulative plot of pore volume vs. pore size (either diameter or radius after assuming a cross section). Usually results are plotted having a dual x-axis, indicating both the pressure and the pore size. Different plots are in common use and the individual experimenter must select the one most useful to his/her purposes.

Not all plots mentioned previously are used to present MIP results. The most often used plots are shown in Figure 3.4 and are described in the following sections.

3.2.1   Cumulative intrusion curves

On a cumulative intrusion curve, there are several choices for the pore volume axis. In common plots, the ordinate includes cumulative intruded volume per unit specimen mass, percent of total intruded volume, or intruded volume per unit bulk specimen volume [3.11]. When the ordinate expresses percentage of total pore volume, an extra measurement is required for independent assessment of the specimen's total pore space. Calculation of the percent porosity requires a value of the apparent volume of the specimen, which can be determined by immersion in mercury at the beginning of the experiment, as it was mentioned in Section 3.1. Similarly, when the ordinate expresses intruded volume per unit bulk specimen volume, the specimen's bulk density also needs to be determined. Attention should be drawn that the plot of percentage intruded volume is different from the plot of cumulative porosity.

The plot of cumulative volume of mercury intruded vs. pore size is called cumulative intrusion curve (see Figure 3.4b). A plot of cumulative porosity vs. pore size is called a cumulative porosity curve (see Figure 3.4c). The cumulative curve usually proceeds from lower right to upper left. Generally, pore sizes span several orders of magnitude, and for this reason it is necessary to use a logarithmic axis in order to avoid crowding the small sizes against one end of the axis. From this curve one readily determines the total volume of mercury intruded, the pore volume in any pore range, the median pore diameter (the diameter for which 50% of the pores is greater and 50% smaller), and the mode pore diameter (the diameter corresponding to the region of steepest slope). The cumulative distribution curve will also readily show the limiting pore size, if there is one, at which there will be no further increase in mercury volume intruded as the pressure is increased [3.14]. From Equation (3.12) it is obvious that in a plot of volume vs. pressure, the total surface area of pores filled by mercury is calculated by the area underneath the curve. The surface area of pores between a pore size interval can be calculated graphically or by numerical methods and is obtained more easily when the pressure and pore size are in linear form.

Some parameters of particular interest that can be determined from the cumulative porosity curve are the total porosity, the critical pore size, and the threshold diameter.

The *total porosity* obtained from the cumulative porosity curve corresponds to the point of highest pressure and the smallest equivalent pore size. This is the point of maximum intruded volume. In Figure 3.4c, the total porosity obtained is 26%.

The *critical pore size* corresponds to the steepest slope of the cumulative porosity curve. For example, the critical pore size obtained from Figure 3.4c is approximately 0.05 μm. The critical pore size is frequently referred to also as the maximum continuous pore radius, and is the pore size through which mercury penetrates the bulk of the specimen, i.e. it is the most frequently occurring pore size in interconnected pores (see Figure 3.5a) [3.23]. The critical pore size controls the transmissivity of the material and this parameter is more often used to examine the effects of factors such as water–cement ratio, temperature etc. on the pore structure change.

All pore size distributions display a *threshold diameter* above which there is comparatively little mercury intrusion, and immediately below which commences the great portion of intrusion. In qualitative terms, the threshold diameter is the largest pore diameter at which significant intruded pore volume is detected. Its determination always requires some judgment as to exactly what level of intrusion is considered "significant." It has been observed that the threshold diameter decreases steadily as the age increases, as the water–cement ratio decreases, and as the silica fume content in cement paste increases [3.23,3.24,3.25,3.26]. According to D. Winslow and S. Diamond, 1970, the threshold diameter represents the minimum radius that is geometrically continuous throughout hydrated cement paste (see Figure 3.5b) [3.23]. Experimental studies by Winslow and Diamond have associated the threshold diameter with significant changes in the mercury intrusion process. Separate specimens of cement paste were intruded by mercury at pressures slightly smaller and slightly greater than the pressure corresponding to the threshold diameter. The specimens were then depressurized and broken open, and the fracture surfaces were examined using an optical microscope. Significant differences were noted in the color and in the appearance of the specimens under the microscope. The specimen intruded to a pressure smaller than the pressure corresponding to the threshold diameter was light gray, nearly like an unintruded specimen and showed localized streams of mercury along the fractured cross section of the interior, but also many local regions without observable evidence of penetration by mercury. In contrast, the specimen pressured to just beyond the threshold value was dark gray in color and showed mercury uniformly spaced over all regions of the fracture surface. The researchers also observed that the intrusion rate is much slower in the vicinity of the threshold diameter than at

(a)   (b)

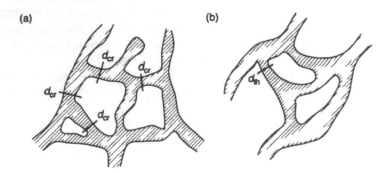

Figure 3.5 (a) The critical pore diameter, d_{cr}, is the most frequently occurring diameter in the interconnected pores that allows maximum percolation. (b) The threshold diameter, d_{th}, corresponds to the narrowest path in the interconnected pores.

any other diameter. When the pressure corresponding to the threshold diameter is reached, mercury begins to flow slowly from the exterior of the specimen to all portions of its interior, probably through long and tortuous paths. Once this flow has been completed, further intrusion consists of filling the smaller pores that are near the already full main channels and is, thus, more rapid.

J. Young, 1974, observed from experiments on hydrating C_3S pastes that there is a very large intrusion in the pressure range corresponding to diameters of 0.5–1.5 μm [3.27]. According to Young, the C_3S pastes show a "threshold diameter" as observed by Winslow and Diamond, 1970, which decreases with hydration in the same way as for cement pastes. Young suggested that above the threshold diameter only the interconnected large capillary pores are intruded, and below the threshold diameter the void space between C–S–H gel "needles" is filled, which represents a large fraction of the intrudable volume. Adding to intrusion at the threshold diameter point will be the filling of larger pores accessible only through intergrowths of needles. Intrusion of the pore space between these needles becomes less marked at later ages, since their total number decreases due to the continued formation of contact points and the growth of calcium hydroxide in the capillary voids occluding the C–S–H gel [3.27].

S. Diamond, 2000, pointed out that capillary pores in cement pastes consist of long percolating chains made up of pores of various sizes [3.28]. Along these pore chains are constrictions, "choke points," where the pore diameter is very narrow. Diamond equates the pore diameter at the choke points with the threshold diameter and suggests that when the mercury pressure is sufficient for mercury to flow through these constrictions, a large proportion of the whole pore system is simultaneously intruded, which produces a very steeply rising in the curve of cumulative pore volume vs. pore diameter. These choke points will gradually narrow with age as the cement hydrates, due to C–S–H gel formation and autogenous shrinkage, and some of the pore volume may become totally isolated by gel. Hence, the threshold radius will decrease as the curing time increases and as the water–cement ratio decreases [3.28].

Even though sometimes the critical pore size is referred to as the threshold or characteristic pore size, the terms critical and threshold pore size can refer to different sizes and should not be used interchangeably. The critical pore size is related, but is not identical, to Diamond's threshold diameter, which is a characteristic feature of the pore size distributions of cement pastes. D. Roy, 1989, has also pointed out attention to the difference between the threshold diameter and critical diameter, and has indicated that the threshold diameter occurs at a larger radius [3.29]. S. Mindess *et al.*, 2001, have clearly identified the difference between a critical diameter and threshold diameter using cumulative intruded volume curves [3.30]. The first inflection point (point on the curve at which the slope of the volume-diameter curve increases abruptly), indicates the threshold diameter, which is the minimum diameter of pores that form a continuous network through the cement paste (see Figure 3.4c) and marks the onset of percolation. The point of steepest slope represents the mean size of pore entryways that allows maximum percolation throughout the pore system, and is called the "continuous (or critical) pore diameter" (d_{cr}). According to this differentiation, in Figure 3.4c the threshold diameter is 0.7 μm, compared to the critical diameter of 0.05 μm. In a similar way, other researchers also associate the threshold diameter to the point where the initial rapid increase in the cumulative porosity curve occurs [3.31,3.32].

Several researchers have referred to the critical diameter and threshold diameter and have defined ways to characterize and use the parameters for their studies [3.33,3.25,3.31,3.26]. P. Mehta and D. Manmohan, 1980, refer to the threshold diameter as the diameter of the largest pore present at which mercury begins to penetrate into the pores of the specimens [3.26]. They found experimentally that with increasing water–cement ratio the cumulative volume of pores at any given diameter and the threshold diameter increased significantly.

Researchers have also found that the permeability of cement paste is more sensitive to the threshold diameter than to the total porosity, with both parameters being determined using the MIP technique [3.33,3.26]. B. Nyame and J. Illston, 1980, studied the relationship between the permeability of hardened cement paste and its porosity and pore size distributions with the particular purpose of finding the significance of the threshold diameter in relation to permeability [3.25]. The increased chloride permeability of pastes with high water–cement ratio is most closely correlated with the increased threshold diameter; this is probably because the threshold pore size is the widest, most accessible, path for fluid transport in the cement paste [3.34].

3.2.2 *Incremental and differential distribution curve*

Differential distributions differ mainly in whether a linear or logarithmic axis/calculation is being applied.

The *incremental volume curve* is obtained by plotting the change in intruded mercury volume in each pressure step that is measured during the experiment vs. the pore diameter. The incremental volume curve is presented as a histogram prepared by dividing the cumulative distribution curve into a number of contiguous pore size ranges and plotting the volume of mercury in each range as the ordinate vs. the log pore size range. The result is a distribution curve showing the pore volume in arbitrary pore ranges. Some advantages of this method of presentation are the fact that it is very simple to obtain, it makes comparisons between samples very straightforward, and that the total porosity and pore fraction in each range may be simply computed. Like most differential distribution methods, however, it is restrained by the fact that the distribution curve is dependent upon the size of the pore intervals used.

The *differential distribution curve* is obtained by differentiating the cumulative distribution curve and is essentially a plot of $dV/d(\log d)$ vs. $\log d$ (see Figure 3.4d). This type of plot has the advantage of showing a spectrum of pore sizes and is especially revealing when the sample has two or more unique peak pore sizes. It is, however, somewhat tedious to obtain the differential curve and the precise shape of the curve depends upon the interval between slope measurements. In addition, it is possible to estimate the pore volume in any pore range from the differential distribution curve as the area under the curve, as it is shown in Figure 3.4d [3.35].

The differential distribution curve presents in a more clear way important and useful information on the pore size distribution than the cumulative porosity curve. Differential MIP curves of hardened cement pastes exhibit a sharp peak that corresponds to the critical pore size (see Figure 3.4d) [3.29,3.36,3.37]. The peak shifts to smaller pore sizes and diminishes in size with increased curing time and for lower water–cement ratios of the cement paste [3.36]. It has been found that a higher water–cement ratio leads to a less pronounced peak in a pore size distribution curve. Sometimes, the distribution curve shows a second usually rounded peak exhibited at larger pore sizes that appears to correspond to the pressure required to break through blockages in the capillary pore network [3.37]. The rounded peak does not shift as much and becomes more dominant with increasing curing time and with lower water–cement ratio [3.37].

Another parameter that can be obtained from the differential distribution curve is the *median pore size*, at which 50% of the pore volume is observed on the pore size range considered. In most cementitious materials, the critical pore size and the median pore size have similar values. Properties of different cement-based materials may be contrasted by comparing the total porosity, median pore size, and the critical pore size.

There are several difficulties associated with the use of differential plots, and for this reason the cumulative plots are greatly preferred by researchers in presenting a pore size distribution. One difficulty refers to the selection of a pore size interval, because the increment must be selected so as to be neither too narrow nor too wide. Different selections of the interval can result in plots of differing appearance generated from the same set of intrusion data. It has been the opinion of S. Diamond, 1970, that the derivative distribution function, even though preferred by many researchers, is severely distorted for data covering several orders of magnitude of pore sizes [3.38]. Furthermore, the cumulative curve has the advantage that the fraction of pore volume between any arbitrary pore sizes can be assessed at a glance. Another problem associated with the differential plot is that the total intruded pore volume is not readily apparent from the plot and the plot does not convey any information that is not also available on the cumulative plot: e.g. a peak on the differential plot is a steep region on the cumulative plot.

Several researchers have reported that the distribution of pore sizes in hardened cement pastes and mortars, as it is determined by MIP, could be described by a mixture of log-normal distribution functions (see also Section 1.4.1.5) [3.39,3.40,3.41]. Pores of sizes ranging from about 1 nm to about 10 μm can be included in a compound log-normal model that can be analyzed in three sub-distributions [3.40]. This observation is valid for ordinary cement pastes, blended cement pastes, and mortars, which have different compositions and degrees of hydration. The first sub-distribution is associated with coarse pores that may extend to air voids, the third sub-distribution is associated with fine pores that may extend to gel pores, and the middle distribution includes capillary pores. Selected properties of the log-normal distribution, such as the median pore diameter, the inflection point, and the mean squared pore diameter, can be used to calculate characteristics of the pore size distribution.

3.2.3 Surface area

If cylindrical pores are assumed, it is also possible to calculate a surface area distribution from the MIP data, as it was expressed by Equation (3.12) [3.42,3.43]. However, it should be kept in mind that mercury porosimetry analysis does not include the very small pores; therefore, the surface areas calculated from MIP data are smaller than the surface areas calculated from other techniques, e.g. gas adsorption. In addition, because of hysteresis phenomena (described in Section 3.3), surface areas obtained from mercury porosimetry are not reliable. However, the surface area calculation can be used to determine and adjust the contact angle of mercury on a specific material.

3.2.4 Range of sizes determined

MIP can measure pores over a wide range of diameters. The volume of large pores (with sizes between 17 and 100 μm) is measured by volume changes in the mercury as the pressure applied to mercury is increased from the evacuated state to atmospheric pressure (low pressure part). The largest pore diameter determined may in theory be up to approximately 1 mm by applying a vacuum pressure on the filling device of the instrument. Currently commercially available equipment is capable of exerting pressures on mercury as high as 405 MPa (4000 atm), which determines a cylindrical pore with radius as small as 2 nm.

D. Winslow and S. Diamond, 1970, using MIP on cement-based materials having different water–cement ratios found that the pore volume left unintruded by mercury at a pressure of 100 MPa was significantly less than 28% [3.23]. The researchers believed that this volume represents gel pores. MIP has been reported to be able to determine surface areas as low as 25 cm^2/g [3.44]. According to D. Winslow, 1984, the MIP technique is well suited for materials with surface areas smaller than 10 m^2/g [3.11].

The range of pores present in hydrated cement paste is greater than what can be measured by MIP. The method shows the problem of "*lost porosity*", i.e. the pores that mercury seems to be unable to reach at any pressure. Research studies clearly demonstrate that MIP is intrinsically limited and that, even if high pressures are applied to the system, mercury is unable to intrude the entire pore volume of the hydrated cement paste. It has been stated that at the early hydration states of cement pastes MIP can characterize the pore structure well, but at later hydration states, MIP underestimates the total pore volume and the specific surface due to the presence of micropores, which MIP cannot determine [3.45]. According to S. Diamond, 1971, this phenomenon can be explained by the presence of pore spaces too fine to be intruded by mercury or completely isolated from the exterior (e.g. encapsulated pockets of gel) [3.46]. J. Beaudoin, 1979, suggested that the lost porosity could rather be explained by the existence of microspace between aggregations of C–S–H gel sheets that are accessible to gas molecules but not to mercury [3.47]. N. Alford and A. Rahman, 1981, concluded that closed spherical voids or macropores that cannot be intruded by mercury contribute significantly to the total porosity of hardened cement paste, affect the pore size distribution, modify surface areas determined by mercury intrusion porosimetry, and thus contribute to the lost porosity [3.48].

3.3   Hysteresis and entrapment of mercury

It had already been observed by H. Ritter and L. Drake in 1945 that, when at the end of mercury penetration the pressure is reduced, the pressure–volume relationship obtained during retraction is different from the pressure–volume relationship during penetration [3.2]. In addition, a certain amount of mercury remains trapped in the porous material after complete reduction of the pressure. S. Lowell and J. Shields, 1981, have divided the total mercury intruded into a specimen into two kinds: (a) mercury that is irreversibly trapped in the pores and never extrudes, and (b) mercury that cycles into and out of the pores as the applied pressure is increased and decreased [3.7]. Therefore, the extrusion of mercury from the pores as the pressure is lowered to atmospheric pressure after intrusion reveals two general phenomena, the *entrapment or retention* of mercury inside the pores of the material and *hysteresis* between intrusion and extrusion curves.

All cumulative porosity curves exhibit hysteresis between the intrusion and extrusion paths (see Figure 3.6). At a given pressure, the volume indicated on the extrusion curve is greater than the volume on the intrusion curve. In a similar way, at a given volume, the pressure indicated on the intrusion curve is greater than the pressure on the extrusion curve. It is obvious that pore size distribution plots derived from volumes and pressures obtained from the mercury intrusion curve will be different than the corresponding plots derived using values from the extrusion curve at the end of the experiment. During the experiment, the time required to reach thermal and mechanical equilibrium is different during intrusion and extrusion: e.g. it could take 2–3 minutes to obtain a mercury intrusion data point, and 20–30 minutes to obtain a mercury extrusion point from a material containing macropores [3.49].

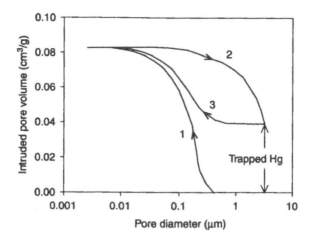

Figure 3.6 Intrusion–extrusion hysteresis and entrapment of mercury in a cement paste specimen. The arrows indicate the directions of intrusion (curve 1), extrusion (curve 2), and reintrusion (curve 3).

Source: Reprinted from *Pore Structure and Permeability of Cementitious Materials*, L.R. Roberts, J.P. Skalny (eds), *MRS Symposium Proceedings Vol. 137*, D. Winslow, "Some experimental possibilities with mercury intrusion porosimetry," pp. 93–103, copyright 1989, with permission from the Materials Research Society, Warrendale, PA.

After the pressure is reduced to atmospheric pressure, various amounts of mercury remain irreversibly trapped in the specimen, depending on the porous material. The amount of trapped mercury can vary from a negligible fraction up to nearly the total volume of the pores, usually in the range of 80–85%, depending of the pore structure of the material [3.2]. Typically, it is estimated that $\frac{1}{3}$ to $\frac{1}{2}$ of the intruded mercury does not spontaneously exit the pores of the material upon depressurization [3.50]. R. Feldman, 1984, reported that the residual mercury in cement pastes depends on their composition, and is higher for plain cement pastes and lower for blended cement pastes that had been distilled for a long time [3.51]. As an example, he reported that the residual mercury for 1-year hydrated cement distilled for 13 days was 4.8% of the initial mass of the sample, and 12.6% for plain cement distilled for 9 days. The corresponding values for blended cement pastes were lower.

3.3.1 Theories proposed to explain hysteresis

Hysteresis between the intrusion of mercury with increasing pressure and extrusion of mercury with decreasing pressure is a universal feature of MIP, observed in nearly all specimens during mercury porosimetry measurements. R. Smithwick and E. Fuller, 1984, showed that even during a blank test with mercury only and no specimen present, a large apparent hysteresis appears between the pressurization and depressurization data [3.49]. Many different theories have been proposed to explain the appearance of intrusion–extrusion hysteresis; however, there is still no single theory that can provide a definite explanation of all the experimentally observed effects. The most commonly used theories are the presence of ink-bottle pores and differences in contact angle. Additional explanations include pore potential, percolation/connectivity model (network effects), surface roughness of the pore walls, and contamination of the material's surface by mercury. These theories are briefly discussed in the following sections.

3.3.1.1   Ink-bottle pores and trapped mercury

Many authors have interpreted intrusion–extrusion hysteresis with reference to the presence of ink-bottle pores, i.e. large pores into which mercury intrudes through a much smaller constriction (neck) (see Figure 3.7a) [3.2,3.17, 3.13,3.6,3.9,3.8,3.7,3.50,3.52]. The "throat" or entrance opening to an ink-bottle pore is usually smaller than the actual cavity. Mercury will enter to fill the pore cavity at a pressure that is determined by the entrance opening and not by the actual cavity size. However, on depressurization, the wide body of the ink-bottle pore cannot empty through the smaller pore entrances during retraction of mercury until a low pressure is reached, leaving entrapped mercury in the wide inner pore. If this mechanism is true, the amount of mercury that is not expelled by the specimen after retraction might be a measure for the pore volume of ink-bottle pores. The entrapment of mercury inside the porous material has been explained by the breaking of the mercury column in the narrowed neck of an ink-bottle pore in the course of its rapid emptying, so that mercury remains trapped in the wide ink-bottle cavity.

It has been shown experimentally that hysteresis and mercury retention appear even in the case of glass with controlled cylindrical pores of constant radius [3.16]. S. Lowell, 1980, suggested that hysteresis in mercury porosimetry cannot be attributed solely to the ink-bottle pore effect commonly offered

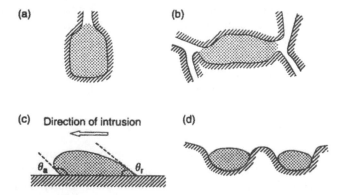

Figure 3.7 Some of the theories proposed to explain hysteresis during MIP. (a) Ink-bottle pores, (b) entrapment of mercury, (c) contact angle hysteresis showing different values for the advancing, θ_a, and the receding, θ_r, contact angle, (d) surface roughness of the pore wall.

as the explanation [3.52]. According to the researcher, the explanation involving ink-bottle pores ignores several factors, which include the following:

- All porous specimens exhibit hysteresis; if the assumption about ink-bottle pores were true, it would require that all porous materials contain pores that are ink-bottle in shape. In addition, experiments involving intrusion and extrusion of mercury between packed spheres, where the openings are wider than the interior space, also show hysteresis.
- Regardless of the maximum pressure attained, depressurization always results in hysteresis, which implies that ink-bottle pores are distributed over the entire range of pore sizes. Therefore, pores with very wide entrances would have to possess even wider inner cavities. Furthermore, cumulative porosity curves exhibit various shapes. If hysteresis were caused by ink-bottle pores, only one shape of hysteresis curve should be observed.
- The ink-bottle pore model ignores the question of the energy required to break the mercury column in the pore in order for the narrow entrance to empty, while the inner cavity remains filled.

A similar theory involves the connectivity of pores and is usually referred to as the *connectivity theory*. In order for a pore to become filled with mercury, the pore entrance must be equal to or larger than the corresponding pore size at the applied pressure, but it is also necessary that a continuous path of mercury leads to that pore. Large internal voids, which are surrounded by small pores will not be filled by mercury unless the pressure is sufficient to fill the small external pores. During mercury extrusion, the reverse process may occur and mercury is trapped in the wide voids causing hysteresis (see Figure 3.7b) [3.8,3.10]. In general, hysteresis is less evident when the porosity is higher, such as in some ceramics; however, the details of hysteresis will depend on the particular geometry of the pore system. In a detailed analysis of structural hysteresis and irreversible trapping of mercury, it has been found that the rate of pressure built-up affects the kinetics of the moving mercury menisci and the places where menisci merge or are newly formed by snap-off [3.22,3.53].

3.3.1.2 Contact angle hysteresis

Calculations using mercury intrusion and extrusion data are conventionally made using the same values for the contact angle. However, the concept of markedly different contact angles for advancing and receding liquids on solid surfaces is a well-documented phenomenon. If the advancing contact angle, θ_a, is different from the receding contact angle, θ_r, the retraction mercury curve will be different from the penetration curve. The advancing contact angle, θ_a, is as a rule greater than the receding contact angle, θ_r (see Figure 3.7c). The difference between the contact angle measured for a liquid advancing across a surface and the smaller contact angle measured for a liquid receding from the surface is called *contact angle hysteresis* [3.54].

Although the reported difference in values between the advancing and receding contact angles are not very large, the effect of different contact angles may be quite important. In conical pores, or diverging–converging pores the effect of contact angle is exaggerated.

S. Lowell and J. Shields, 1981, showed experimentally that appropriate adjustments in the contact angle can significantly reduce the intrusion–extrusion hysteresis on porous glass and alumina [3.55,3.52]. When the value of contact angle used for analysis of data is increased for the intrusion curve and decreased for the extrusion curve, the two curves can be brought closer to each other. By using a contact angle of 170° for intrusion and 107° for extrusion, they replotted intrusion and extrusion curves. The researchers pointed out that since some mercury is permanently retained inside the porous material, the intrusion and extrusion curves cannot be brought into coincidence along their entire path by means of contact angle adjustments. One can choose a contact angle for intrusion and then find a corresponding contact angle for extrusion, which will result in superimposition of the curves. It should be kept in mind though that the true contact angles for intrusion and extrusion remain unknown, since an infinite number of choices are possible.

Even though results in the literature suggest that hysteresis due to difference in contact angle cannot be neglected, it seems that several observations cannot be resolved only by this theory. Some of these observations include the following [3.10]:

- The intrusion and extrusion curve should theoretically be shifted parallel to each other when using a logarithmic scale for the pressure axis or the pore size axis.
- When scanning is carried out between the hysteresis branches in the extrusion and intrusion curve, no change in volume should be observed.
- The contact angle hysteresis cannot explain sufficiently why some mercury remains trapped in the pore system after complete depressurization.

3.3.1.3   Pore potential theory

S. Lowell, 1980 proposed an additional theory to explain hysteresis, which treats pores as wells of potential energy in a manner similar to the gas adsorption theory described in Chapter 4 [3.52]. Mercury atoms are about 3.14 Å in diameter and are larger and more polarizable than non-polar adsorbate atoms such as nitrogen (which is about 1.10 Å in diameter) or argon, which are known to condense into pores with diameter several hundred Angstroms. Therefore mercury, once forced into a pore, interacts with the pore walls and is trapped in a potential well of energy. During pressurization, the mercury column is under compression as it intrudes into the pore. During depressurization the mercury column is under tension due to the pore potential and can break if the pore potential is high and the pore radius very small [3.8]. The pores with small radii represent deep potential wells due to the close proximity of the pore walls and their small cross-sectional area, and require less energy to break the mercury column than large pores. Accordingly, narrow pores can permanently trap mercury, while large pores act as shallow potential wells and will be less able to trap mercury. During mercury intrusion and extrusion, the pore potential causes mercury to extrude from a pore at a pressure lower than the pressure at which mercury intruded.

The pore potential theory is also referred to as the energy barrier theory and explains intrusion–extrusion hysteresis and mercury retention independent of pore shape [3.52]. The pore potential model does not require the presence of ink-bottle pores. In addition, the pore potential theory is not inconsistent with the theory about differences in contact angle during intrusion and extrusion. It is possible that the mechanism by which the pore potential exerts its influence is presented by changes in contact angle.

3.3.1.4 *Surface roughness*

In repeated cycles of intrusion and extrusion, the first penetration curve is different from the subsequent curves. Part of the difference may be caused by a difference in contact angle that is the result of contamination of the surface by mercury for a fresh surface and a surface that has already been in contact with mercury [3.53]. Different contact angle may also be caused by a chemical or geometric heterogeneity (roughness) of the solid surface. Roughness of the surface of pore walls can cause the mercury to stick on the pore walls so that the mercury thread is broken (see Figure 3.7d). It has been observed that after retraction, part of the mercury is distributed as a film on the walls of large pores [3.8]. Therefore, it has been suggested that at high pressures mercury wetting can be irreversible due to adhesion to the pore walls.

In addition, the roughness of the pores' surface tends to increase the effective advancing contact angle. More information on the effect of surface roughness on the contact angle is given in Section 3.4.4.1 later.

3.3.1.5 *Compression of the solid*

M. Ternan and L. Mysak, 1987, tested experimentally the contribution to hysteresis from dimensional changes in pores (compression of the solid) [3.56]. Their results indicated that at high pressures and in solids with high porosities and low moduli of elasticity, dimensional changes make significant contributions to hysteresis. In addition, some materials will fracture when subjected to high pressure, which will change the specimen's pore dimensions and also contribute to hysteresis.

G. Johnston *et al.*, 1990, proposed correction of the pore size distributions to compensate for specimen compression because pore collapse and compression may lead to inaccurate and misleading pore size information [3.57]. They performed intrusion experiments on silica gel samples sealed in an impermeable membrane, and found that the volume change associated with pore collapse compression was proportional to $P^{-1/3}$, where P is the applied pressure. They accomplished partial correction of intrusion data by subtracting $aP^{-1/3}$ from the intrusion data, where a is a constant that relates to the solid's strength.

In addition to the theories mentioned here, other theories for hysteresis are possible. For example, at high pressure, mercury atoms might be pushed into the compound's crystal lattice. However, for samples that exhibit large amounts of sample compression (as determined from isostatic compaction) such as amorphous, high porosity materials, it does not appear possible to accurately obtain the pore size distribution via mercury porosimetry [3.57]. Another theory is thermal hysteresis due to increase in temperature of the solid sample during penetration and decrease during retraction [3.56]. It is obvious that none of the theories proposed explains by itself the phenomena of hysteresis and entrapment sufficiently. The various mechanisms could be interacting and the phenomena of hysteresis and entrapment could be a combination of more than one mechanisms.

3.3.2   *Entrapment of mercury and second intrusion method*

When going from the maximum pressure reached during the experiment down to the final atmospheric pressure, part of the mercury is expelled again from the specimen and part of the mercury in pores of certain geometry cannot be discharged. Hence, in materials containing pores with a variety of shapes, the ink-bottle pores can be differentiated from the uniform pores of capillary shape, i.e. perfect cylinders or pores that contain small contractions and enlargements along their lengths. This differentiation can be achieved by re-intruding the specimens after the first intrusion–depressurization cycle is completed to permanently trap mercury into the ink-bottle pores; this important development in MIP has been called the *second intrusion method*. It has been found that in repeated experimental intrusion–extrusion cycles the second penetration curve can follow the first penetration curve closely, and the second retraction curve can be completely identical to the first retraction curve [3.9]. If during the second intrusion cycle the hysteresis loop closes, it is an indication that no additional mercury is retained inside the pores.

R. Feldman, 1984, has devised a technique by which mercury trapped in a porous specimen can be removed by distillation, which is done by heating specimens at 105°C in vacuum for 1–15 days [3.51]. When the final mass of the specimen approximates its original mass before mercury intrusion, it indicates that mercury has been removed. Complete removal of mercury allows the study of a second intrusion to be made in order to provide additional information about the material's pore structure.

Researchers have used the second intrusion method and the entrapment of mercury to determine the porosity in air-entrained cement pastes, and to determine the pore structure parameters that affect permeability of cement paste. The second intrusion method has also been used to determine damage to the microstructure of cement pastes and in this aspect, it is further discussed in Section 3.4.6.

Air-entraining agents introduce mainly large spherical air voids in the cement paste without significantly affecting the cement paste's microstructure. Air voids in cement paste show behavior of ink-bottle pores. After the first intrusion and depressurization, the air voids containing mercury behave as solid portions of the cement paste and do not show up in further intrusion–depressurization cycles [3.58]. In this way, the second intrusion curve of air-entrained pastes is similar to the second intrusion curve of pastes containing no entrapped or entrained air, provided that the water–cement ratio and age of the two sets of pastes are equal.

S. Diamond and M. Leeman, 1995, measured the pore size distribution of an air-entrained paste using both MIP and SEM image analysis and found significant differences in the results [3.59]. The volume and distribution of air voids, which ranged in diameter from 10 to 300 μm, was correctly quantified using image analysis. However, the researchers observed that the volume of air voids was incorrectly correlated to a pore size of 0.2 μm when using MIP, a difference in pore size of 3 orders of magnitude.

Z. Liu and D. Winslow, 1989, 1995, used first and second intrusion and introduced the idea of dividing the pore size distribution into reversible and irreversible sub-distributions [3.34,3.50]. The first intrusion curve gives the total intrudable pore size distribution. The second intrusion is the reversible portion of the total intrusion distribution, and may be called the "reversible pore size distribution". When the reversible distribution is subtracted from the total intrusion, the result is a sub-distribution that represents the part of the pore system that is irreversibly filled and may be called the "irreversible pore size distribution". The reversible sub-distribution, and not the total nor the irreversible, is the pore structure parameter that provides the best correlation with chloride diffusion. The researchers expressed the opinion that other properties of cement paste might be able to correlate to the reversible sub-distribution.

R. Hill, 1960, attempted to quantify the fraction of mercury retained in the specimen after the pressure has been reduced to atmospheric pressure [3.60]. Hill regards the volume of mercury retained by the specimen at atmospheric pressure as giving the volume of ink-bottle pores, and the volume of mercury ejected as giving the volume of pores of approximately uniform diameter. In order to make a distinction between these two types of pores, he introduced the *retention factor*, which is determined by Equation (3.13):

$$R = \frac{V_{ret}}{V_{max}} \tag{3.13}$$

where R is the retention factor (unitless), V_{ret} is the retained volume of mercury after the first cycle of intrusion–extrusion is completed (mm^3), V_{max} is the total volume of mercury intruded at the maximum applied pressure (mm^3).

The retention factor can range between 0 and 1. For example, in Figure 3.6 the retention factor is calculated to be 0.48 (=0.04/0.084). Small retention factors indicate pores with approximately uniform cross sections, and large retention factors indicate the presence of ink-bottle pores. Hill also proposed correcting the conventional pore size distributions by multiplying the volumes by the factor $(1 - R)$. A high retention factor means that the mercury intrusion curve can be biased towards small pores. Hill considered intrusion curves with retention factor R greater than about 0.5 to be misleading. Therefore, he proposed that comparisons of pore size distributions be made between specimens that have about the same retention factor.

A. Auskern and W. Horn, 1973, studied the change in retention factor as the degree of hydration of cement pastes increases [3.14]. They obtained mercury extrusion curves and determined the relative amount of mercury retained at atmospheric pressure. The retention factor increased almost linearly with degree of hydration and was independent of the water–cement ratio; it was about 0.20 when the degree of hydration was 20%, and 0.55 when the degree of hydration was 80%. The fact that the retention factor is independent of water–cement ratio and dependent upon the maturity of the pastes shows that the retention factor is affected by the accumulation of hydration products in the capillary pores in such a manner as to make the pores more nonuniform. Hydration is accompanied by a decrease in capillary porosity, a net filling of large pores, and a relative change in pore shape from cylindrical with more or less uniform cross section, to pores with considerable variation in cross section.

3.4   Parameters affecting results

A review of the practical applicability and reproducibility of MIP results with possible sources of error has been summarized in the past [3.22,3.9,3.61]. The effect of microstructural changes on the pore structure parameters of cement pastes measured using MIP have been studied by several researchers. Microstructural effects usually include water–cement ratio, degree of hydration, curing conditions, and addition of mineral admixtures. As the degree of hydration increases, the pore size distribution is shifted towards smaller average pore diameters [3.23,3.62,3.64,3.36,3.63]. The proportion of total porosity of hardened cement paste lying in small pores (less than 10 nm) increases dramatically with hydration time [3.62]. Total porosity measured can vary from 16% to 56% depending on the water–cement ratio and length of curing time: long curing times and low water–cement ratios result in low porosity values [3.26,3.65,3.40,3.36,3.64]. The water–cement ratio affects both the microstructure of mature cement pastes and the change in microstructure during maturing. At low water–cement ratios very few pores occur having a radius in the range 0.1–7.5 μm, and the distribution of pores with radii between 10 and 100 nm change from a maximum at about 50 nm at early ages to 30 nm at maturity [3.36]. The addition of mineral admixtures (such as fly ash and silica fume) also modifies the microstructure [3.63,3.24,3.66].

The results of measurements are not only influenced by the microstructure of the specimen, but by experimental parameters as well. The effect of experimentally related factors have been studied by researchers in the past as well, such as pretreatment technique, specimen size, rate of pressure build-up during the experiment, contact angle between mercury and the pore wall, surface tension of mercury, purity of mercury, compressibility of mercury and specimen, and specimen damage. These factors are discussed in more detail in the following sections.

3.4.1   Specimen pretreatment

Before the MIP experiment, the cement paste specimen has to be dried using one of the techniques described in Chapter 2. The effect of pretreatment on pore structure parameters determined using MIP has been studied by several researchers by comparing the MIP results obtained after using different drying techniques [3.67,3.68]. The drying procedure does not seem to have a significant effect on total pore volume determined by MIP (see also Figure 2.3) but seems to be responsible for differences in pore size distribution, particularly in the fine pore region. C. Gallé, 2001, tested the effect of oven-drying, vacuum-drying, and freeze-drying on the cumulative intruded volume and the pore size distribution of cement pastes with w/c = 0.40 [3.68]. His results indicate that oven-drying results in increased total porosity and a larger critical pore size, with freeze-drying resulting in the smallest critical pore size. An example of the effect of drying techniques on pore size distribution curves is shown in Figure 3.8.

Oven-drying at 105°C is usually selected as a drying technique mainly because it gives a more complete water removal and because it has been reported that a single value of mercury contact angle could be used for all oven-dried specimens [3.67]. Oven-dried specimens exhibited an increase in the volume of pores between 0.2 and 0.1 μm and this alteration of the pore structure appears to be more severe for cement pastes with low water–cement ratios [3.69]. Oven-drying treatment affects also the value of threshold radius measured and it appears that oven-drying opens some pores and creates continuity in the cement paste matrix through irreversible changes in the cement paste microstructure.

Experimental studies have shown that P-dried and oven-dried cement paste specimens yield pore size distribution curves in which the coarse pore regions roughly coincide [3.23]. However, for pore diameters smaller than about 0.1 μm, the oven-dried pastes have increased pore volume compared to pastes that are P-dried. Pore size distributions of D-dried cement pastes are more nearly like the pore size distributions of P-dried cement pastes. It is believed that D-drying results in more complete removal of pore water compared to P-drying and probably provides a truer assessment of the actual pore spaces present with mercury intrusion. In addition, the thickness of the layer of adsorbed water remaining in D-dried and P-dried specimens can significantly reduce the effective diameter of the small pores measured by the mercury intrusion technique, compared to the oven-dried specimens.

Experiments have shown that *solvent-replacement* results in less damage to the pore structure of cement pastes during drying than other techniques [3.67,3.70,3.71]. The pore size distribution of cement paste specimens was unaffected by soaking vacuum-dried specimens in methanol, for periods up to 90 days. Therefore, it has been concluded that solvent exchange techniques are suitable for the preparation of specimens for pore structure determinations [3.71]. Differences have been observed between the pore size distributions of cement pastes that have been solvent replaced and those that are not solvent replaced [3.72]. It is possible that the sorbed *solvent* remaining in the cement paste specimens after vacuum-drying exists in the gel porosity and that its presence leaves the pore structure in the range above 3.7 nm unaltered [3.71]. R. Feldman and J. Beaudoin, 1991, studied the effect of solvent replacement pretreatment on MIP measurements [3.32]. Solvent replacement allows water

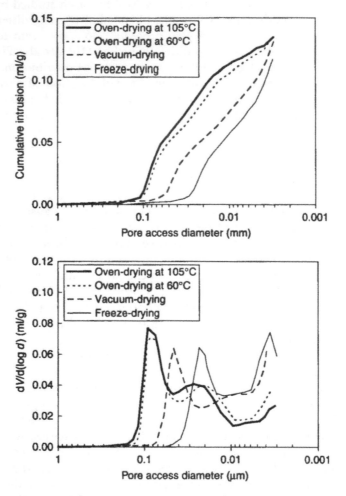

Figure 3.8 Influence of the drying technique on the cumulative intruded pore volume and the differential pore size distribution curves of Portland cement pastes 4 months old with w/c = 0.40.

Source: Reprinted from *Cement and Concrete Research*, Vol. 31, C. Gallé, "Effect of drying on cement-based materials pore structure as identified by mercury intrusion porosimetry – a comparative study between oven-, vacuum-, and freeze-drying," pp. 1467–1477, copyright 2001, with permission from Elsevier Science, Amsterdam, The Netherlands.

removal from cement pastes without major stress application to the cement paste matrix. Length of subsequent evacuation time and temperature of heating are in themselves relatively unimportant to the resultant pore size distribution. Solvent replacement by methanol increases the threshold pore diameter of the pore size distribution curve by changing surface characteristics of the hydrated cement paste; however, this was not observed with isopropanol. The researchers recommended as the most suitable technique for preparation for

MIP experiments replacement using isopropanol and immediate evacuation and heating to 100°C for 20 hours. It has also been found that *freeze-dried* specimens contain residual water after drying and that subsequent oven-drying of the freeze-dried specimens leads to the same calculated pore volume [3.69].

Attention should be given to any differences in contact angles used by researchers when different pretreatment techniques are used. More information on how pretreatment affects the contact angle is given in Section 3.4.4.

3.4.2 Specimen size

The selection and preparation of specimens for MIP tests needs to be done with great care. One must know, from other sources or experiments, something about the pore structure and the response of the specimen during testing. Failure to consider these matters may lead to inaccurate and/or nonreproducible results. Often, at the outset of an experiment, the porosity of the specimen is not known. In such cases, an initial intrusion test should be made in order to use the results to adjust the specimen size to utilize fully the maximum available intrusion volume. Frequently it is the penetrometer's size that controls the specimen size. However, the specimen's bulk volume may be the controlling factor when specimens with a small porosity are to be tested, and then an instrument that can accept a large specimen can be advantageous. The geometrical properties of the specimen may affect the reproducibility of measured values and may cause difficulty in giving an unambiguous interpretation of the results [3.22]. A primary consideration in selecting the size of the specimen to be tested in a single pore size determination is that the volume of intrudable pores within the specimen must not exceed the range over which the instrument can detect intrusion. Typical specimen volumes acceptable to commercially available instruments range from several cm^3 up to 15 cm^3 with maximum intrusion volumes of 5–10% of the specimen volume [3.11].

Large specimens are also advantageous when their pore structure is heterogeneous, because a large specimen tends to "smooth out" local heterogeneities and produces a more representative result of the pore structure [3.11,3.37]. Furthermore, using a single large specimen in a porosimetry experiment minimizes the surface area to volume ratio of the specimen, thus minimizing boundary effects. The use of large specimens that almost fill the penetrometer assures that less mercury is needed to fill the penetrometer, which has several important advantages: less waste mercury is created, density measurements are more accurate, and mercury compressibility effects are reduced, making the entire analysis more accurate. When a sufficiently large specimen cannot be tested, an alternative is to make several replicate tests and then combine the results to obtain the intrusion that would have been measured, had all the material been tested at once.

The effect of particle size of the test specimen on the results has been tested by several researchers [3.65,3.35,3.73,3.11,3.22]. It has been found that the use of a single specimen eliminates the possibility that the intrusion of mercury between the particles, i.e. the inter-particle voids, of a porous specimen will be confused with the intrusion corresponding to the porosity of the particles themselves, i.e. the intra-particle voids [3.74,3.75,3.73]. These two pore systems must be separated before a comparison between the distribution functions for different particle sizes can be made.

N. Hearn and D. Hooton, 1992, carried out experiments to evaluate the effect of specimen critical (minimum) dimensions and specimen mass on mercury porosimetry results [3.73]. They found that large specimens give consistently larger porosities compared to small specimens from the same material. Minimum specimen dimensions of up to 6.5 mm do not influence total intruded porosity, but influence pore size distribution; therefore it might be useful to make sure that the dimensions of specimens are similar when comparing results. In addition, they determined that the specimen's mass and not its critical dimension influences total intruded porosity and pore size distribution. There was however, some uncertainty because the critical dimension varied in their experiments between 1.4 and 2 mm. Use of crushed specimens instead of prisms can lead to introduction of an external porosity peak. On statistical grounds, one could argue that the larger the specimen mass, the more likely it is that faults will be present. However, their results showed very consistent changes in porosity with changes in specimen mass, and not the random effects that would be expected if faults were present.

Preparation for MIP measurements usually involves crushing, which may change the internal pore space. If closed pores are present in the material, some of these will be broken open. In addition, particles will tend to break along large pores; hence the relative volume of large pores will decrease [3.22].

D. Bager and E. Sellevold, 1974, 1975, showed the effect of specimen size on the results using powders and chunks of oven-dried cement pastes of the same maturity and water–cement ratio [3.65,3.35]. They found that there is overlapping between the external porosity between the particles (inter-particle pore space) and internal porosity inside the specimens (intra-particle porosity). They made the separation by simply defining the maximum internal pore radius as the radius where the differential curve has a minimum value. Even though the separation procedure was somewhat arbitrary, the accuracy of their data was not great enough to justify a more elaborate procedure. Their choice seemed reasonable because the curves came together at a relatively high pressure and maximum penetration was about the same for all particle sizes. Their results showed a systematic change in pore structure parameters as the particle size decreases: the threshold radius becomes less well-defined, and the pore size distributions show an increasing concentration of large pores as particle sizes decrease.

Another study has shown the effect of crushed fragments and small cores drilled out from concrete on porosity and pore size distribution [3.76]. Small cores drilled out from large concrete specimens exhibit less specimen-to-specimen variation than crushed fragments, since small cores drilled from concrete are likely to preserve the specimen-to-specimen uniformity more reliably. From the point of view of similarity in specimen mass, specimen dimensions, nature of concrete contained in specimen, and ease of obtaining it, the cored specimens of concrete is preferable.

3.4.3 Rate of pressure build-up

It has been shown experimentally that there is not much difference in the pore size distributions obtained using a continuous or a stepwise pressuring instrument, provided that the pressure is not allowed to decrease during the stepwise pressuring and that the total elapsed time between initial pressurization and final pressure for the tests in the two modes is the same [3.11]. When the pressure is applied in a continuous mode, the total test time depends upon the selected pressurization rate. When the pressure is applied in a step-wise mode, the total test time depends upon both the rate of pressure rise between the steps and the length of the pause selected at the end of each step.

The rate of pressure build-up during MIP is a parameter that requires attention. In order to reduce the analysis time, it is preferable to apply pressure build-up at a high rate; however, this rate may not be too high, because a certain time is needed for the transport of mercury through the porous system, and for the system to reach intrusion equilibrium. It has been shown experimentally using a continuous mode that the scanning speed does not affect total pore volume values, but affects pore size distributions [3.73]. With fast scanning (achieving a pressure of 400 MPa in approximately 10 minutes), the measured volume of the small pores is low, because mercury does not have sufficient time to intrude into the pores properly. With low scanning speed (reaching a pressure of 400 MPa in 4–6 hours), the small pores are determined more accurately, which is evident in the pore size distributions and the high surface areas. No clear effect on the median pore size has been observed, because this parameter emphasizes differences in the large pore diameter range.

For some specimens, the rate of mercury intrusion into the specimen is so slow that the pressure generator must be run at a low speed, two or more hours from ambient pressure to 400 MPa, in order to avoid running ahead of the intruded volume. This phenomenon is called *delayed intrusion* and has been attributed to the inability of liquid mercury to rapidly penetrate the narrower regions of a matrix of interconnecting pores [3.23]. An alternative explanation for delayed intrusion has been proposed based on the inability to thermodynamically distinguish between liquid and vapor transport of mercury into a pore [3.77]. It is thermodynamically possible for vapor transport to continue filling a pore once mercury stops flowing, provided that the inner portion of the pore is wider than the point at which the flow stopped. In porosimeters that pressurize in incremental mode, there are necessarily pores that are in an incipient state of filling after each pressure increment. The mercury has developed a curved meniscus and has just commenced filling the pore when the pressurization ceases. In such instances, the pores can continue filling by vapor transport, as evidenced by a slow rate of intrusion. When the pressure is applied in a continuous mode, this incipient condition is not possible and the pores fill by the faster liquid transfer mechanism.

3.4.4 Contact angle

The selection of the appropriate contact angle for a specific application has received considerable attention. The largest error in the determination of the pore radius may be caused by the use of an incorrect wetting angle because the value of constant angle is entered as the $\cos\theta$, thus small differences are magnified [3.63,3.78,3.16]. For example, an error of even 1° at 140° would introduce an error of slightly more than 1.4% in $\cos\theta$ and thus in the pore radius. The choice of contact angle has no effect on the results obtained for total porosity, but has been found to have an impact on pore size distribution and the threshold diameter determined [3.63]. The contact angles for mercury on various solids may differ substantially. For example, it has been reported that for concrete aggregates the contact angle can range from 118° to 129°, with an average value of 124°. Similarly, for clays the contact angle can range between 139° and 147° [3.64].

Hydrated Portland cement pastes do not have a unique contact angle with mercury and researchers have stated values as low as 117° [3.23] and as high as 175° [3.79]. Originally, L. Edel'man *et al.*, 1961, used a contact angle of 145° for their application of MIP on cement pastes [3.5]. The most frequently quoted values for the contact angle of mercury are 130° and 140°. It is recommended that the true value of the contact angle be determined for each material studied (with reference to the cement paste's age, composition, pretreatment method etc.), and the corrected value of contact angle be used during analysis of results. Before mercury enters the penetrometer, the penetrometer and the specimen are outgassed. Evacuation of the specimen changes the contact angle, which is higher when measured in air under atmospheric conditions than in vacuum [3.53]. In addition, it is not known whether the contact angle remains constant as the applied pressure on mercury increases.

3.4.4.1 Parameters affecting contact angle

The contact angle depends on several parameters such as the composition properties of the cement paste, the characteristics of the pores, and the mercury itself. Some of these parameters have been studied and are discussed in the following paragraphs.

Cement paste characteristics The contact angle has been found to vary for cement pastes with maturity, presence of mineral admixtures, drying method, and whether the specimens were previously intruded and the mercury subsequently removed [3.63]. Summarized values for the contact angle depending on the type of cement paste, the age, and the pretreatment technique are reported in Table 3.1.

Pastes of different *maturity* have different porosities and chemically dissimilar surfaces, and it has been found that the contact angle for plain cement pastes becomes smaller as the maturity of the pastes increases [3.63,3.34,3.14]. The water–cement ratio has no significant effect on the contact angle of mercury with cement paste [3.78]. Cement pastes of approximately the same maturity, but containing *fly ash*, have chemically dissimilar surfaces and a larger contact angle than plain cement pastes [3.63]. The contact angle for polymer-impregnated cement pastes has been determined to be 155° [3.62]. In addition, heating a cement paste for a prolonged period of time increases the contact angle

Table 3.1 Values of contact angle between mercury and different types of cement pastes, at different ages, and dried by different techniques

Type of cement paste	Contact angle (degrees)	Source
Plain cement paste, 1 day old	129	[3.63]
Plain cement paste, 28 days old	123	[3.63]
Cement paste with 20% fly ash, 28 days old	132	[3.63]
Plain cement paste, re-intruded after mercury distillation	135	[3.63]
Oven-dried cement paste	117	[3.23]
D-dried cement paste	117–130	[3.23]
P-dried cement paste	130	[3.23]
Freeze-dried cement paste	141	[3.23]

[3.63]. The contact angle has been found to vary with relative humidity and range between 123° and 136° over the entire humidity range above 50% RH [3.80]. Thus, the contact angle cannot be regarded as a constant property.

Pore size One assumption commonly made during MIP is that both the contact angle and the surface tension have constant values and are independent of the radius of the capillary. It is possible that the contact angle varies within a porous solid, having a high value for the large pores and a small value for the small pores. It has been suggested that both the advancing contact angle, θ_a, during mercury intrusion, and the receding contact angle, θ_r, during mercury extrusion are pore size dependent, having a value of 180° for pores larger than 2 nm, and lower values 135°–140° for smaller pores [3.8,3.53,3.79,3.78].

In the presence of ink-bottle pores, since the total volume of the ink-bottle pore is assigned to the size of its entry way using a small contact angle, it is obvious that MIP consistently underestimates the true pore sizes of ink-bottle pores; this has been verified by experimental results [3.65].

Surface roughness It has been theoretically and experimentally proven that roughness of the solid surface has an effect on the contact angle: the contact angle of a non-wetting liquid such as mercury on a rough or porous surface will be greater than the contact angle on a microscopically smooth surface of the same composition [3.79,3.63,3.10,3.53]. The cement paste is microscopically rough and porous, even when prepared by molding against a smooth solid surface.

Mercury purity It is often recommended that double- or triple-distilled mercury be used, because of the implication that mercury impurities affect the contact angle and the surface tension of mercury. However, it is expensive and impractical to flush a porosimeter with fresh mercury in-between test runs; therefore, mercury is commonly re-used in a porosimeter over an extended period of time. Studies with chemical analysis of the original mercury and the used mercury have shown that only minor pollution of the mercury occurs during the test [3.9]. Comparison of experimental results obtained on cement pastes using fresh triple-distilled mercury with those obtained using mercury that had been used in MIP tests for several months indicated that the re-use of mercury has little effect on the results [3.78]. However, in a similar study, it has been found that the contact angle between the pore walls of a hardened cement paste and mercury that has been distilled is greater than when using non-distilled mercury [3.63].

Since there are so many factors affecting the value of the contact angle, it is apparent that a unique correct value cannot be determined. An obvious solution to the uncertainty of contact angle is to select a conventional value for the contact angle and to recognize that when that contact angle is used, the pore size distribution data may contain a constant error. Comparison with results published in the technical literature is only possible when full information is given about the parameters examined and the values selected for analysis of results. Whatever contact angle is used during an experiment, the value that has been selected needs to be mentioned. One should also describe the water-removing procedure employed.

3.4.4.2   Determination of contact angle

The determination of the exact contact angle between mercury and a solid is not easy and a variety of ways to determine the contact angle have been developed and are described in detail elsewhere [3.21,3.81]. Contact angle measurements have been reported using both *static* and *dynamic techniques*. In static techniques, mercury is in a non-mobile condition, and in dynamic techniques mercury is advancing through an opening. Some of the most commonly used techniques, also in concrete science, include the sessile drop technique and the use of the mercury contact anglometer.

The *sessile drop method* is used to measure mercury contact angles on flat surfaces and determines the maximum vertical height that a drop or small pool of liquid can attain without spreading on a horizontal flat surface [3.8]. A small drop of mercury is placed on the surface of a smooth bed of powder, which assumes a spherical shape, except for the portion in contact with the surface. As additional mercury is added to the drop, the height increases until it reaches a maximum value. Further additions of mercury increase the drop diameter without any additional change in the drop height (see Figure 3.9) [3.7]. The maximum drop height that can be supported regardless of increasing drop size is related to the contact angle by Equation (3.14) [3.38]. The values of density and surface tension of mercury are given in Table 2.2. The sessile drop technique is a reliable and practical method for non-porous or low-porous materials [3.9]:

$$\cos \theta = 1 - \frac{g \rho_m h^2}{2 \gamma_m} \tag{3.14}$$

where θ is the contact angle (degrees), g is the acceleration of gravity ($=9.81$ m/s^2), ρ_m is the density of mercury (kg/m^3), h is the maximum height of the sessile drop (m), γ_m is the surface tension of mercury (N/m).

Early use of the sessile drop method consisted of placing a drop of mercury on the horizontal surface of the solid to be studied and measuring the heights of sessile drops of mercury (or any other non-wetting liquid) as a function of

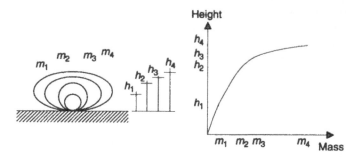

Figure 3.9 Change of the shape of a mercury drop with size. The height h of the drop of mercury increases as its mass m increases, and reaches an asymptotic value after a certain mass.

drop mass. From a graph of height of the drop of mercury vs. mercury mass, the maximum height can be determined and the contact angle is determined using Equation (3.14).

Modern equipment allows the measurement of the contact angle between a drop of mercury and the solid surface using drop contours and modern image processing. Usually, an image of the drop on a flat surface is digitized and the contact angle is measured as the tangent at the point of contact. Optical devices are sometimes employed for the measurement of contact angle, where the operator must attempt to establish the tangent to the contact angle of a drop of mercury resting on a plane surface. This method has not proven sufficiently accurate because of its inherent subjectivity: different experimenters will measure substantially different contact angles and even the same person will observe different angles on the same material on different occasions [3.7].

The maximum height method measures an equilibrium contact angle and not the advancing angle. It is known that liquids exhibit contact angle hysteresis on solids, i.e. the advancing contact angle differs from the receding angle (see also Section 3.3.1.2). Therefore, it is important that the technique used for measuring contact angles for mercury porosimetry studies, matches the actual intrusion conditions as nearly as possible. The angle measured should be the advancing contact angle in an evacuated cylindrical pore in order to conform to the assumptions implicit in the Washburn equation. For this reason, J. Shields and S. Lowell, 1982, developed the *mercury contact anglometer*, which they have found to provide a constant contact angle for a given powder over a range of pore radii from about 1.8 nm to 100 μm [3.81].

In order to measure the contact angle of a given material with the mercury contact anglometer, the material must first be crushed into a powder. A press is used to compact the powder around a precision bore pin, which upon removal produces a cylindrical open-ended pore of known radius. A fixed volume of mercury is placed above the powder at a known height and pressure in a specimen holder. After placing the powder specimen and the mercury into the anglometer, the apparatus is evacuated. Air is slowly allowed into the upper chamber of the anglometer and mercury pressure is constantly monitored by a sensitive transducer. When the breakthrough pressure is reached, the pressure is recorded. The corresponding value of $\cos \theta$, which is proportional to the mercury pressure can be calculated from the known radius of the cylindrical hole, using the Washburn equation. The mercury contact anglometer has been reported to give highly reproducible results [3.7,3.8].

A slightly different procedure was first applied by D. Winslow and S. Diamond, 1970, on cement pastes and has been used to measure directly the contact angle of mercury with cement pastes [3.23]. The technique was developed and previously applied on a block of nickel, by drilling holes of known diameter [3.82]. The particularity of the method developed by the researchers is that cylindrical holes of known diameter, e.g. 200 μm, are drilled directly in a hardened cement paste specimen, avoiding in this way the crushing operation into powder. The contact angle is measured by determining the pressure required to intrude the drilled, cylindrical holes of known diameter and then calculating the contact angle that must have been in effect during the intrusion. The pressure required to intrude these artificial "pores" is measured and the contact angle is calculated from the known values of pressure and pore diameter using Equation (3.3). The technique has been used for cement pastes by several researchers [3.22,3.63,3.64]. A modified technique for calculating the contact angle of mercury with hardened cement pastes has also been proposed [3.78]. Instead of drilling holes into the cement paste, researchers produced artificial cylindrical pores of 0.15–1.05 mm in diameter by casting cement paste around small diameter wires.

The difference in contact angle values can be pronounced, depending on the technique used to measure it and it is not easy to decide which technique gives the correct value. The surface of a drilled hole can be expected to have a higher surface energy than that of a crystal formed by a growth process such as occurs during Portland cement hydration and this would lead to a measured contact angle lower than the actual. However, the interior of a drilled hole in a brittle solid will be rough, which would lead to a measured contact angle higher than the actual, counteracting in this way the former effect [3.83].

3.4.5   Surface tension of mercury

The surface tension of mercury is usually quoted to be $0.485 \pm 0.001 \, \text{N/m}$ even though for pure mercury, values as different as 410 and 515 N/m have been quoted [3.79,3.22]. The surface tension of mercury is influenced by temperature, mercury purity, and the curvature of the solid's surface [3.61]. Even when one starts experiments using clean mercury, mercury will soon be contaminated by components that were adsorbed on the surface of the specimen. It is known that impurities may change the surface tension of mercury by as much as 30%; however, it seems that in mercury porosimetry this is not such a serious problem since such variations do not affect the results significantly [3.22]. For very strongly curved surfaces, the surface tension is a function of the surface curvature. It has also been shown that the surface tension of mercury is affected by pore size: during both intrusion and extrusion, as the pore radius increases, the surface tension increases [3.53]. Changes in surface tension values due to temperature have minimal effect on measured pore volume and pore size distribution.

3.4.6 Alteration of pore structure

The extent of pore deformation under high pressure was studied by S. Brown and E. Lard, 1974, on porous silicas [3.43]. Their experiments have shown that the material tested is compressed under pressure, which reduces the pore volume, an alteration that becomes greater in materials with high pore volume. The compressibility of the material reduces significantly the average pore diameter of the material, but does not change the surface area of the sample, since only pore walls are compressed and the number of pores are not altered [3.43]. It is generally expected that specimens under test will be damaged only if the porosity is very high, or if there is a significant number of closed pores [3.22].

Two techniques commonly used for checking the alteration of pore structure are the second intrusion of specimens that was mentioned in Section 3.3.2, and the microscopic examination of specimens intruded by mercury.

3.4.6.1 Reintrusion of mercury

The effect of the mercury intrusion process on the microstructure of cement pastes has been investigated experimentally by reintruding specimens after the first intrusion and after the mercury was removed by extended distillation. Distillation of mercury and reintrusion has shown that damage is caused in the pore structure of cement pastes during intrusion of mercury. However, distillation of mercury itself from a cement paste has been found to alter the pore structure, and this effect must be considered separately before alterations due to intrusion are discussed. Pore size distributions obtained during a first and second intrusion of mercury in cement pastes always show discrepancies [3.63,3.51]. Comparison of the corrected second intrusion curve with the first intrusion curve indicates the creation of additional large pores and an increase in the threshold diameter. The observed discrepancies can be explained by a combination of three factors: (a) damage induced to the pore structure of the specimen by the intruding mercury during the first intrusion, (b) modification of the contact angle induced by the high-temperature distillation treatment, and (c) presence of residual mercury on the pore wall during the second intrusion that could not be removed by distillation.

Results published in the technical literature for the first and second intrusion of blended cement pastes containing fly ash or blast furnace slag display a coarser pore size distribution during second intrusion compared to the pore size distribution during first intrusion [3.51,3.84,3.15,3.85,3.63]. Blended cement pastes include relatively large, but discontinuous pores, into which mercury enters by breaking through the pore structure. It has been suggested that the high pressure applied on the mercury necessary for the intrusion of the smaller pores can result in temporary or permanent alteration in the microstructure of cement paste. During mercury intrusion, the discontinuous thin-walled pores are disrupted and the damage is greater at high pressures, or in small pores. It has been found that for pressures lower than about 70–98.1 MPa there is no measurable alteration to the specimen, which was indicated by complete coincidence of the pore size distribution of an unintruded specimen and a specimen intruded for the second time [3.51,3.85]. It is possible that damage may occur at even lower pressures, but the intrusion curves cannot reveal damage because of the increase in the volume of large pores. During the second intrusion, some pores are created between 40 nm and 100 nm in diameter at the expense of smaller pores between 20 and 40 nm in diameter. Although threshold diameters change from about 40 to 100 nm after distillation, the volume of pores with diameter greater than 88 nm increases from 1.0% to only 2.4% [3.51,3.86].

The heat required to remove trapped mercury after a first intrusion also alters the pore structure and may distort one's impression of the alteration that is produced solely by the pressure involved in intrusion [3.63]. For cement-based materials, compression of the specimen cell affects the pore size distribution curve only in the region below the pore diameter of 50 nm [3.76]. It has been reported that the error due to alteration of the pore structure is not more than 3% [3.8]. Thus, for the study of pore size distribution of concrete, effect of these factors can be neglected without introducing significant error [3.76].

3.4.6.2   Microscopic examination

Several researchers have examined microscopically damage to the microstructure of cement pastes after mercury intrusion [3.59,3.87,3.88]. Specimens before and after MIP were viewed using an environmental scanning electron microscope, which allowed imaging at low vacuum conditions in the chamber and eliminated constant out-gassing of the residual mercury in the specimen. Specimens were intruded to a pressure just below the critical threshold pressure, removed for observation, then intruded to a pressure well above the critical threshold pressure. Significant damage caused by relatively low pressures of 10–20 MPa was found in the interior of the specimen. The resulting porosity is far more interconnected within the material after the first MIP, and any slice through the intruded specimen reveals these interconnected networks. It has been found that after intrusion, the number of pores observed microscopically was lower and the average pore size was greater than the companion values measured before intrusion, indicating that damage was inflicted by the intrusion process.

3.4.7 Alternative intrusion liquids

J. Davidson, 1979, proposed a new technique in which mercury is replaced by a low melting metal or alloy, and intrusion experiments offer a convenient and simple technique to aid in interpretation of mercury intrusion curves [3.89]. By substituting mercury with a low melting alloy in an intrusion experiment, it is possible to "freeze" the intrusion process at any given pressure by cooling the metal or alloy below its melting point. The "frozen" intrusion specimen may then be sectioned and examined under the microscope to determine how the metal is entering the pore structure (see also Chapter 8). Such metals and alloys include gallium and Wood's metal, which were mentioned in Chapter 2.

D. Winslow, 1989, proposed the use of a solution of mercury and 5% by mass sodium as an alternative liquid to run the MIP test because it has a lower contact angle, lower surface tension, and allows intrusion at lower pressures than pure mercury [3.50]. Pure mercury has values of contact angle between 120° and 140°. From an experimental viewpoint, this is not an optimum situation since the degree to which the contact angle exceeds 90°, merely increases the required pressure for intrusion. In addition, mercury also has a reasonably high surface tension, about 0.485 N/m, and this contributes to the high pressure required for intrusion. An optimum intruding liquid would have a contact angle barely in excess of 90° and a low surface tension. Such a liquid would allow intrusion into the pores at much lower pressures, which provides several advantages. One advantage is that the intrusion instrument can be less massive and expensive. More importantly, the corrections for compression and heating increase greatly at high pressures and errors owing to inaccuracies in the corrections needed become larger at high pressures. Therefore, intrusion volumes measured at low pressures are likely to be more accurate than the volumes measured at high pressures. Another advantage is that low pressures reduce the possibility of crushing the pore walls during intrusion. Since virtually all amalgams have reduced surface tensions and, possibly, reduced contact angles, an intruding liquid that is a solution of another metal in mercury will allow intrusion at a lower pressure than will pure mercury.

From experiments on porous alumina, Winslow found that using a solution of mercury with 5% by mass sodium results in intrusion at pressures that are less than one-half of those required for intrusion with pure mercury [3.50]. He used the intrusion factor $[-4\gamma\cos\theta]$ for intrusion of pure mercury, and from the intrusion factor he calculated the contact angle assuming a surface tension. Then, by assuming that the pore size distribution of the alumina is unaffected by the intruding liquid, one can also calculate a new intrusion factor for the sodium–mercury solution. This latter factor can be used to find a contact angle, if one assumes a surface tension for the solution. These intrusion factors, along with assumed surface tensions and calculated contact angles, are given in Table 3.2. The results showed that both intruding liquids give essentially the same pore size distribution over the range of pore sizes common to both fluids. The distribution obtained with the sodium–mercury solution also

Table 3.2 Intrusion factors for Al_2O_3 calculated with pure mercury and with a solution of mercury with 5% by mass sodium [3.50]

Liquid	Intrusion factor $[-4\gamma \cos\theta]$ (psi μm)	Assumed surface tension (mN/m)	Calculated contact angle (degrees)
Pure Hg	167.6	480	127
Na–Hg solution	74.1	390	109

allows one to explore smaller pores at the maximum available pressure. In this case, the diameter of smallest intrudable pore has been reduced from 28 Å to 12 Å. Using such a solution, however, presents problems because sodium tends to leave the solution and oxidize. More research is needed to find the best intruding liquid. However, Winslow's study indicated that it is possible to make and use an improved intruding liquid that will allow intrusion measurements to be made at lower pressures.

3.5 Advantages and limitations

The comparative advantages and limitations of mercury porosimetry over other techniques have been detailed in studies, and the appropriateness of MIP for pore size analysis has been questioned by several researchers [3.47,3.36,3.90,3.28]. Despite its perceived fundamental and practical limitations, mercury porosimetry will continue to be regarded as a common pore structure characterization technique for cement pastes for years to come. This is so because the technique is conceptually simpler, experimentally much faster, and has the ability to evaluate a much wider range of pore sizes, than any alternative method practised currently. Some of the issues that need to be kept in mind when using MIP and which impose restrictions to the method are mentioned in the following paragraphs.

The method is relatively easy and quick to perform, and the time needed for running the experiment can be as low as 30–45 minutes. A fully automated procedure is possible with currently commercially available equipment and a variety of pore structure parameters can be determined. In addition, the technique has been used extensively in the past and plenty of experimental data is available for reference and comparison.

Extreme care and safety precautions are needed during the test because mercury is a hazardous substance that can be absorbed through the skin and cause illness or even death. Several safety precautions can be used to minimize any possible exposure to mercury vapor. The operator should wear rubber gloves whenever handling mercury. Mercury is slightly volatile at room temperature, mercury vapor is extremely poisonous, and strict rules should govern any work with mercury, such as proper ventilation of the workplace to avoid accumulation of mercury vapor. The intruded specimens must be handled with great care and disposed in a safe and environmentally acceptable manner immediately after completion of the test.

The specimen must be dried before the mercury intrusion process, because mercury cannot intrude into pores that are filled with another liquid. Different drying techniques might lead to different results, therefore the method of drying needs to be reported for comparison of results to be meaningful.

The dimensions of the specimen cell (penetrometer) limit the size of the specimen to be tested. The specimen is only a fraction of the original specimen and precautions are required to ensure that these specimens are representative of the bulk volume of the material. Therefore, the results are not necessarily representative of an infinite pore space as larger pore openings at the surface are more easily accessible in smaller specimen pieces. The use of dynamic porosimeters, in which the pressure changes continuously, might generate different pore size distributions than static porosimeters, where the pressure increases in steps allowing time for the system to come to equilibrium between steps.

At high pressures, when the pressure is raised to several hundreds of atmospheres, a number of new problems develop in measuring intrusion such as compression of the specimen, compression of mercury, and the penetrometer, and rise in the system's temperature. Corrections can be made for the compressibility of the dilatometer and the mercury by doing a blank run. Due to the slight compressibility of mercury, the measured total pore volume of a porous material appears larger than its actual pore volume. Consequently, the larger the specimen and pore volume in comparison to the amount of mercury in the penetrometer, the smaller the error is.

Compression of the specimen leads to false values for the pore volume and pore size. This effect can be shown especially in specimens that contain closed pores and is observed as a too large volume of small- or medium-sized pores. The specimen's compressibility depends both upon the specimen and upon the degree to which its pores have been filled with mercury; thus, the specimen's compressibility changes as the intrusion experiment progresses. There does not appear to be any practical way to account for the changing compressibility during intrusion. However, an approximate correction can be made for the compressibility of the nonporous, solid portion of the specimen. In addition, due to the high pressure involved, permanent structural changes could occur by breakdown of the porous structure during mercury intrusion.

When most substances are compressed, their temperature rises. Due to the compression of the mercury, the temperature in the penetrometer may rise up to as much as 15°. The temperature rise is not a difficulty in porosimetry, if the experiment is conducted slowly enough for the pressured system to remain in thermal equilibrium with the laboratory environment. Under conditions of steadily increasing temperature during pressuring, an apparent extrusion will be observed. The compressive heating of hydraulic oil and mercury causes the mercury to expand in both the penetrometer stem and filled pores. Compression of mercury introduces an additional error opposite to the heating effect.

The pore size distributions obtained by MIP are based on numerous assumptions. The Washburn equation models the individual pores as cylinders of uniform diameter, a model that grossly departs from the reality of cement paste pore systems, which have pores of different sizes and shapes. Furthermore, much of the porosity may be composed of narrow, tortuous pathways, as evidenced by the time required for intrusion to be complete after the applied pressure is increased. Another source of inaccuracy is the presence of ink-bottle pores, which leads to hysteresis and mercury retention in the pores. Large pores with a small opening are thus filled at high pressures, and detected as smaller pores than they actually are. In this way, the population of large pores is systematically misrepresented and the pore size distribution curve is biased in favor of small pore sizes. In this way, MIP does not measure the true pore size distribution but rather the accessibility to mercury of the overall porosity as a function of pore size. Practical mercury porosimetry is limited to the study of pores with a diameter larger than about 4 nm.

Usually, constant surface tension and contact angle values are assumed for mercury. However, the value of contact angle used in calculations differs due to many parameters and more data should be collected. Failure to consider changes in the operant contact angle may produce significant distortions in pore size distributions.

Large pores due to entrained air or accidental voids and microcracks are mostly not interconnected and therefore can only be reached via smaller capillary pores; this leads to a systematic error.

When intruding a number of pieces of cement paste, a distinction must be made between inter-particle void space and the beginning of true intra-particle porosity.

One problem at low pressures is the apparent reluctance of mercury to enter into pores that it should enter. This problem is particularly prevalent at pressures below about 0.2 atm and gradually disappears as the pressure is raised. Even when sufficient pressure is available to intrude a large pore, mercury may not intrude the pore readily unless some mechanical vibration is available to encourage it. Perhaps, this is because the extremely low driving force is insufficient to move the mercury past regions of local roughness along the pore walls; at such places, the contact angle is greater than elsewhere.

Residual air in the penetrometer and in particle pores may lead to minor errors in reported values. Depending on the volume of the bulb holding the test specimen, errors in pore volume distribution due to entrapped air are generally less than 10%.

References

3.1 Washburn E.W. "Note on a method of determining the distribution of pore sizes in a porous material," *Proceedings of the National Academy of Sciences*, Vol. 7, No. 4, 1921, pp. 115–116.

3.2 Ritter H.L., Drake L.C. "Pore-size distribution in porous materials – pressure porosimeter and determination of complete macropore-size distributions," *Industrial and Engineering Chemistry Analytical Edition*, Vol. 17, No. 12, 1945, pp. 782–786.

3.3 Drake L.C. "Pore-size distribution in porous materials," *Industrial and Engineering Chemistry*, Vol. 41, No. 4, 1949, pp. 780–785.

3.4 Joyner L.G., Barrett E.P., Skold R. "The determination of pore volume and area distributions in porous substances. II. Comparison between nitrogen isotherm and mercury porosimeter methods," *Journal of the American Chemical Society*, Vol. 73, No. 7, 1951, pp. 3155–3158.

3.5 Edel'man L.I., Sominskii D.S., Kopchikova N.V. "Pore size distribution in cement rocks," *Colloid Journal* (in English Translation), Vol. 23, No. 2, 1961, pp. 192–196 (pp. 228–233 in original text in Russian).

3.6 Svatá M. "The applicability of mercury porosimetry to materials wetted by mercury," *Powder Technology*, Vol. 29, No. 1, 1981, pp. 145–149.

3.7 Lowell S., Shields J.E. *Powder Surface Area and Porosity*, 3rd Edition, Chapman and Hall, London, 1991.

3.8 Allen T. *Particle Size Measurement*, 5th Edition, Chapman and Hall, New York, 1997.

3.9 Moscou L., Lub S. "Practical use of mercury porosimetry in the study of porous solids," *Powder Technology*, Vol. 29, No. 1, 1981, pp. 45–52.

3.10 Giesche H. "Mercury Porosimetry," *Handbook of Porous Solids*, Vol. 1, F. Schüth, K.S.W. Sing, J. Weitkamp (eds), Wiley-VCH, Weinheim, Federal Republic of Germany, pp. 309–351.

3.11 Winslow D.N. "Advances in experimental techniques for mercury intrusion porosimetry," in *Surface and Colloid Science*, Vol. 13, E. Matijevic, R.J. Good (eds), 1984, Plenum Press, New York, pp. 259–282.

3.12 León y León C.A. "New perspectives in mercury porosimetry," *Advances in Colloid and Interface Science*, Vol. 76–77, 1998, pp. 341–372.

3.13 Orr C. Jr. "Application of Mercury Penetration to Materials Analysis," *Powder Technology*, Vol. 3, No. 1, 1969/70, pp. 117–123.

3.14 Auskern A., Horn W. "Capillary porosity in hardened cement paste," *Journal of Testing and Evaluation*, Vol. 1, No. 1, 1973, pp. 74–79.

3.15 Winslow D.N. "The validity of high pressure mercury intrusion porosimetry," *Journal of Colloid and Interface Science*, Vol. 67, No. 1, 1978, pp. 42–47.

3.16 Liabastre A.A., Orr C. "An evaluation of pore structure by mercury penetration," *Journal of Colloid and Interface Science*, Vol. 64, No. 1, 1978, pp. 1–18.

3.17 Rootare H.M., Prenzlow C.F. "Surface areas from mercury porosimeter measurements," *Journal of Physical Chemistry*, Vol. 71, No. 8, 1967, pp. 2733–2736.

3.18 Rootare H.M., Nyce A.C. "The use of porosimetry in the measurement of pore size distribution in porous materials", *International Journal of Powder Metallurgy*, Vol. 7, No. 1, 1971, pp. 3–11.

3.19 Jenkins R.G., Rao M.B. "The effect of elliptical pores on mercury porosimetry results," *Powder Technology*, Vol. 38, No. 2, 1984, pp. 177–180.

3.20 Cook R.A., Hover K.C. "Mercury porosimetry of cement-based materials and associated correction factors," *ACI Materials Journal*, Vol. 90, No. 2, 1993, pp. 152–161.

3.21 Hiemenz P.C., Rajagopalan R. *Principles of Colloid and Surface Chemistry*, 3rd Edition, Marcel Dekker Inc., New York, 1997.

3.22 Van Brakel J., Modrý S., Svatá M. "Mercury porosimetry: state of the art," *Powder Technology*, Vol. 29, No. 1, 1981, pp. 1–12.

3.23 Winslow D.N., Diamond S. "A mercury porosimetry study of the evolution of porosity in Portland cement," *Journal of Materials*, Vol. 5, No. 3, 1970, pp. 564–585.

3.24 Feldman R.F., Cheng-yi H. "Properties of Portland cement-silica fume pastes I. Porosity and surface properties," *Cement and Concrete Research*, Vol. 15, No. 5, 1985, pp. 765–774.

3.25 Nyame B.K., Illston J.M. "Capillary pore structure and permeability of hardened cement paste," in *Proceedings of the Seventh International Congress on the Chemistry of Cement*, Paris, 1980, Vol. III, pp. VI-181–VI-186.

3.26   Mehta P.K., Manmohan D. "Pore size distribution and permeability of hardened cement pastes," in *Proceedings of the Seventh International Congress on the Chemistry of Cement*, Paris, 1980, Vol. III, pp. VII-1–VII-5.

3.27   Young J.F. "Capillary porosity in hydrated tricalcium silicate pastes," *Powder Technology*, Vol. 9, No. 4, 1974, pp. 173–179.

3.28   Diamond S. "Mercury porosimetry: an inappropriate method for the measurement of pore size distributions in cement-based materials," *Cement and Concrete Research*, Vol. 30, No. 10, 2000, pp. 1517–1525.

3.29   Roy D.M. "Relationships between permeability, porosity, diffusion and microstructure of cement pastes, mortar, and concrete at different temperatures," in *Symposium Proceedings Vol. 137: Pore Structure and Permeability of Cementitious Materials*, L.R. Roberts, J.P. Skalny (eds), Materials Research Society, Warrendale, PA, 1989, pp. 179–189.

3.30   Mindess S., Young J.F., Darwin D. *Concrete*, 2nd Edition, Prentice Hall, Upper Saddle River, NJ, 2003.

3.31   Katz A.J., Thompson A.H. "Quantitative prediction of permeability in porous rock," *Physical Review B*, Vol. 34, No. 11, 1986, pp. 8179–8181.

3.32   Feldman R.F., Beaudoin J.J. "Pretreatment of hardened hydrated cement pastes for mercury intrusion measurements," *Cement and Concrete Research*, Vol. 21, Nos 2/3, 1991, pp. 297–308.

3.33   Winslow D.N., Cohen M.D., Bentz D.P., Snyder K.A., Garboczi E.J. "Percolation and pore structure in mortars and concrete," *Cement and Concrete Research*, Vol. 24, No. 1, 1994, pp. 25–37.

3.34   Liu Z., Winslow D. "Sub-distributions of pore size: a new approach to correlate pore structure with permeability," *Cement and Concrete Research*, Vol. 25, No. 4, 1995, pp. 769–778.

3.35   Bager D.H., Sellevold E.J. "Mercury porosimetry of hardened cement paste: the influence of particle size," *Cement and Concrete Research*, Vol. 5, No. 2, 1975, pp. 171–178.

3.36   Midgley H.G., Illston J.M. "Some comments on the microstructure of hardened cement pastes," *Cement and Concrete Research*, Vol. 13, No. 2, 1983, pp. 197–206.

3.37   Cook R.A., Hover K.C. "Mercury porosimetry of hardened cement pastes," *Cement and Concrete Research*, Vol. 29, No. 6, 1999, pp. 933–943.

3.38   Diamond S. "Pore size distributions in clays," *Clays and Clay Minerals*, Vol. 18, No. 1, 1970, pp. 7–23.

3.39   Diamond S., Dolch W.L. "Generalized log-normal distribution of pore sizes in hydrated cement paste," *Journal of Colloid and Interface Science*, Vol. 38, No. 1, 1972, pp. 234–244.

3.40   Shi D., Brown P.W., Ma W. "Lognormal simulation of pore size distributions in cementitious materials," *Journal of the American Ceramic Society*, Vol. 74, No. 8, 1991, pp. 1861–1867.

3.41   Shi D., Brown P.W., Kurtz S. "A model for the distribution of pore sizes in cement paste," in *Symposium Proceedings Vol. 137: Pore Structure and Permeability of Cementitious Materials*, L.R. Roberts, J.P. Skalny (eds), Materials Research Society, Warrendale, PA, 1989, pp. 23–34.

3.42   Spitzer Z. "Mercury porosimetry and its application to the analysis of coal pore structure," *Powder Technology*, Vol. 29, No. 1, 1981, pp. 177–186.

3.43   Brown S.M., Lard E.W. "A comparison of nitrogen and mercury pore size distributions of silicas of varying pore volume," *Powder Technology*, Vol. 9, No. 4, 1974, pp. 187–190.

3.44   Mikijelj B., Varela J.A., Whittemore O.J. "Equivalence of surface areas determined by nitrogen adsorption and by mercury porosimetry," *American Ceramic Society Bulletin*, Vol. 70, No. 5, 1991, pp. 829–831.

3.45 Abdel-Jawad Y., Hansen W. "Pore structure of hydrated cement determined by various porosimetry and nitrogen sorption techniques," in *Symposium Proceedings Vol. 137: Pore Structure and Permeability of Cementitious Materials*, L.R. Roberts, J.P. Skalny (eds), Materials Research Society, Warrendale, PA, 1989, pp. 105–118.

3.46 Diamond S. "A critical comparison of mercury porosimetry and capillary condensation pore size distributions of Portland cement pastes," *Cement and Concrete Research*, Vol. 1, No. 5, 1971, pp. 531–545.

3.47 Beaudoin J.J. "Porosity measurements of some hydrated cementitious systems by high-pressure mercury – microstructural limitations," *Cement and Concrete Research*, Vol. 9, No. 6, 1979, pp. 771–781.

3.48 Alford N.McN., Rahman A.A. "An assessment of porosity and pore sizes in hardened cement pastes," *Journal of Materials Science*, Vol. 16, No. 11, 1981, pp. 3105–3114.

3.49 Smithwick R.W., Fuller E.L. Jr. "A generalized analysis of hysteresis in mercury porosimetry," *Powder Technology*, Vol. 38, No. 2, 1984, pp. 165–173.

3.50 Winslow D. "Some experimental possibilities with mercury intrusion porosimetry," in *Symposium Proceedings Vol. 137: Pore Structure and Permeability of Cementitious Materials*, L.R. Roberts, J.P. Skalny (eds), Materials Research Society, Warrendale, PA, 1989, pp. 93–103.

3.51 Feldman R.F. "Pore structure damage in blended cements caused by mercury intrusion," *Journal of the American Ceramic Society*, Vol. 67, No. 1, 1984, pp. 30–33.

3.52 Lowell S. "Continuous scan mercury porosimetry and the pore potential as a factor in porosimetry hysteresis," *Powder Technology*, Vol. 25, No. 1, 1980, pp. 37–43.

3.53 Kloubek J. "Hysteresis in porosimetry," *Powder Technology*, Vol. 29, No. 1, 1981, pp. 63–73.

3.54 Adamson A.W., Gast A.P. *Physical Chemistry of Surfaces*, 6th Edition, John Wiley & Sons, Inc., New York, 1997.

3.55 Lowell S., Shields J.E. "Influence of contact angle on hysteresis in mercury porosimetry," *Journal of Colloid and Interface Science*, Vol. 80, No. 1, 1981, pp. 192–196.

3.56 Ternan M., Mysak L.P. "Hysteresis caused by dimensional changes of porous solids during mercury porosimetry," *Powder Technology*, Vol. 52, No. 1, 1987, pp. 29–34.

3.57 Johnston G.P., Smith D.M., Melendez I., Hurd A.J. "Compression effects in mercury porosimetry," *Powder Technology*, Vol. 61, No. 3, 1990, pp. 289–294.

3.58 Cebeci Ö.Z. "Pore structure of air-entrained hardened cement paste," *Cement and Concrete Research*, Vol. 11, No. 2, 1981, pp. 257–265.

3.59 Diamond S., Leeman M.E. "Pore size distributions in hardened cement paste by SEM image analysis," in *Symposium Proceedings Vol. 370: Microstructure of Cement-Based Systems/Bonding and Interfaces in Cementitious Materials*, S. Diamond, S. Mindess, F.P. Glasser, L.W. Roberts, J.P. Skalny, L.D. Wakeley (eds), Materials Research Society, Warrendale, PA, 1995, pp. 217–226.

3.60 Hill R.D. "A study of pore-size distribution of fired clay bodies II. An improved method of interpreting mercury penetration data," *Transactions of the British Ceramic Society*, London, Vol. 59, 1960, pp. 198–212.

3.61 Metz F., Knöfel D. "Systematic mercury porosimetry investigations on sandstones," *Materials and Structures*, Vol. 25, 1992, pp. 127–136.

3.62 Whiting D., Kline D.E. "Pore size distribution in epoxy impregnated hardened cement pastes," *Cement and Concrete Research*, Vol. 7, No. 1, 1977, pp. 53–60.

3.63 Shi D., Winslow D.N. "Contact angle and damage during mercury intrusion into cement paste," *Cement and Concrete Research*, Vol. 15, No. 4, 1985, pp. 645–654.

3.64 Winslow D.N., Lovell C.W. "Measurements of pore size distributions in cements, aggregates and soils," *Powder Technology*, Vol. 29, No. 1, 1981, pp. 151–165.

3.65 Sellevold E.J. "Mercury porosimetry of hardened cement paste cured or stored at 97°C," *Cement and Concrete Research*, Vol. 4, No. 3, 1974, pp. 399–404.

3.66 Day R.L., Marsh B.K. "Measurement of porosity in blended cement pastes," *Cement and Concrete Research*, Vol. 18, No. 1, 1988, pp. 63–73.

3.67   Marsh B.K., Day R.L., Bonner D.G. "Pore structure characteristics affecting the permeability of cement paste containing fly ash," *Cement and Concrete Research*, Vol. 15, No. 6, 1985, pp. 1027–1038.

3.68   Gallé C. "Effect of drying on cement-based materials pore structure as identified by mercury intrusion porosimetry – a comparative study between oven-, vacuum-, and freeze-drying," *Cement and Concrete Research*, Vol. 31, No. 10, 2001, pp. 1467–1477.

3.69   Moukwa M., Aitcin P.-C. "The effect of drying on cement pastes pore structure as determined by mercury porosimetry," *Cement and Concrete Research*, Vol. 18, No. 5, 1988, pp. 745–752.

3.70   Hansen W., Almudaiheem J. "Pore structure of hydrated Portland cement measured by nitrogen sorption and mercury intrusion porosimetry," in *Symposium Proceedings Vol. 85: Microstructural Development During Hydration of Cement*, L.J. Struble, P.W. Brown (eds), Materials Research Society, Warrendale, PA, 1987, pp. 105–114.

3.71   Thomas M.D.A. "The suitability of solvent exchange techniques for studying the pore structure of hardened cement paste," *Advances in Cement Research*, Vol. 2, No. 5, 1989, pp. 29–34.

3.72   Konecny L., Naqvi S.J. "The effect of different drying techniques on pore size distribution of blended cement mortars," *Cement and Concrete Research*, Vol. 23, No. 5, 1993, pp. 1223–1228.

3.73   Hearn N., Hooton R.D. "Sample mass and dimension effects on mercury intrusion porosimetry results," *Cement and Concrete Research*, Vol. 22, No. 5, 1992, pp. 970–980.

3.74   Palmer H.K., Rowe R.C. "The application of mercury porosimetry to porous polymer powders," *Powder Technology*, Vol. 9, No. 4, 1974, pp. 181–186.

3.75   Smith D.M., Schentrup S. "Mercury porosimetry of fine particles: particle interaction and compression effects," *Powder Technology*, Vol. 49, No. 3, 1987, pp. 241–247.

3.76   Kumar R., Bhattacharjee B. "Study on some factors affecting the results in the use of MIP method in concrete research," *Cement and Concrete Research*, Vol. 33, No. 3, 2003, pp. 417–424.

3.77   Lowell S., Shields J.E. "Pore spectra from continuous scan mercury porosimetry," *Powder Technology*, Vol. 28, No. 2, 1981, pp. 201–204.

3.78   Cook R.A., Hover K.C. "Experiments on the contact angle between mercury and hardened cement paste," *Cement and Concrete Research*, Vol. 21, No. 6, 1991, pp. 1165–1175.

3.79   Good R.J., Mikhail R.Sh. "The contact angle in mercury intrusion porosimetry," *Powder Technology*, Vol. 29, No. 1, 1981, pp. 53–62.

3.80   Adolphs J., Setzer M.J., Heine P. "Changes in pore structure and mercury contact angle of hardened cement paste depending on relative humidity," *Materials and Structures*, Vol. 35, 2002, pp. 477–486.

3.81   Shields J.E., Lowell S. "A new instrument for mercury contact angle measurements," *Powder Technology*, Vol. 31, No. 2, 1982, pp. 227–229.

3.82   Winslow N.M., Shapiro J.J. "An instrument for the measurement of pore-size distribution by mercury penetration," *ASTM Bulletin*, No. 236, February 1959, pp. 39–44.

3.83   Good R.J. "The contact angle of mercury on the internal surfaces of porous bodies. A footnote to D.N. Winslow's Review of Porosimetry," *Surface and Colloid Science*, Vol. 13, E. Matijevic, R.J. Good (eds), 1984, Plenum Press, pp. 283–287.

3.84   Hughes D.C. "Pore structure and permeability of hardened cement paste," *Magazine of Concrete Research*, Vol. 37, No. 133, 1985, pp. 227–233.

3.85   Modry S., Hejduk. J. "The limitation of high pressure mercury porosimetry to the study of hardened cement pastes," in *Proceedings of the Seventh International Congress on the Chemistry of Cement*, Paris, 1980, Vol. IV, pp. 387–389.

3.86   Feldman R.F. "A discussion of the paper 'Contact angle and damage during mercury intrusion into cement paste' by D. Shi and D.N. Winslow," *Cement And Concrete Research*, Vol. 16, No. 3, 1986, pp. 452–454.

3.87 Olson R.A., Neubauer C.M., Jennings H.M. "Damage to the pore structure of hardened Portland cement paste by mercury intrusion," *Journal of the American Ceramic Society*, Vol. 80, No. 9, 1997, pp. 2454–2458.

3.88 Diamond S., Bonen D. "Microstructure of hardened cement paste – a new interpretation," *Journal of the American Ceramic Society*, Vol. 76, No. 12, 1993, pp. 2993–2999.

3.89 Davidson J.A. "Mercury porosimetry studies I. Low melting alloy intrusion studies as an aid to the interpretation of mercury porosimetry data," *Powder Technology*, Vol. 23, No. 2, 1979, pp. 233–238.

3.90 Bentur A. "The pore structure of hydrated cementitious compounds of different chemical-composition," *Journal of The American Ceramic Society*, Vol. 63, No. 7–8, 1980, pp. 381–386.

4 Gas adsorption

Gas adsorption has been one of the most popular techniques used for the study of pore structure in materials that contain micropores and mesopores. When a clean solid surface is exposed to a gas, the gas molecules impinge upon the solid and may reside on its surface for a finite time; this phenomenon is called *adsorption*. Adsorption processes may be classified as physical or chemical, depending on the nature of the interaction forces involved between the gas and the solid. For pore structure characterization in cement-based materials, physical adsorption is used and will be considered throughout this chapter. Physical adsorption is a reversible process because it is the result of a relatively weak interaction between the solid surface and the gas molecules. Therefore, almost all the gas adsorbed on the solid surface can be removed by evacuation at the same temperature at which it was adsorbed, a process that is called *desorption*.

The methods developed in the past for analysis of experimental adsorption results usually involve models for condensation of gas in capillaries, such as the Kelvin equation, and models for adsorption of gas molecules on free surfaces such as the Brunauer–Emmett–Teller (BET) equation. The systematic development of gas adsorption as a technique for pore structure characterization started in the early 1900s. In 1911, A. Zsigmondy proposed capillary condensation in small pores as the mechanism for vapor adsorption [4.1]. In 1916–1918 I. Langmuir formulated his theory about coverage of a surface with a monomolecular layer of gas [4.2,4.3]. In 1938, S. Brunauer *et al.*, published their theory about formation of multilayers on a solid surface from vapor adsorption [4.4]. Tremendous further developments in theory occurred from 1930–1980 mostly in the form of analytical equations, and a similar extensive development occurred after 1980 until present times in which microstructure simulations became more feasible than in previous years.

Despite technical improvements and the development of new sophisticated equations for a wide selection of materials, the Kelvin equation and the BET equation still hold their place as the basis of many calculations in gas adsorption because of their simplicity. Theories developed for analysis of gas adsorption data usually assume that the pores are cylindrical and for this reason, the term "pore radius" is frequently mentioned in the technical literature. Gas adsorption as a pore structure characterization technique only measures pores in cement pastes with radii between 1 nm and about 60 nm. In this way, the gas adsorption technique can only characterize gel pores and small and medium capillary pores.

The developments of gas adsorption on porous materials distinguish between micropores and mesopores and these terms are quite often used throughout this chapter, rather than the terms gel and capillary pores. The corresponding sizes to pores in concrete science have already been given in Table 1.1.

The first adsorption experiments on cement pastes were carried out by T. Powers and T. Brownyard, 1946–1947, using water vapor as adsorbate, and by R. Blaine and H. Valis, 1949, using water vapor and nitrogen as adsorbates [4.5,4.6]. Since then, researchers have used adsorption experiments regularly, in particular for determining the surface area of pores in cement-based materials and to investigate the effect of various parameters on the pore structure development.

4.1 Theory and testing procedure

When a porous solid is exposed to a gas of a certain volume and under a finite pressure, it begins to adsorb gas molecules on the outside surface and inside its pores. The solid is referred to as the *adsorbent* and the gas as the *adsorbate*. *Adsorptive* is the general term for the material in the vapor phase that is capable of being adsorbed. The adsorption process is accompanied by an increase in the mass of solid and a decrease in the pressure of gas. Adsorption of gas molecules on the solid's pore walls is considered to take place progressively and can be of different nature, i.e. monolayer or multilayer adsorption, or capillary condensation. First, the surface of the solid's pores adsorbs gas molecules forming a monomolecular layer (monolayer) on the surface of the solid. The monolayer formed on the solid surface will attract further gas molecules, and the adsorbed layer formed becomes several molecules thick, called a multilayer. At a certain stage, the thickness of the adsorbed layers on the pore walls approximates the size of the pore, and then capillary condensation takes place (see Figure 4.1). In micropores where the size of the pore is comparable to the size of the gas molecule, the mechanism is different and is known as *micropore filling*. When the gas pressure is reduced, the gas is evaporated from the pore in a reverse procedure.

For gas adsorption, it is assumed that the gas molecules are strongly attracted and adsorbed on a clean surface of the porous solid, and are evenly distributed throughout the surface of the solid. The pores of the adsorbent are

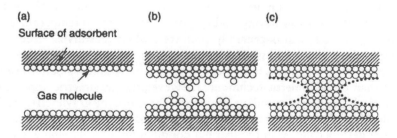

Figure 4.1 Schematic oversimplified process showing various stages during the process of adsorption. (a) Monolayer adsorption, (b) multilayer adsorption, and (c) capillary condensation.

considered to be large compared to the molecular dimensions of the gas. The amount of gas adsorbed on the solid surface depends upon several factors: (a) the surface area of the specimen, (b) the nature of both the specimen and the gas, (c) the pressure of the gas at which adsorption takes place, and d) the temperature. The quantity of physically adsorbed gas at a given pressure increases as the temperature decreases; consequently, most gas adsorption measurements are carried out at low temperatures. The vapor most frequently used for adsorption experiments is nitrogen at its boiling point, 77.3 K (−195.7°C), because nitrogen has low reactivity with most solids, and it is readily available as pure gas or as liquid. Other gases used include water vapor, argon, krypton, and alcohols, and are discussed in Section 4.7.2 later.

In their usual state, all solid surfaces are covered with a film of physically adsorbed vapors from the air (e.g. nitrogen, oxygen, water) and contaminants, which fill or occlude pores and crevices, and change surface characteristics. Vapors and contaminants adsorbed on the surface of a solid must be removed before any quantitative measurements using gas adsorption are made. Physical adsorption is due to van der Waals forces, which are weak; therefore, the physically adsorbed layer can be readily removed from the solid's surface if the specimen is maintained under high vacuum. The exact conditions (temperature and pressure) required for removing the physically adsorbed layer from the specimen's surface before the experiment depend on the particular gas–solid system. For routine analysis it has been found that a vacuum of 0.133–1.33 Pa (10^{-3}–10^{-2} mmHg) is sufficient in order to remove the physically adsorbed vapors from the surface of a material [4.7].

The amount of gas adsorbed during an adsorption experiment can be calculated by using three general methods: gravimetric, volumetric, or thermal conductivity measurements. With the gravimetric method, the increase in mass of the specimen is determined as gas is adsorbed on the pore walls of the solid and is recorded by a high precision balance. During the gravimetric technique, separate instrumentation is used to measure changes in pressure and mass of the adsorbate. The gravimetric method is preferred when water vapor is used as adsorbate. With the volumetric method, the amount of gas removed from the gas–solid system due to adsorption is determined using gas laws, if the volumes of the gas and the specimen are known. The volumetric method requires high precision pressure transducers, and high precision volume measurements. The volumetric method is most often used with nitrogen, because the method is most suitable for gases having boiling points below room temperature. The volumetric method is also generally preferred when reasonable accuracy is required in the region of pressures close to the saturation vapor pressure [4.8]. The gravimetric technique is more precise and accurate and is a better research method than the volumetric technique. However, the gravimetric apparatus is more difficult to construct and maintain, and therefore the gravimetric method is costlier than the volumetric method. The thermal conductivity method uses a cell that detects changes in the thermal conductivity of a flowing gas stream passing over the specimen. Adsorption of gas by the specimen changes the composition, and consequently the thermal conductivity, of the gas. The early types of equipment applied vacuum to the specimen and dosed nitrogen into the vacuum; this procedure was time-consuming, because of the slow heat transfer in a vacuum (sometimes several hours were required to obtain equilibrium at one point). In modern equipment, nitrogen is adsorbed from a mixture of helium and nitrogen, which has high thermal conductivity causing rapid equilibrium.

A variety of forms for volumetric apparatus exist, modified for use with a specific adsorbate. A schematic set-up of equipment for the volumetric gas adsorption experiment using nitrogen is shown in Figure 4.2. To cause sufficient gas to be adsorbed for surface area measurements, the specimen must first be cooled, normally to the boiling point of nitrogen. After degassing under vacuum to remove contaminants from the specimen's surface, helium is admitted into a vessel of known volume and its pressure and temperature are measured. The helium is then allowed to flow into the specimen tube. Helium is not adsorbed by the specimen and the volume in the specimen tube is termed dead space volume being linearly dependent on pressure. The residual amount of gas in the vessel and the amount of gas expanded into the specimen tube are determined. Then the helium is removed and nitrogen is added into the system. First, nitrogen gas is taken into the vessel and its pressure is read on the pressure meter. The valve between the specimen and the calibrated vessel is then opened and, after sufficient time is allowed for equilibrium to be established, the new gas pressure is measured, which is the equilibrium adsorption pressure. The volume of gas admitted in the specimen tube is proportional to the difference in the pressure before and after opening the valve. Nitrogen in the system is split into three parts: residual amount of gas in the calibrated volume vessel and connections, dead space in the specimen tube that can be calculated from the stage involving helium, and amount adsorbed by the specimen. The gas volume adsorbed by the specimen together with the equilibrium pressure gives one adsorption point. Known quantities of nitrogen gas are added stepwise to

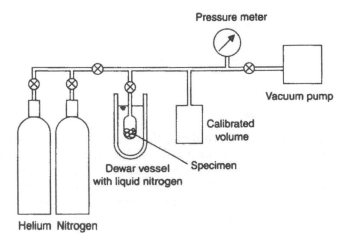

Figure 4.2 Schematic out-of-scale set-up of the equipment for volumetric nitrogen adsorption.

the specimen in such amounts that after about 30 equilibrium points the saturation vapor pressure, P_0, of nitrogen is reached. Each dose of nitrogen is introduced to the specimen only after the foregoing dose of nitrogen has reached adsorption equilibrium with the specimen. By definition, equilibrium is reached when the change in nitrogen pressure is not greater than 13.3 Pa (0.1 mmHg) over a 5 minutes interval [4.9]. The volume of adsorbed gas increases with increasing pressure but not linearly. During desorption, nitrogen is removed from the specimen in a stepwise mode with the same precautions taken to ensure desorption equilibrium in a similar way as it was applied during the adsorption conditions. About 30 equilibrium points are obtained during desorption. During the experiment, the amount of gas adsorbed and the corresponding pressure are recorded, both during adsorption and desorption. It is essential that during adsorption and desorption the experimental points be evenly distributed over the adsorbed volume vs. pressure plot. If too large additions or withdrawals of nitrogen are made, the temporarily too high nitrogen pressure during adsorption or too low pressure during desorption may result in "scanning effects" within the hysteresis loop of the adsorption–desorption branches of the isotherm. The quantity of gas adsorbed may be expressed in many ways, such as mass of gas, volume of gas reduced to standard temperature and pressure, gas moles, and to some extent gas molecules.

4.2   Analysis of data

From the amount of gas adsorbed and the corresponding relative gas pressure recorded during the experiment, a great number of methods have been developed for analysis of data. The most commonly used analysis methods are discussed in the following sections, and include methods for determining the pore volume, pore size distribution, and the specific surface of the material tested. Since the mechanism of adsorption is different in small and large pores, different methods of experimental analysis have been developed with different assumptions. More information on additional methods and equations used for analysis can be found elsewhere [4.7,4.8].

The first steps of analysis include obtaining the adsorption isotherm, determining the thickness of adsorbed film on the pore walls, and determining the pore size from the values of relative pressure measured during the experiment.

4.2.1   Adsorption isotherm

The graph of the volume of adsorbed gas, V_a, vs. the corresponding adsorption pressure, P, at constant temperature is called an *adsorption isotherm* and is the corresponding curve to the porogram obtained using mercury intrusion porosimetry that was mentioned in Section 3.2.1. If the gas is at a pressure lower than the critical pressure, i.e. if the gas is a vapor, the relative pressure P/P_0 rather than the pressure P is preferred to be used for analysis of data and for plots.

The majority of adsorption isotherms obtained for various materials may be grouped into six types, as shown in Figure 4.3. The first five types were

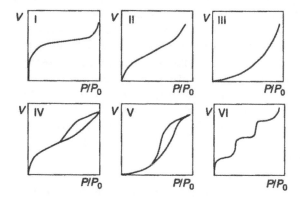

Figure 4.3 Types of adsorption isotherms obtained for various materials given by IUPAC Recommendations. The classification of isotherms of Types I–V were originally proposed by S. Brunauer *et al.*, 1940 [4.10] and Type VI was proposed by Joyner *et al.*, 1945 [4.11]. Type I is obtained from an adsorbent with micropores; Type II from a non-porous adsorbent; Type III from a non-porous adsorbent or an adsorbent with macropores; Type IV from an adsorbent with mesopores and macropores; Type V from an adsorbent with micropores or mesopores; Type VI is observed in carbon blacks and is rather rare.

given by S. Brunauer *et al.*, 1940 [4.10], while Type VI was identified by L. Joyner *et al.*, 1945 [4.11]. From the type of isotherm obtained during a gas adsorption experiment certain conclusions may be drawn about the characteristics of the specimen used. Type I isotherms are referred to as Langmuir-type isotherms and are obtained when adsorption on the specimen's pore walls is restricted to a monolayer. These isotherms are characteristic when chemisorption is occurring, or when the adsorbents have micropores, which are too small to accommodate adsorption in a few layers and subsequent capillary condensation. Type II isotherms are often referred to as sigmoid isotherms, are frequently encountered, and represent multilayer physical adsorption on non-porous solids. Types III and V isotherms are rare, show no rapid initial uptake of gas, and occur when the interaction between adsorbent and adsorbate is rather weak. Type III isotherms are obtained from non-porous solids or solids with macropores. Type V isotherms are obtained from solids having micropores or mesopores and show an inflection point at a fairly high relative pressure. Type IV isotherms are characteristic of solids having mesopores and macropores. Type VI isotherms indicate a non-porous solid with a uniform outside surface, and show the rise known as the Joyner–Emmett step due to continued adsorption in the first layer with simultaneous second layer adsorption. Type VI isotherms are rather rare, observed in carbon blacks and are not usually considered for analysis.

Isotherms of types IV and V are valid for porous materials only and are modifications of isotherms of types II and III respectively: in the low pressure range, there is no distinction between isotherms of types II and IV or between

III and V. They also reflect capillary condensation phenomena, because they level off close to the saturation vapor pressure. For isotherms of types I, II, and III, reducing the pressure during desorption results in following the exact same path as during adsorption. For isotherms of types IV and V, the path followed during desorption is not the same as adsorption, therefore the isotherms possess a hysteresis loop as their characteristic feature.

The Type IV isotherm is an isotherm that has been studied and analyzed extensively. For a porous material that shows a Type IV isotherm, gas is adsorbed inside the pores of the material forming first a monolayer, then a multilayer, and finally capillary condensation takes place. A. Zsigmondy, 1911, put forward the capillary condensation theory, which is still accepted and has served as the basis for virtually all subsequent theoretical treatments of Type IV isotherm [4.1]. Zsigmondy made use of the principle established earlier by Lord Kelvin on thermodynamic principles, that the vapor pressure over a concave meniscus of liquid in a pore of capillary shape is less than the saturation vapor pressure at the same temperature. His theory implies that a vapor will be able to condense to a liquid inside the pores of a solid, even when the saturation vapor pressure is not reached. The Type IV isotherm includes the following features (see also Figure 4.4): at low pressures, the isotherm follows the same path as a Type II isotherm, but at a certain point it begins to deviate upwards, until at high pressures its slope decreases. Along the initial part of the isotherm (segment AB in Figure 4.4), adsorption is restricted to a thin layer of adsorbate on the pore walls. The "knee," point B in Type II and Type IV isotherms, occurs at a relative pressure somewhere between

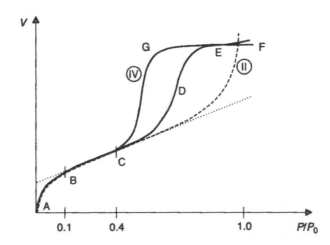

Figure 4.4 Type IV adsorption isotherm and its characteristic features. The dotted line corresponds to Type II isotherm. Point B is at the beginning of the linear segment and corresponds to the completion of a monolayer on the pore walls. The segment CDE corresponds to adsorption and EGC to desorption. Point C corresponds to commencement of capillary condensation in the fine pores. The segment EF corresponds to filling of all pores with liquid adsorbate. The relative pressures corresponding to points B and C are approximate values.

$P/P_0 = 0.05$ and 0.10, and has been interpreted as indicative of the completion of a continuous monomolecular layer of gas on the pores' surface. After point B, the first steep part of the isotherm ends and a relatively straight and less steep part begins at which the thickness of the film adsorbed on the pore walls increases, and is identified to correspond to multilayer adsorption. At point C, which is the inception point of the hysteresis loop, capillary condensation commences first in pores of small radii, and continues progressively into large pores. As the saturation vapor pressure $P/P_0 = 1$ is approached towards point E, wide pores are filled by gas, until at the saturation vapor pressure the entire pore system is full of condensate. Curve EF in Figure 4.4 represents the filling of all pores with liquid adsorbate. As the pressure is decreased, thinning of the multilayer film adsorbed on the pore walls occurs progressively, starting at pores with large radii and progressing into pores with small radii. The path followed during desorption, segment EGC, is not the same as the path followed during adsorption, segment CDE, and as a result a hysteresis loop is formed, the shape of which depends both on the adsorbent and the adsorbate. The amount of gas adsorbed at any given relative pressure is always greater along the desorption branch than along the adsorption branch. The point where the hysteresis loop closes, point C, corresponds to a relative pressure of approximately 0.4 [4.8].

The adsorption isotherm obtained for hardened Portland cement paste is a Type IV isotherm and an example of it is shown in Figure 4.5 obtained using nitrogen as adsorbate on cement pastes pretreated in different ways [4.12]. More

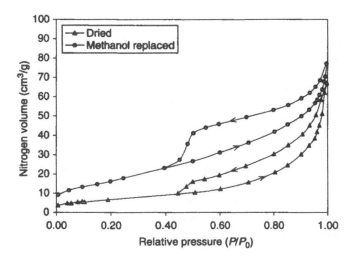

Figure 4.5 Nitrogen adsorption isotherm of hardened cement pastes with water–cement = 0.40, pretreated with oven-drying and methanol replacement.

Source: Reprinted from *Pore Structure and Permeability of Cementitious Materials*, L.R. Roberts, J.P. Skalny (eds), *MRS Symposium Proceedings Vol. 137*, Y. Abdel-Jawad, W. Hansen, "Pore structure of hydrated cement determined by mercury porosimetry and nitrogen sorption techniques," pp. 105–118, copyright 1989, with permission from the Materials Research Society, Warrendale, PA.

examples on isotherms obtained for cement pastes using different adsorbates, and discussion of their features are given in Section 4.6 later.

4.2.2   Thickness of adsorbed film

The average thickness of a layer of adsorbate on the pore walls of a material at a certain relative pressure can be calculated theoretically or by empirical experimental extrapolations. The thickness of the adsorbed film is used in analyses for pore size distributions mentioned in Section 4.4, and to calculate the specific surface area mentioned in Section 4.5.

Theoretically, by assuming that for any value of relative pressure the thickness of the adsorbed film on the pore walls is the same as the thickness of the adsorbed film on a plane surface, the thickness of the adsorbed film is given by Equation (4.1) [4.13,4.14]:

$$t = \frac{V_a}{V_m} \tau \qquad\qquad (4.1)$$

where t is the thickness of adsorbed film on the pore wall (mm), V_a is the volume of adsorbed gas (mm^3 of adsorbate per gram of adsorbent), V_m is the volume of gas adsorbed when the entire surface is covered with a mono-molecular layer (mm^3 adsorbate per gram of adsorbent), τ is the thickness of a monomolecular layer of adsorbate (mm).

Equation (4.1) simply states that the thickness of the adsorbed film on a pore wall is equal to the thickness of a monomolecular layer times the number of layers formed. The volume of adsorbed vapor, V_a, can be measured during the experiment, and the thickness of a monomolecular layer of adsorbate, τ, can be calculated theoretically by assuming a certain packing of gas molecules. The volume of adsorbate that would cover the surface with a monomolecular layer, V_m, can be determined using several approaches, such as the Langmuir equation, the BET equation, or the Dubinin–Kaganer equation, which are described in Section 4.5.

As an example, the thickness of a monomolecular layer of nitrogen can be calculated from the area occupied by one molecule of nitrogen and from the volume and surface occupied by one mole of nitrogen. Nitrogen adsorption is governed by adsorbate–adsorbate interaction, which pulls the gas molecules together, particularly near the completion of a monolayer. The area occupied by one molecule of nitrogen has been calculated from the density of liquid nitrogen using Equation (4.2) [4.15]:

$$\sigma = 1.091 \left(\frac{M_r}{\rho_L N_A} \right)^{2/3} = 16.2 \cdot 10^{-20} \, m^2 \qquad\qquad (4.2)$$

where σ is the area occupied by one molecule of nitrogen (m^2), M_r is the molecular weight of nitrogen (=28.01), ρ_L is the density of liquid nitrogen (=0.81 $\cdot$ 10^6 g/m^3), N_A is the Avogadro constant (=6.023 $\cdot$ 10^{23} molecules/mol).

The type of packing of nitrogen molecules has an effect on the thickness of the monomolecular layer. For an assumed cubical packing of nitrogen molecules, the thickness of a monomolecular layer has been calculated to be $\tau = 0.402$ nm [4.15]. However, the most commonly assumed packing of nitrogen molecules is hexagonal close-packing, which is the densest packing of molecules. In the case of hexagonal close-packing, the thickness of a monomolecular layer of nitrogen can be calculated from Equation (4.3) [4.16]:

$$\tau = \frac{V_{mol}}{S_{mol}} = \frac{V_{mol}}{\sigma \cdot N_A} = \frac{34.6 \cdot 10^{-6}}{(16.2 \cdot 10^{-20}) \cdot (6.023 \cdot 10^{23})} = 3.54 \times 10^{-10} \, m$$

(4.3)

where τ is the thickness of a monomolecular layer of adsorbate (m), V_{mol} is the volume occupied by one mole of liquid nitrogen (m^3), S_{mol} is the area occupied by one mole of liquid nitrogen (m^2), σ is the area occupied by one molecule of nitrogen (m^2), N_A is the Avogadro constant ($=6.023 \cdot 10^{23}$ molecules/mol).

Using the thickness of a calculated monomolecular layer, τ, the volume of adsorbed gas, V_a, measured during the experiment, and the volume of adsorbate, V_m, needed to cover the pores' surface with a monomolecular layer determined by one of the methods described in Section 4.5, one can calculate at any point during the experiment the thickness of the adsorbed film, t, on the pore walls using Equation (4.1).

Alternatively, the thickness, t, of the adsorbed layer can be determined *empirically* from experimental curves. Researchers have found that the amount of nitrogen adsorbed per unit of surface area of the specimen is a unique function of relative pressure and can be represented by a single curve, which is called a *t-curve* (see Figure 4.6). With the aid of the *t*-curve, the thickness of the adsorbed layer can be calculated as a function of P/P_0. Composite curves for a variety of adsorbents for nitrogen at 77.3 K have been published by several researchers, but the most commonly used curves are those published by R. Cranston and F. Inkley, 1957 [4.17] and by J. de Boer *et al.*, 1965 [4.18]. The curve reported by Cranston and Inkley was obtained from nitrogen adsorption measurements on several non-porous materials, including zinc oxide, titanium oxide, tungsten powder, precipitated silver etc. The curve reported by de Boer *et al.*, was obtained from nitrogen adsorption experiments on titanium oxide, aluminum oxide, magnesium oxide, zirconium oxide, and barium sulfate. It should be noted that the values from the *t*-curve cannot be used for all materials and the *t*-curve should be used with caution. In addition, the *t*-curve is considered inappropriate at pressures below the closure pressure of the hysteresis loop (approximately 0.4 of relative pressure), because at these low pressures the mechanism of adsorption includes also micropore filling that has not been completed [4.19].

As it is seen from Figure 4.6a, the *t*-curves for nitrogen provided by de Boer *et al.*, and by Cranston and Inkley compare very well to each other. The

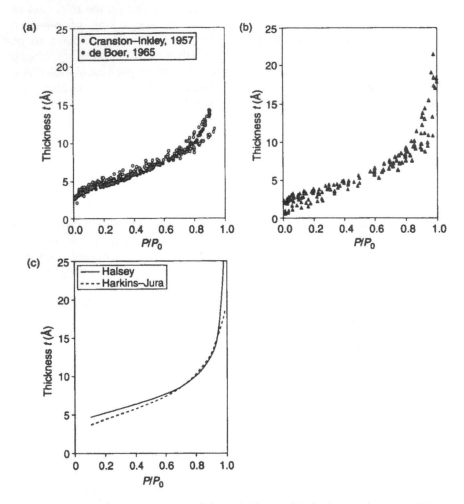

Figure 4.6 Examples of *t*-curves used for calculating the thickness of adsorbed film on an adsorbent surface. (a) Curves obtained from nitrogen adsorption on materials by de J. Boer *et al.*, 1965 [4.18] and by R. Cranston and F. Inkley, 1957 [4.17]; (b) curves obtained from water vapor adsorption on materials by J. Hagymassy *et al.*, 1969 [4.20]; (c) curves drawn from empirical equations given by W. Harkins and G. Jura, 1944 [4.21,4.22], and G. Halsey, 1948 [4.23] (Equations (4.4) and (4.5), respectively, mentioned in the text).

t-curve published by de Boer gives smaller values for the thickness of adsorbed film at relative pressures below 0.7 than the curve published by Cranston and Inkley; this is probably because the adsorbents used for obtaining the de Boer *t*-curve had lower heats of adsorption than the adsorbents used by Cranston and Inkley. At a given relative pressure, more gas is adsorbed by the specimen if the heat of adsorption is high, a phenomenon that is especially noticeable at low relative pressures. The effect of the adsorbent's surface is much weaker in the formation of the second adsorbed layer and hardly noticeable in the third and higher layers [4.20]. The *t*-curve by Cranston and Inkley gives lower values for the thickness of adsorbed film than the de Boer curve at $P/P_0 > 0.75$; this is probably because some capillary condensation occurred in de Boer's adsorbents at high relative pressures. De Boer and co-workers have also expressed doubts about the correctness of their *t*-curve above a relative pressure of $P/P_0 = 0.75$, which is a vitally important region in the analysis of wide pores.

J. Hagymassy *et al.*, 1969, have published t-curves using water vapor as adsorbate on various materials including zirconium silicate, rutile, silica, silica gel, zinc oxide, quartz, and calcite (see Figure 4.6b) [4.20]. The original curve published by the researchers was determining values of the number of layers formed (V/V_m); these values have been multiplied by a factor of 3, which is estimated to be the thickness of a monomolecular layer of water, for the curve presented in Figure 4.6b.

Several equations have been proposed for determining the thickness of the adsorbed film from measured P/P_0 values during the gas adsorption experiment. For the derivation of these equations, it is assumed that the adsorbed nitrogen in the monolayer has the same density and packing of molecules as for liquid nitrogen. The most commonly used equations are the Harkins–Jura equation and the Halsey equation.

W. Harkins and G. Jura, 1944, proposed the empirical Equation (4.4) assuming hexagonal close-packing for nitrogen molecules, which results in thickness of a monomolecular layer of nitrogen of 0.354 nm, as it was calculated in Equation (4.3) [4.21,4.22]:

$$t = \sqrt{\frac{13.99}{0.034 - \log(P/P_0)}} \tag{4.4}$$

where *t* is the thickness of adsorbed film on the pore wall (Å), P/P_0 is the relative pressure of vapor.

Later, G. Halsey, 1948, proposed Equation (4.5) for determining the thickness of the adsorbed film [4.23]. The values of -5 and 3.54 used in Equation (4.5) are empirical:

$$t = 3.54 \sqrt[3]{\frac{-5}{2.303 \log(P/P_0)}} \tag{4.5}$$

where *t* is the thickness of adsorbed film on the pore wall (Å), P/P_0 is the relative pressure of vapor.

The Halsey equation is convenient to use and results in a similar relationship between the thickness of adsorbed film on the pore walls and the relative pressure P/P_0 as the *t*-curve obtained by Cranston and Inkley, 1957. The Halsey equation has been proposed for correcting the Kelvin equation and obtaining pore size distributions, as it is described later in Section 4.4. As it is shown in Figure 4.6c, the curves obtained using the Harkins–Jura equation, and the Halsey equation compare relatively well to each other.

For the *t*-curve published by de Boer *et al.*, the Harkins–Jura equation (Equation 4.6a) and an empirical equation (Equation 4.6b) can be used for describing the curve [4.7]:

$$\text{For } \frac{P}{P_0} < 0.75 \quad \log\frac{P}{P_0} = 0.034 - \frac{0.1399}{t^2} \tag{4.6a}$$

$$\text{For } 0.4 < \frac{P}{P_0} < 0.96 \quad \log\frac{P}{P_0} = -\frac{0.1611}{t^2} + 0.1682 \exp\left(-1.137t\right)$$

$$\tag{4.6b}$$

where t is the thickness of adsorbed film on the pore wall (Å), P/P_0 is the relative pressure of vapor.

An experimental isotherm can be analyzed by plotting the volume adsorbed, V_a, at a given relative pressure vs. the value of thickness of adsorbed film, t, at the same relative pressure taken from the t-curve. The obtained plot of V_a vs. t is called a *t-plot* from which the specific surface can be obtained graphically, as it is described in Section 4.5.5.

4.2.3   Pore size (Kelvin equation)

The theory of condensation effects in pores can be traced back to the work of P. Laplace in 1806 and T. Young in 1855. In 1871, W. Thomson (later named Lord Kelvin) published an equation that relates the size of a pore to the relative pressure, P/P_0, of the gas at which capillary condensation occurs inside the pore [4.24]. The Kelvin equation has been the basis of all procedures used over the last seventy years for determining the pore size distribution of a material from the Type IV isotherm. The Kelvin equation does not apply well to macropores, which are too big for capillary condensation, applies well to mesopores, which are typical for capillary condensation, and is inapplicable to micropores in which the pore size is comparable to the size of the gas molecules. The Kelvin equation can be applied to either the adsorption or the desorption branch of an isotherm.

The Kelvin equation determines the *Kelvin radius* or *critical radius* of a cylindrical pore into which capillary condensation occurs at a specific relative pressure. For the derivation of the equation, Kelvin assumed that the gas is ideal, the condensed liquid is incompressible, and that the molar volume of the liquid phase is negligible compared to the molar volume of the gas. These assumptions lead to significant errors at high temperatures, especially as the critical point is approached. The Kelvin equation for cylindrical pores is expressed by Equation (4.7):

$$r_K = \frac{-2\gamma V_{mol} \cos\theta}{RT \ln(P/P_0)} = \frac{k}{\ln(P/P_0)} \quad \text{where } k = \frac{-2\gamma V_{mol} \cos\theta}{RT} \tag{4.7}$$

where r_K is the radius of the pore in which condensation occurs (m), γ is the surface tension of the liquid adsorbate (N/m), V_{mol} is the volume occupied by one mole of condensate at temperature T (m^3/mol), θ is the contact angle between the liquid and the pore wall (usually assumed to be 0), R is the gas constant = 8.314 J/(Kmol), T is the temperature (K), k is a constant, function of the adsorbate and the experimental conditions (m), P is the pressure (N/m²), P_0 is the saturation vapor pressure (N/m²).

For *nitrogen*, the surface tension is 8.85×10^{-3} N/m and the molar volume is 3.467×10^{-5} m³/mol, so at nitrogen's boiling point ($T = 77.3$ K), the Kelvin equation may be written as it is expressed by Equation (4.8):

$$r_K = \frac{-2 \cdot (8.85 \cdot 10^{-3}) \cdot (3.467 \cdot 10^{-5})}{(8.314)(77.3) \ln(P/P_0)} = \frac{-9.59}{\ln(P/P_0)} \times 10^{-10} \text{ (m)}$$

or   (4.8)

$$r_K = \frac{-4.15}{\log(P/P_0)} \times 10^{-10} \text{ (m)}$$

The upper limit in pore radius is limited by the term $\ln(P/P_0)$ in the denominator of the Kelvin equation and approaches infinity as the relative pressure approaches 1.0. The largest pore radius determined using Equation (4.8) is about 95 nm, at a relative pressure $P/P_0 = 0.99$. An increase in pore size from 50 to 60 nm corresponds to a negligible change in relative pressure P/P_0 from 0.981 to 0.984. It is therefore impossible to allocate the condensate that fills pores with radius greater than about 50 nm to specific diameter classes [4.12,4.25].

The lower limit in pore size associated with capillary condensation theories depends on the particular isotherm, as well as on the calculational procedure. Some researchers have determined pore size distributions with pore diameters as low as 1.5 nm. It is generally conceded that the Kelvin equation tends to break down for micropores. O. Kadlec and M. Dubinin, 1969, presented data suggesting that the Kelvin equation does not apply for pore diameters as small as 3–4 nm [4.19]. They also concluded that the Kelvin equation is inapplicable at relative pressures slightly higher than those at which the adsorption–desorption hysteresis loop closes. Pores with a size smaller than the critical value do not exhibit capillary condensation, but instead fill continuously as the pressure is increased. The analysis of the micropore size distribution is even more uncertain, because these sizes are too small for thermodynamic techniques to be applicable.

The Kelvin equation ignores the fact that adsorption inside the pores is not caused only by capillary condensation, but by adsorption on the pore walls as well. In a similar way, during desorption, an adsorbed film remains on the pore walls while evaporation of the gas at the center of the pore takes place. A. Wheeler, 1945, was the first to point out the necessity of taking into account the thickness of the adsorbed multilayer of gas when applying the Kelvin equation, especially in small pores, where the thickness of the adsorbed multilayer is a significant fraction of the pore radius [4.26]. Wheeler assumed that at any point during desorption, all pores are covered with an adsorbed film of thickness t, and that all pores smaller than the critical radius r_K are completely filled by adsorbed or condensed gas. The approach proposed by Wheeler to

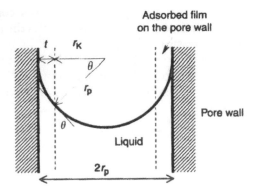

Figure 4.7 Schematic graph showing the Kelvin radius, r_K, the pore radius, r_p, and the thickness t of the film adsorbed on the pore wall of a cylindrical pore. The contact angle θ is also indicated at the contact with the adsorbed layer.

determine the pore radius is satisfactory for measurements on pores ranging in radius from about 2 to 30 nm [4.26]. When a cylindrical pore model is assumed, the radius of the pore can be determined by addition of the core radius (Kelvin radius) and the thickness of the adsorbed film when condensation or evaporation occurs, as is given by Equation (4.9) (see also Figure 4.7). When a slit-pore model is used, the pore width is taken to be $r_K = 2t$:

$$r_p = r_K + t \tag{4.9}$$

where r_p is the pore radius (m), r_K is the radius of the pore in which condensation occurs (m), t is the thickness of adsorbed film on the pore wall (m).

The thickness t of the adsorbed layer as a function of relative pressure may be determined from empirical measurements or theoretical calculations, as it was described in Section 4.2.2. Wheeler suggested the use of Halsey's equation for determining the thickness of the adsorbed layer, in association with Kelvin's equation to calculate pore size using Equation (4.9). The analysis methods used for pore size distribution described in Section 4.4 use the Kelvin equation and the thickness of adsorbed film.

4.3 Total pore volume

The total pore volume of the specimen may be calculated from the gas adsorbed at saturated vapor pressure. Due to capillary condensation, all the pores are filled with liquid condensate when the saturated vapor pressure is reached. If the amount of gas adsorbed on the pore walls is neglected, then the volume of gas adsorbed would be equal to the volume of pores; this, however, can only be true when the uptake at saturation vapor pressure is definite.

Furthermore, for the case of micropores the Dubinin–Radushkevich equation is used, as it is described in the following section.

4.3.1 *Dubinin–Radushkevich equation*

Micropores have pore dimensions that are only few times greater than the gas molecule diameter, and the use of the Kelvin equation is generally regarded as inappropriate. Various methods have been put forward to obtain information about micropores from relative pressure values. In 1960, M. Dubinin derived the Dubinin–Radushkevich (DR) equation for determining the volume of micropores based on the researchers' earlier theory for the mechanism of micropore filling [4.27,4.28]. The DR equation uses the low and medium pressure parts of the adsorption isotherm. According to Dubinin, if the pore diameter is comparable to the size of the gas molecule, micropores are filled by volume rather than by layer-to-layer adsorption on the pore walls. Dubinin also considered that bringing a molecule from the gas phase to a given point in the adsorbed film depends on both the adsorbate and the adsorbent. In order to obtain a relation that is characteristic of the adsorbent only and independent of the adsorbate, Dubinin introduced coefficient β, a constant that is characteristic of the adsorbate, and parameter k_a, which is determined by the pore size distribution. On the assumption that the pore size distribution is Gaussian, Dubinin and Radushkevich derived Equation (4.10), which determines the volume of micropores in a material from values of measured relative pressure:

$$\frac{V}{\rho V_0} = \exp\left[-D\log^2\left(\frac{P_0}{P}\right)\right] \quad \text{with} \quad D = 2.303\frac{k_a}{\beta^2}(RT)^2 \qquad (4.10)$$

where V is the volume of micropores that have been filled at relative pressure P/P_0 and at temperature T (m^3 per gram of adsorbent), V_0 is the total micropore volume (m^3/g), ρ is the density of liquid adsorbate (g/m^3), R is the gas constant = 8.314 J/(Kmol), T is the temperature (K), k_a is a characteristic parameter relating to the adsorbent only, β is the affinity (or similarity) coefficient that depends on the adsorbate (unitless), P is the pressure (N/m^2), P_0 is the saturation vapor pressure (N/m^2).

For plotting purposes, Equation (4.10) can be written in linear form as it is expressed by Equation (4.11), with all the parameters defined in the same way as for Equation (4.10):

$$\log V = \log(\rho V_0) - D\left[\log^2\left(\frac{P_0}{P}\right)\right] \qquad (4.11)$$

According to Equation (4.11), a plot of log V vs. [$\log^2(P_0/P)$], which is known as the DR-plot, should be a straight line with an intercept that is equal to the total micropore volume V_0 (see Figure 4.8). From the slope of the line, the value of D can be obtained. The DR plot deviates from linearity and displays an upward turn as the saturation vapor pressure is approached; this

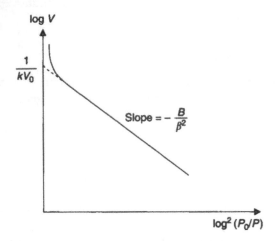

Figure 4.8 DR-plot showing the extrapolation of the volume of micropores V_0 and the constant B/β^2 for calculation of the specific surface.

feature can be understood considering multilayer adsorption and capillary condensation in mesopores.

The DR equation has been found to apply for the adsorption of nitrogen, and organic adsorbents over the range of relative pressure 10^{-5}–10^{-1} on a variety of microporous materials [4.7]. It has been widely used in Materials Science for determining the micropore volume of carbons and for determining the thermodynamic parameters of gas adsorption. The method has been used recently by E. Robens *et al.*, 2002, to calculate the micropore volume in cement pastes made from ordinary Portland cement and blast furnace slag cement using nitrogen as adsorbate [4.29]. The researchers reported the micropore volume to be 0.46 mm³/g for hydrated ordinary Portland cement paste and 0.92 mm³/g for cement paste containing blast furnace slag. Further information on their work is given in Section 4.7.3 (Table 4.7).

4.4   Pore size distribution

From the obtained adsorption isotherm and after converting the relative pressure P/P_0 to pore radius, it is possible to obtain a pore size distribution, which is a plot of the cumulative pore volume vs. the pore radius. Pore size distributions of cement pastes having different water–cement ratios obtained using nitrogen as adsorbate are shown in Figure 4.9 [4.30]. The distributions demonstrate the increase in number of pores with large diameters as the water–cement ratio increases.

Although obtaining accurate volume and pressure measurements during a gas adsorption experiment is a matter of routine, the conversion of data to yield a reliable pore size distribution is rather difficult. In order to calculate the pore size distribution of a material, a model is usually selected, e.g. cylindrical pores, ink-bottle pores, slit-shaped pores, etc. Procedures commonly used

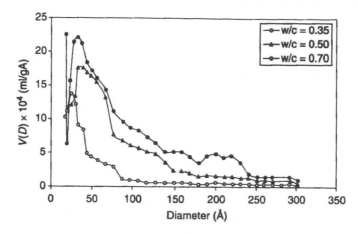

Figure 4.9 Pore size distributions obtained using nitrogen adsorption on hardened Portland cement pastes having different water–cement ratios.

Source: Reprinted from the *Canadian Journal of Chemistry*, Vol. 42, No. 2, R.Sh. Mikhail, L.E. Copeland, S. Brunauer, "Pore structures and surface areas of hardened Portland cement pastes by nitrogen adsorption," pp. 426–438, copyright 1964, with permission from the National Research Council of Canada.

for determining the pore size distribution from capillary condensation measurements are the Barrett–Joyner–Halenda method [4.31], the Cranston–Inkley method [4.17], and the micropore method and modelless method proposed by S. Brunauer *et al.* [4.32,4.33]. The methods used for pore size distribution make use of the combined effect of capillary condensation in the inner pore volume, which is the pore core, and physical adsorption on the pore walls.

All pore size distribution procedures describe an algorithm that gives the volume and surface area contained in a certain group of pore sizes. By summing up the calculated partial volumes and surface areas, one can calculate the total pore volume and surface area of the material tested. During analysis, it is necessary to decide which branch of the isotherm to use, the adsorption or the desorption branch. It has been suggested that in most cases of porous materials, thermodynamic equilibrium is more nearly complete on the desorption branch of the isotherm, and therefore pore size distributions should be determined using the desorption branch of the isotherm [4.34]. However, in the case of ink-bottle pores, the wide portion of the pore is unable to evaporate until the narrow neck empties during desorption; therefore, for ink-bottle pores equilibrium is believed to be achieved in the adsorption branch of the isotherm [4.13].

For pore size distribution, it is of great assistance if regularly spaced experimental points are available. Since these are difficult to obtain in practice, it may be preferable to fit the experimental measurements to an analytical curve and employ numerical interpolation, especially if computing facilities are available [4.35].

4.4.1   *The Barrett–Joyner–Halenda method*

The Barrett–Joyner–Halenda (BJH) method was proposed in 1951 and was originally developed for relatively coarse porous adsorbents having a wide range of pore sizes [4.31]. However, the procedure proved to be applicable to almost all types of porous materials.

For their analysis the researchers assumed that the pores are cylindrical in shape and that the radius of the capillary is equal to the sum of the Kelvin radius and the thickness of the adsorbed film on the pore walls. For determining the thickness of the adsorbed layer, t, as a function of P/P_0, the researchers used a t-curve provided by C. Shull, 1948 [4.36]. They also assumed that the thickness of the adsorbed layer inside the pores is the same as the thickness of the adsorbed layer on a flat surface and that all pores are filled at a relative pressure of 1.0. They divided the pore sizes into groups, and for simplicity they assumed that all pores in each group of capillaries have an average radius r_p. In general, they recommended that the average pore radius increases by 0.5 nm from one pore size group to the other up to a pore radius of about 6 nm, and increases by 1 nm above 6 nm.

The desorption branch of the isotherm obtained experimentally is used for analysis. Computations start at a relative pressure of 0.967 down to a relative pressure of approximately 0.4; in this way, pores with a radius greater than 30 nm are ignored. The large pores could contain up to 10% of the total volume of condensate, but their number is generally small compared to the number of smaller pores; hence they do not contribute appreciably to the pore size distribution. The BJH method considers the isotherm as a series of steps as the pressure is lowered. The amount of adsorbate removed during each pressure lowering step is divided between inner pore emptying from liquid adsorbate and film thinning processes. When the relative pressure is lowered from $(P/P_0)_1$ to $(P/P_0)_2$, the largest pores empty of their capillary condensate and a reduction in thickness of the adsorbed layer occurs by the amount Δt_1. Further reduction of pressure will not only result in evaporation and thinning of the adsorbed film in the second group of pore sizes, but also from a second thinning of the physically adsorbed layer from the first group of pore sizes. For each group of pores, the pore core is calculated, as well as the change in thickness of the adsorbed film during each step of pressure reduction. The total pore volume and the surface area of the pores can be easily calculated by adding the values calculated in each group of pores.

The researchers also showed that the previous assumption for Gaussian distribution of pore sizes is inadequate for many adsorbents, and that the thickness of the physically adsorbed layer on the pore walls is not constant. Even though they assumed cylindrical pores for their analysis, their method is suitable also when slit-shaped pores are assumed.

Several researchers have used the method for determining the pore size distribution of cement-based materials [4.29,4.37,4.38]. E. Robens *et al.*, 2002, determined the total surface area and pore volume of cement pastes using the BJH method by analyzing both the adsorption and desorption branch of the nitrogen isotherm. They found that the pore volume and surface area values determined using the desorption branch of the isotherm were higher than the values using the adsorption branch. Their experimental results are listed in Table 4.7 and are discussed in Section 4.7.3.

4.4.2 The Cranston–Inkley method

R. Cranston and F. Inkley, 1957, proposed a method similar to but more precise than the BJH method for deriving pore size distributions from adsorption isotherms [4.17]. The pore structure analysis they proposed can be made for either the adsorption or the desorption branch of the isotherm. The researchers found that for most adsorbents, the analysis of the adsorption isotherm gave more reliable results than the desorption isotherm, as judged by the agreement between the cumulative surface area and the surface area calculated using the BET method (see Section 4.5.2). The Cranston–Inkley analysis starts like the BJH method with pores having a radius of 30 nm and ignores larger pores.

Initially, it is assumed that pores are cylindrical with one end closed. They divided the pore sizes into groups with predetermined average radius r_p for each group from 1 to 30 nm and they provided the corresponding critical relative pressure. At any relative pressure P/P_0, all pores with radii larger than r_p contain an adsorbed film of thickness t on their walls and all pores smaller than r_p are filled due to joint multilayer adsorption and capillary condensation. The critical pore radius is taken as the sum of the calculated Kelvin radius and the approximately determined thickness of the multilayer existing on a flat surface at the same relative pressure. For determining the thickness of adsorbed film, the researchers provided a t-curve from published isotherms on 15 non-porous materials (shown in Figure 4.6a) and used the Halsey equation (Equation 4.5) for describing their t-curve. The volume adsorbed in the pores is obtained from the experimental isotherm using the P/P_0 values. They corrected the volumes of pores taking into consideration the gaseous and liquid phase of nitrogen. When the pressure increases from P_1 to P_2, the total volume of nitrogen is used to fill the pores whose critical pressures have been reached and to increase the thickness of the adsorbed layer on the walls of larger pores.

The cumulative pore volume and surface area can be calculated at the end of the analysis. The Cranston–Inkley method provides an estimate of surface area almost independently from the BET method. The differences in the surface area determined by the Cranston-Inkley method compared to the values obtained by the BET method are not experimental errors, but may give an indication of the character of the pores in materials.

4.4.3 The modelless method and micropore (MP) analysis method

S. Brunauer *et al.*, 1967, proposed that a complete pore structure analysis can be performed in two parts using a combination of the MP method for micropores and the corrected modelless method for wide pores (mesopores and macropores) [4.39]. The lower part of the *t*-curve, up to a relative pressure of about $P/P_0 = 0.5$, is used in the analysis of the pore volume and pore surface distributions of micropores and is called the micropore analysis method [4.32]. The large pores are analyzed from the *t*-curve for values of relative pressure P/P_0 between 0.4 and 1.0 using the modelless method, which is based on capillary condensation in pores having widths greater than 1.5 nm [4.33]. The criterion for selecting the correct *t*-curve for analysis is that the heat of adsorption of nitrogen on the adsorbent to be analyzed should be approximately the same as the heat of adsorption of the non-porous material on which the *t*-curve is based. An adequate test for the applicability of the *t*-curve from an adsorbent is a comparison of its BET constant C with that of the adsorbent under test (see Section 4.5.2).

Both the micropore system and the wide pore system are arbitrarily divided into pore groups, and the surface area and volume of pores in each group is determined. After the volume and surface area in each pore group are determined, one can sum up the results and obtain a cumulative pore volume and cumulative pore surface area. Two criteria are used for the correctness of the analysis: the cumulative pore volume must agree with the total pore volume, V_s, and the cumulative surface area must agree with the total surface area obtained using the BET method, S_{BET}.

Modelless method   The corrected modelless method can be used for the analysis of volume and surface area of wide pores of any adsorbent, using any adsorbate for which a *t*-curve is available. The analysis uses the hysteresis region of an adsorption isotherm, the region in which multilayer adsorption and capillary condensation occur simultaneously. S. Brunauer *et al.*, 1967 reported that the desorption branch rather than the adsorption branch represents capillary condensation equilibrium [4.33]. For the modelless method, the researchers assumed no shape for the pores. They used the hydraulic radius, and not the Kelvin radius, as a measure of the average size of a group of pores regardless of their shapes. The volumes, surface areas, and hydraulic radii of the pores are obtained using the modelless method for the cores of each group of pores.

For their analysis, they divided the cores into 16 groups, from $P/P_0 = 1.0$ to $P/P_0 = 0.40$, the relative pressure range for each group being 0.05. If the relative pressure ranges are chosen to be sufficiently small, the plot of log P/P_0 vs. *V* can be considered linear for each group of pores. The hydraulic radius of the core of each pore group is determined assuming either parallel plate pores or cylindrical pores. Conversion from core to pore is done using the thickness of the adsorbed film, taken from the *t*-curve. In their analysis, the researchers used the *t*-curve provided by Cranston and Inkley; however, they stated that other *t*-curves may be used in analysis.

When the pressure is lowered from $(P/P_0)_1$ to $(P/P_0)_2$, the first group of pores is emptied by capillary evaporation, but a multilayer adsorbed film still remains on the pore walls. The core is the part of the pore that fills up by capillary condensation or remains empty after capillary evaporation. The core volume is the volume desorbed between $(P/P_0)_1$ and $(P/P_0)_2$. The method of analysis proposed gives the distribution of the core volumes and the distribution of surface areas as functions of the core hydraulic radii. When P/P_0 is lowered from $(P/P_0)_2$ to $(P/P_0)_3$, the volume desorbed is not only the volume of the cores of the second group, but includes an adsorbed film on the first group of pores as well. The researchers used thinning correction factors that were calculated from the t-curve used.

One proceeds in a similar manner to obtain the volumes and the surfaces of all core groups. The thinning corrections for the adsorbed layer cannot be made without assuming a shape for the pores and, in this respect, the method proposed cannot be regarded as entirely modelless. The correction factors are the same for cylindrical and parallel plate models, increase as one progresses from group to group, and eventually the volume correction becomes equal to the volume desorbed. This occurs near the closure of the hysteresis loop, indicating that at low pressures only adsorption takes place in narrow pores and not capillary condensation. Corrections add little, if any, significant information to that obtainable from the uncorrected values. If no corrections are applied, the volume obtained for each group of pores is somewhat larger than the true core volume of those pores, the uncorrected surface is somewhat larger than the true core surface, and the hydraulic radius is almost the same whether corrections are applied or not.

The modelless method proposed by Brunauer involves an integration, which in the original work was performed graphically. However, an analytical integration is possible, which is faster and can carry out the entire procedure of pore size analysis [4.40].

Micropore analysis method   The corrected modelless method is inapplicable for analyzing the structure of the micropores. S. Brunauer and colleagues, 1968, presented a method for the analysis of pore volume and pore surface distributions of micropores, which they named the micropore analysis method or MP-method [4.32]. The method is a further development of the t-method proposed by B. Lippens and J. de Boer, 1965, described in Section 4.5.5 later [4.41]. The volumes, surfaces, and hydraulic radii of groups of pores are calculated from the downward deviations from the straight line in the de Boer t-plot. In the analysis of micropore structure, it is absolutely necessary that a correct t-curve be employed, because in micropores the adsorbed film constitutes the entire pore dimension. The use of the correct t-curve is not quite so critical in the analysis of the large pores, because the adsorbed film constitutes only part of the pore dimension [4.32].

For the analysis, the researchers first converted the adsorption isotherm into a V–t-plot using the method of Lippens and de Boer. The value of thickness of adsorbed film, t, for the corresponding relative pressure P/P_0 is read from the t-curve and the adsorbed volume, V_t, is plotted against the thickness of adsorbed film, t. The points are taken from the smooth isotherm curve at relative pressure intervals of 0.05. The slope of the first curve gives the surface area S_t. From the slope of the next line, S_2, the surface area, hydraulic radius and volume of the first group of pores are calculated.

The analysis is continued until there is no further decrease in the slope of the t-plot, which means that no blocking of pores occurs due to multilayer adsorption [4.32].

The researchers showed that the MP-method results in the same volume calculated for the micropores whether the pores are assumed to be cylindrical or slit-shaped. Correct values for the volumes of the groups of pores can be obtained only if the surface area and adsorbed film thickness values are correct. The correctness of the surface area values depends on the correctness of the starting surface area, S_t, which should be the same as the surface area calculated from the BET method, S_{BET}, if the appropriate t-curve is used.

The MP-method may fail to detect micropores in certain cases, which can be due to the simultaneous occurrence of micropore filling and capillary condensation, the so-called "compensation mechanism" [4.44]. If micropore filling and capillary condensation occur at the same time, i.e. at the same values of t on the t-plot, they will cancel each other out and yield a t-plot with a zero or positive deviation; such a plot makes the analysis of micropores impossible. The reason for the possible occurrence of the compensation mechanism in specimens remains unclear.

J. Hagymassy *et al.*, 1972, have shown that the MP method and the "corrected modelless method" originally used for nitrogen isotherms can be used to analyze the pore structure of cement pastes when water vapor is used as the adsorbate [4.42]. In their experiments, shown in Table 4.1, cumulative volumes and surface areas as well as surface area and volume distributions in the pores were obtained assuming parallel plates and cylindrical pores. They found that if the real shape of the pores is unknown, the cylindrical pore model gives the best approximation [4.39]. O. Kadlec and M. Dubinin, 1969, have reported that it is impossible to calculate micropore distribution from the desorption branches of water isotherms, but the adsorption branch should be used [4.19]. I. Odler and J. Skalny, 1971 showed that micropores could be detected from water vapor isotherms only, not from nitrogen isotherms when the modelless method is used [4.43].

Agreement between the cumulative pore surface area, S_t, and the BET surface area, S_{BET}, frequently cited as the criterion of the correctness of the analysis is not correct. The micropore analysis leaves out the large pores that fill or empty in the vicinity of the saturation vapor pressure, P_0, as well as all the pores that fill or empty at low pressures, i.e. below the region of capillary condensation. Therefore, the cumulative pore volume should always be smaller than the total pore volume, and the cumulative pore surface area should be smaller than the surface area calculated using the BET method, and this is observed in the results presented in Table 4.1. The total surface area of an adsorbent, S_t cannot be greater than S_{BET}; at best it can be equal to S_{BET}. The great importance of the correct value of the surface area S_t lies in the fact that it constitutes the starting point of the MP-method for pore structure analysis. If the surface area S_t is close to the surface area S_{BET}, it can be concluded that the t-curve on which the analysis is based is reliable for the given adsorbent.

Table 4.1 Pore structure parameters (surface area S, pore volume V, and hydraulic radius r_h) of regular cement pastes that have been D-dried [4.42]

| w/c | Time in years | Water vapor | | | | | | | | | | | | | | | Nitrogen | | |
| | | Micropores | | | Mesopores and macropores | | | | | | Cumulative | | | | | | | | |
		S m^2/g	V m^2/g	r_h (Å)	S_{pp} m^2/g	V_{pp} ml/g	r_{h-pp} (Å)	S_{cp} m^2/g	V_{cp} ml/g	r_{h-cp} (Å)	S_{cp} m^2/g	V_{cp} ml/g	V_s ml/g	r_h (Å)	S_{BET} m^2/g	S_t m^2/g	S_{BET} m^2/g	V_s ml/g	r_h (Å)
0.40	2	30	0.014	4.4	70.1	0.152	21.7	88.3	0.157	17.8	118	0.171	0.184	11.8	156	157	53	0.118	22.2
0.45	7	70	0.023	4.2	70.6	0.180	25.5	88.7	0.186	21.0	151	0.209	0.225	11.8	191	200	89	0.136	15.3
0.70	2	67	0.024	3.7	79.5	0.353	44.4	97.3	0.359	36.9	165	0.383	0.385	19.2	200	200	—	—	—

Notes
The BET-method and the t-plot method have been used for analysis. pp = parallel plate and cp = cylindrical pore model. The values were obtained using water vapor and nitrogen as adsorbates.

4.5   Specific surface

A variety of methods is available for determining the specific surface from gas adsorption measurements. The methods can be categorized into two groups: analytical and graphical, even though a strict distinction is not possible. Analytical methods are based on an analytical equation fitted to the isotherm data from which a parameter can be extracted representing the specific surface. These methods include the Langmuir equation, the BET equation, and the Dubinin–Kaganer equation. In graphical methods the specific surface is determined graphically from a plot, and this category includes the Harkins–Jura method, the t-plot method, and the α_s-plot method.

A commonly used analytical method for determining the specific surface of a porous material is to deduce from the isotherm the amount of gas required to form a monomolecular layer, which is called the *monolayer capacity* [4.16,4.45,4.46]. The number of gas molecules required to form a monomolecular layer may be determined, and since the area occupied by one gas molecule is known, or may be estimated, the specific surface area of the material may be calculated from Equation (4.12):

$$S = n_m A_m N_A \quad \text{or} \quad S = \frac{V_m N_A A_m}{V_M} \cdot 10^{-20} \tag{4.12}$$

where S is the specific surface area (m^2/g), V_m is the volume of gas adsorbed when the entire surface is covered by a monomolecular layer (m^3 per gram of adsorbent), A_m is the average area occupied by one molecule of adsorbate in the completed monolayer (m^2/molecule), V_M is the molar volume (m^3/mol), N_A is the Avogadro constant ($=6.023 \cdot 10^{23}$ molecules/mol), n_m is the amount of gas adsorbed when the entire surface is covered by a monomolecular layer (moles per gram of adsorbent).

When *nitrogen* is used as the adsorbate, the specific surface area is given by Equation (4.13), where 4.35 m^2 is the area occupied by 1 cm^3 of nitrogen. When the volume of gas V_m is expressed in m^3/g, the surface area is obtained in m^2/g:

$$S = \frac{(6.023 \cdot 10^{23})(16.2 \cdot 10^{-20})V_m}{22.4 \cdot 10^{-3}} \quad \text{or} \quad S = 4.35 V_m \times 10^6 \,(\text{m}^2/\text{g}) \tag{4.13}$$

The volume of gas, V_m, required for the completion of a monomolecular layer can be obtained by the Langmuir equation, the BET equation, and the Dubinin–Kaganer equation, which are described in Sections 4.5.1, 4.5.2, and 4.5.3, respectively.

Alternatively, the specific surface may be calculated from Equation (4.14), which is obtained after combining Equation (4.1) mentioned earlier and Equation (4.12):

$$S = \frac{V_a N_A A_m}{t V_M} \tau \tag{4.14}$$

where S is the specific surface area (m²/g), V_a is the volume of adsorbed vapor (m³ per gram of adsorbent), N_A is the Avogadro constant ($=6.023 \cdot 10^{23}$ molecules/mol), A_m is the average area occupied by one molecule of adsorbate in the completed monolayer (m²/molecule), t is the thickness of adsorbed film on the pore wall (m), V_M is the molar volume (m³/mol), τ is the thickness of a monomolecular layer of adsorbate (m).

Equation (4.14) determines the specific surface area from the measured volume of adsorbed gas, V_a, and the calculated thickness, t, of the adsorbed film. For nitrogen, Equation (4.14) becomes Equation (4.15) for calculating the specific surface:

$$S = 15.47 \frac{V_a}{t} \times 10^{-4} \tag{4.15}$$

where S is the specific surface area (m²/g), V_a is the volume of adsorbed vapor (m³ per gram of adsorbent), t is the thickness of adsorbed film on the pore wall (m).

Equation (4.15) is the basis of the t-plot, which is another way of calculating the specific surface and is discussed in Section 4.5.5.

4.5.1 The Langmuir theory

The Langmuir equation is perhaps the most important of all equations in the field of gas adsorption and establishes a relationship between the pressure of a gas in equilibrium with a flat surface and the volume of gas adsorbed on the surface. I. Langmuir studied extensively the solid–gas interactions and won the Nobel prize in Chemistry in 1932 for his discoveries and investigations in surface chemistry. Langmuir's equation describes well Type I isotherms and provides the basis for subsequent developments. Some basic assumptions in deriving the Langmuir equation are that the adsorbate behaves as an ideal gas and the amount of gas adsorbed is confined to a monomolecular layer. In addition, it is assumed that as a monomolecular layer of the adsorbed gas forms on the pore walls, the repulsion from the already adsorbed gas molecules hinders the formation of a second and subsequent layers. This assumption is not very realistic, because most gas molecules do not adsorb into a single layer, but rather pile up into multilayers at some sites, even before other regions are covered by the first layer. It is also assumed that the energy of adsorption is constant, which implies that the adsorbent's surface is entirely uniform, although this is not supported by experimental evidence. Other assumptions are that the ability of a molecule to bind on the solid's surface is independent of whether or not nearby sites are occupied, i.e. there is no adsorbate–adsorbate interaction. For his approach, I. Langmuir, 1915, 1918, equated the number of gas molecules evaporating from the surface with the number of gas molecules condensing on the surface and derived Equation (4.16) [4.2,4.3]:

$$\frac{P}{V} = \frac{1}{k_L V_m} + \frac{P}{V_m} \quad \text{or} \quad \frac{P}{n} = \frac{1}{k_L n_m} + \frac{P}{n_m} \tag{4.16}$$

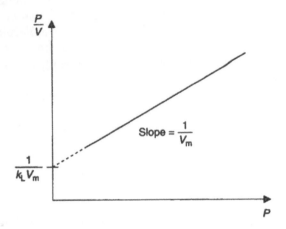

Figure 4.10 Langmuir plot used to calculate the volume of gas adsorbed at a monolayer, V_m, from the slope, and from that the surface area of the porous
material.

where P is the pressure (N/m²), V is the volume of adsorbed vapor at pressure
P (m³ per gram of adsorbent), V_m is the volume of gas adsorbed when the entire
surface is covered by a monomolecular layer (m³ per gram of adsorbent), k_L is
an empirical constant (m²/N), n is the amount of gas adsorbed at pressure P
(moles per gram of adsorbent), n_m is the amount of gas adsorbed when the entire
surface is covered by a monomolecular layer (moles per gram of adsorbent).

A plot of P/V vs. P will give a straight line of slope $1/V_m$ and intercept
$1/k_L V_m$, from which both k_L and V_m can be calculated (see Figure 4.10). Having
established V_m graphically, the specific surface area of the specimen can be
calculated using Equation (4.12), or in the case of nitrogen using Equation (4.13).

At relatively low temperatures, and at pressures approaching the saturation
vapor pressure, agreement of experimental data with Langmuir's theory is
generally poor. The Langmuir theory ignores the possibility that the initial
monomolecular layer may act as a substrate for further adsorption. It is
possible that at low temperatures and high pressures adsorbed molecules can
hold other gas molecules so that multimolecular layers are possible [4.34].

The Langmuir theory has been used for analysis on ordinary Portland cement
pastes by J. Adolphs and M. Setzer, 1998 [4.47]. The researchers determined
the Langmuir constant, k_L, to be 23 when nitrogen was used as adsorbate, and
3152 when water vapor was used as adsorbate in their experiments. Their
results are presented in Table 4.8 and discussed in Section 4.7.3.

4.5.2   The Brunauer–Emmett–Teller (BET) theory

In 1938, S. Brunauer *et al.* extended Langmuir's theory by introducing the
adsorption of several layers of gas on a solid surface [4.4]. Their theory assumes
that the solid surface is energetically homogeneous, i.e. that all adsorption
sites are energetically identical. It is also assumed that there is no

adsorbate–adsorbate interaction, and there is no variation in properties of adsorbed layers after the first layer. Another assumption is that the heat of adsorption of the second and higher layers is the same and equal to the heat of liquefaction, i.e. to the heat of bulk liquid. At saturation, the number of layers is assumed to become infinite, therefore, from a theoretical point of view, the theory is only valid for a non-porous material. At dynamic equilibrium, the number of molecules evaporating from a layer is equal to the number of molecules condensing on the layer below. The researchers equated the rate of condensation of gas molecules onto an adsorbed layer to the rate of evaporation of gas molecules from the layer, and by summing for an infinite number of layers beneath, they arrived at the expression given by Equation (4.17). This equation is known as the BET equation after the initials of the names of the three researchers, and is applicable to isotherms of types I through V, depending on the value of the BET constant C [4.7]. In linear form, Equation (4.17) is written as Equation (4.17a):

$$V = \frac{V_m CP}{(P_0 - P)\left[1 + (C-1)P/P_0\right]} \quad \text{with } C = \exp\left[\frac{q_1 - q_L}{RT}\right] \tag{4.17}$$

$$\frac{P}{V(P_0 - P)} = \frac{1}{V_m C} + \frac{(C-1)P}{V_m CP_0} \tag{4.17a}$$

where V is the volume of adsorbed vapor at pressure P (m^3 per gram of adsorbent), V_m is the volume of gas adsorbed when the entire surface is covered by a monomolecular layer (m^3 per gram of adsorbent), C is the BET constant, function of the net heat of adsorption ($q_1 - q_L$) of the monomolecular layer (unitless), P is the pressure (N/m^2), P_0 is the saturation vapor pressure (N/m^2), q_1 is the heat of adsorption of the first layer (J), q_L is the heat of liquefaction of the liquid adsorptive (J).

A plot of $P/[P_0 V(1 - P/P_0)]$ vs. P/P_0 is the BET plot, and generally gives a straight line in the region of relative pressures near completed monolayers ($0.05 \leq P/P_0 \leq 0.35$), as it is shown in Figure 4.11 [4.8]. When the experimental data is plotted, the line should be drawn as the best fit to the data. The slope and intercept of the BET plot are given by Equation (4.18). From the slope and intercept of this line, the volume of gas adsorbed in a monomolecular layer and the BET constant are determined from Equation (4.19):

$$s = \frac{C-1}{V_m C} \quad i = \frac{1}{V_m C} \tag{4.18}$$

$$V_m = \frac{1}{s + i} \quad C = \frac{s}{i} + 1 \tag{4.19}$$

where s is the slope (g/m^3), i is the intercept (g/m^3), C is the BET constant (unitless), V_m is the volume of gas adsorbed when the entire surface is covered by a monomolecular layer (m^3 per gram of adsorbent).

After determining the volume of gas adsorbed in a monomolecular layer, V_m, the specific surface can then be calculated from Equation (4.12) mentioned in Section 4.2.4, in a similar way as it is done for the Langmuir method.

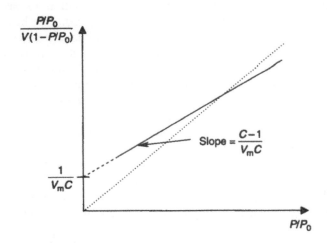

Figure 4.11 BET-plot used for determining the surface area of a porous material using parameters V_m and C determined from the slope and intercept of the linear fit to the data. V_m is the volume of gas adsorbed at a monomolecular layer and C is the BET constant. The dotted line corresponds to the single point BET-method, which is carried out by forcing the fit line through the origin of the plot.

It is worth noting that at very low relative pressures, i.e. when $P \ll P_0$, Equation (4.17) reduces to Equation (4.20), which is Langmuir's equation (Equation (4.16)) with $k_L = C/P_0$ and the parameters defined as for Equation (4.17) [4.34]:

$$V = \frac{V_m C \, (P/P_0)}{1 + C \, (P/P_0)} \tag{4.20}$$

The value of the BET constant C is most frequently between 50 and 300, when using nitrogen at 77.3 K as adsorbate, and is frequently lower when using water vapor as adsorbate [4.20]. A relatively low C value indicates low heat of adsorption during the formation of the first adsorbed layer. For high C values, a Type II isotherm with a well-defined point B is obtained, and for $C \le 2$, a Type III isotherm is obtained. An example of the change of the isotherm curves using Equation (4.17) for different values of the constant C, is shown in Figure 4.12. It is believed that low C values are improbable with microporous materials, therefore the validity of the technique is questionable [4.7].

Although derived over 60 years ago, the BET theory continues to be extensively used for gas adsorption analysis, because of its simplicity and its ability to accommodate each of the five isotherm types. The BET theory is mostly applicable to Type II and Type IV isotherms, provided that the value of the constant C is not too high and that the BET plot is linear in the region of relative pressures near completed monolayers $(0.05 \le P/P_0 \le 0.35)$ [4.8]. The BET equation may not be used generally outside the relative pressure ranges from 0.05 to 0.35; the theory underestimates the extent of adsorption at low

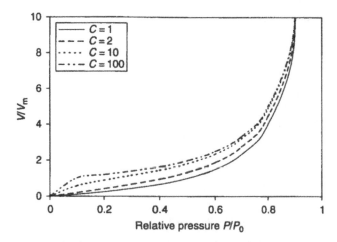

Figure 4.12 Change of the isotherm curves using Equation (4.19) mentioned in the text for different values of the BET constant C.

pressures $(P/P_0 < 0.05)$ and overestimates the extent of adsorption at high pressures $(P/P_0 > 0.35)$ [4.13,4.48]. The reason for overestimating adsorption at high pressures is the assumption that at $P/P_0 = 1$, an infinite number of layers are adsorbed. Experimental data on non-porous adsorbents show that the number of adsorbed layers close to P/P_0 is not infinite, but about 5 or 6 [4.48]. S. Brunauer *et al.*, 1940, extended the BET theory by considering capillary effects near the saturation vapor pressure [4.10]. The theory and derivation of their equation has been discussed in Brunauer's book and his original papers [4.4,4.10,4.49]. However, the extended BET theory is complex and not suitable for surface area determinations. Derivations of modified BET equations have been proposed by other researchers as well [4.23,4.50].

The technique has been used extensively by researchers on cement-based materials [4.30,4.38,4.51–4.60]. The BET constant C has been found to be around 80 when using nitrogen as adsorbate on cement pastes [4.62]. The constant C has been found to depend on the method of drying of the cement pastes and is significantly reduced when specimens are dried by solvent replacement procedures. The constant C is 76.7 ± 21.7 for specimens that have been vacuum-dried for short periods, and 82.0 ± 23.3 for specimens that have been vacuum-dried for long periods [4.61]. The values of constant C are low when using butane as adsorbate, and range from 3.6 to 30, averaging 11.6 with a standard deviation of 4.8 [4.62]. When using water vapor as adsorbate, the constant C has a value of about 8 for silica gel, and for cement pastes its value is considerably higher, about 35 [4.63].

A development of the BET method is the *single point* BET method and has the advantages of simplicity and speed, often with little loss in accuracy. In the BET plot, the intercept term is generally small compared to the slope of the line, so the plot may be approximated by forcing the fit line through the origin, after slightly changing the slope. This is equivalent to assuming that $1/(V_m C)$ is zero, or that $C \gg 1$. If $C \gg 1$, then $C-1 \approx C$ and this assumption yields the

BET single point relationship, given by Equation (4.21) with the parameters defined in the same way as for Equation (4.17) [4.13,4.45]:

$$\frac{P}{V(P_0-P)} = \frac{1}{V_m} \cdot \frac{P}{P_0}$$
(4.21)

T. Powers, 1955, has advised against the use of the single point BET method for research purposes and considered the method more useful for quality control [4.63]. Instead of drawing the line through the origin of the BET plot, he proposed to draw the line to a point on the negative x-axis, corresponding to an assumed finite value for the constant C [4.5]. Powers further suggested that the researcher should know the average value of constant C for a particular kind of material before applying the single point method.

4.5.3 The Dubinin–Kaganer equation

M. Kaganer, 1959, extended Dubinin's theory for micropores and developed a method for the calculation of specific surface of micropores [4.64]. The Dubinin–Kaganer (DK) equation is applicable to relative pressures below 0.01. The DK equation has the same form as the DR equation (Equation (4.10)) in which fractional filling of the pore volume, V/V_0 has been replaced by the fractional coverage, V/V_m, as it is shown by Equation (4.22):

$$\ln\left(\frac{V}{V_m}\right) = -D_1\left[\ln\left(\frac{P_0}{P}\right)\right]^2 \quad \text{with } D_1 = 2.303k_1(RT)^2$$
(4.22)

where V is the volume of micropores that have been filled at relative pressure P/P_0 and at temperature T (m^3 per gram of adsorbent), V_m is the volume of gas adsorbed when the entire surface is covered by a monomolecular layer (m^3 per gram of adsorbent), P_0 is the saturation vapor pressure (N/m^2), P is the gas pressure (N/m^2), k_1 is a constant that characterizes the pore size distribution, R is the gas constant ($=8.314\,J/(K \cdot mol)$), T is the temperature (K).

A plot of $\log V$ vs. $[\log (P_0/P)]^2$ will yield a straight line with an intercept of $\log V_m$, from which the specific surface can be calculated using Equation (4.12). The plot deviates from linearity at high pressures. Using nitrogen as adsorbate, Kaganer showed good agreement between specific surfaces measured using the DK equation and the BET-method on a variety of adsorbent/adsorbate systems at relative pressures between 0.0001 and 0.01 [4.64]. However, it should be noted that the DK equation is very different from the BET equation and the surface area is determined from data obtained at very low relative pressures, orders of magnitude below the BET region.

J. Adolphs and M. Setzer, 1998, have used the technique for analysis of pore structure of hardened cement pastes and have determined the parameter D_1 in Equation (4.22) to be -0.08 when using nitrogen as adsorbate, and -0.01 when using water vapor as adsorbate [4.47]. The surface area they determined using the DK method compares very well to the surface area determined using the BET method, when using nitrogen and water vapor as adsorbates (see also Table 4.8 in Section 4.7.3).

4.5.4 The Harkins–Jura (HJ) relative method

The method developed by W. Harkins and G. Jura, 1944, for determining the surface area is an empirical method [4.22]. The HJ-method does not need the volume V_m of gas required to form a monomolecular layer in order to calculate the surface area, like the methods described in the previous sections do. The researchers provided Equation (4.23), which involves only the gas pressure, P, and the volume of adsorbed gas, V_a, that are measured during the experiment:

$$\log\left(\frac{P}{P_0}\right) = B - \frac{A_1}{V_a^2} \quad \text{with } A_1 = \frac{a10^{20}S^2V_M^2}{2\,RTN_A} \tag{4.23}$$

where P is the gas pressure (N/m^2), P_0 is the saturation vapor pressure (N/m^2), B is a constant, V_a is the volume of adsorbed vapor (m^3 per gram of adsorbent), a is a constant, S is the specific surface area (m^2/g), V_M is the molar volume of the gas (m^3/mol), N_A is the Avogadro constant ($=6.023 \cdot 10^{23}$ molecules/mol), R is the gas constant ($=8.314$ J/(K·mol)), T is the temperature (K).

A plot of $(1/V_a^2)$ vs. $\log(P/P_0)$ gives a straight line with a negative slope A_1, which according to Equation (4.23) is proportional to the square of the specific surface as it is shown in Equation (4.24), with the parameters defined as for Equation (4.23):

$$S = k\sqrt{A_1} \quad \text{with } k = \sqrt{\frac{2RTN_A}{a10^{20}V_M^2}} \tag{4.24}$$

The proportionality constant, k, depends only upon the adsorbate and the temperature, and is independent of the nature of the adsorbent. The constant k must be determined by calibration using an independent method for surface area determination and this can be done from adsorption measurements on an adsorbent with known surface area. Table 4.2 lists values for constant k for some gases, so that when the volume of adsorbed gas, V_a, is expressed in cm^3/g at standard temperature and pressure, the surface area is expressed in m^2/g [4.34]. Equation (4.23) is analogous to Equation (4.4) provided by the same researchers for determining the thickness of the adsorbed layer on the pore wall, in which the empirical values 13.99 and 0.034 are substituted for the slope A_1 and intercept B, respectively. The Harkins–Jura equation is applicable in the high pressure range, and yields surface areas similar to the BET surface areas, if the BET constant C is in the approximate range from 50 to 250 [4.65]. For low values of the BET constant

Table 4.2 Values for the parameter k in the HJ method [4.34]

Gas	Temperature (°C)	k
Nitrogen	−195.8	4.06
Argon	−195.8	3.56
Water vapor	25	3.83

C ($C = 2$, 5, and 10), and at low pressures ($P/P_0 < 0.4$), there is no linear relationship in the plot. For $C = 100$, the range of mutual validity of the BET equation and the equation provided by Harkins–Jura is limited to the region $0.01 < P/P_0 < 0.13$ [4.7].

Harkins and Jura also proposed another method for surface area determination based on high precision calorimetric measurements, which is referred to as the HJ-absolute method, as opposed to the HJ-relative method used to refer to the method presented in this section [4.21].

4.5.5   The t-plot

The analysis of micropores has been made possible by the studies of J. de Boer *et al.*, in 1964–1965. In a series of seven papers, they developed a method that is suitable for analysis of pore structure [4.16,4.18,4.41,4.66–4.69]. The researchers proposed a method to separate the pore systems in the adsorbents they tested into narrow and wide pores, and to determine the surface areas of micropores. The *t*-plot method (also called the *t*-method) has attracted great attention as a simple and direct means of interpreting nitrogen isotherms. The values of relative pressure, P/P_0, measured during the experiment can be transformed to values of thickness t of the adsorbed layer, as it was explained in Section 4.2.2. A plot of the volume adsorbed, V_a, vs. the thickness t of the adsorbed layer can then be obtained, and is called a *V–t curve* or a *t-plot* (curve (a) in Figure 4.13). The *t*-plot should not be confused with the *t*-curve, which is a plot of the thickness of adsorbed layer, t, vs. the relative pressure, P/P_0, and is not linear. The method of analysis using a *t*-plot is based on the

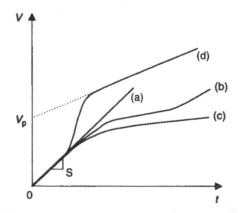

Figure 4.13 Different types of *t*-plots used for determining the specific surface S of porous materials from the slope of the line. (a) A straight line indicates multilayer adsorption without vapor condensation. (b) The curve will drop below the straight line, when micropores become filled with vapor gas. (c) When slit-shaped pores are completely filled with vapor, no further adsorption can occur. (d) In mesopores, capillary condensation can occur in addition to ongoing multilayer adsorption and the curve deviates above the straight line.

BET surface area as a primary standard and can give valuable information concerning the distribution of pores in the specimen.

The *t*-plot should be a straight line in the pressure range where multilayer adsorption occurs which by extrapolation to $t = 0$ goes through the origin. However, the *t*-plot obtained experimentally is not a straight line and deviations from linearity usually occur. Three different characteristic types of *t*-plots can be identified [4.69,4.70,4.71].

1. The pores' surface is available for adsorption up to high relative pressures, therefore as the volume of adsorbed vapor increases, the thickness of the film adsorbed on the pore walls increases. A straight line indicates multilayer adsorption without vapor condensation inside the pores, as well as the absence of micropores (curve (a) in Figure 4.13). In this case the surface area of the adsorbent can be calculated from the slope of the *t*-plot.

2. When some of the micropores become filled with adsorbate, then part of the pores' surface becomes unavailable for further adsorption and, even though the volume of gas adsorbed increases, there is no change in the thickness of adsorbed film. Therefore, the points of the plot should begin to drop below the straight line (curve (b) in Figure 4.13).

Commonly, a second straight part in the *t*-plot is found at intermediate relative pressures, corresponding to the unhindered growth of the adsorbed layer either in mesopores or at the external surface of the adsorbent. Therefore, the second slope corresponds to the surface area of mesopores or to the external surface area. The micropore surface area cannot be estimated from the *t*-plot, and the initial slope of the line probably leads to a minimum estimate of the micropore surface area.

In slit-shaped pores, capillary condensation cannot occur, but at a certain point the pores may be completely filled by the adsorbed layers on both parallel walls. The surface area in such pores is no longer accessible above a certain relative pressure; therefore, the *t*-plot will get a smaller slope corresponding to the surface area that is still accessible to the adsorbent (curve (c) in Figure 4.13).

3. In mesopores, capillary condensation starts in addition to ongoing multilayer adsorption. In this case, the amount of gas adsorbed increases, while there are only small increases in the adsorbed thickness and, consequently, the *t*-plot rises above the straight line (curve (d) in Figure 4.13). After complete filling of the pores with capillary condensate, a second straight line can be found in the *t*-plot, corresponding to unhindered growth of the adsorbed layer at the external surface area of the adsorbent. In this case, the surface area of the mesopores may be estimated from the initial slope only.

Variable information may be obtained from a *t*-plot: the specific surface area, the size, and shapes of pores, the setting in of reversible capillary condensation, and of complete filling of pores [4.16]. The *t*-plot, together with the experimental adsorption and desorption isotherm can give a very good picture of the whole pore system. It should be noted that the *t*-plot does not furnish an independent value of the specific surface than the BET plot, since the thickness of the adsorbed layer is itself calculated from V/V_m. The *t*-plot rather provides a different, and sometimes more convenient way of arriving at the same value and gives more information than the BET-method. The specific surface area obtained using the *t*-plot is comparable to the values obtained using the BET-method, provided that multilayer adsorption, and not capillary condensation, takes place inside the pores [4.7,4.13]. The *t*-plot applies between relative pressures $P/P_0 = 0.08$ and $P/P_0 = 0.75$, whereas the BET equation covers the experimental curve only between $P/P_0 = 0.05$ and 0.35. The dependence of the *t*-plot on the BET-method demands that at the relative pressure range between 0.08 and 0.35 the values for the surface area from the two methods coincide, which is generally observed. The *t*-plot has also the advantage over the BET-method that departure from a linear relation is more easily detected and quantitatively assessed than in the deviation from the BET plot.

The *t*-plot as a method for determining the surface area of cement pastes has been used by several researchers [4.12,4.30,4.53,4.72–4.74]. Some of these plots are shown in Figure 4.14 on cement pastes having different water–cement ratios using nitrogen as adsorbate [4.30]. The researchers used the *t*-plots to discuss the increase in size of the small pores with increasing water–cement ratio. The curve for w/c = 0.35 in Figure 4.14 deviates from straight line already at the thickness of adsorbed film of 0.4 nm, which indicates that pores with radii of 0.4 nm are already filled with adsorbate. The thickness value at which deviation from linearity is observed increases as the water–cement ratio increases. If the small (gel) pores have the same size distribution in all pastes with different water–cement ratios, then the curves would show a deviation from linearity at the same point. It has been stated that the *t*-method yields incorrect values for the specific surface of mesopores at pressures below the closure pressure of the hysteresis loop, since at these pressures filling of micropores has not been completed [4.19]. In other nitrogen

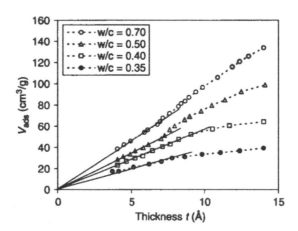

Figure 4.14 Typical *t*-plots for hardened cement pastes with different water–cement ratios.

Source: Reprinted from the Canadian Journal of Chemistry, Vol. 42, No. 2, R.Sh. Mikhail, L.E. Copeland, S. Brunauer, "Pore structures and surface areas of hardened Portland cement pastes by nitrogen adsorption," pp. 426–438, copyright 1964, with permission from the National Research Council of Canada.

adsorption experiments the *t*-plots obtained on cement pastes containing silica fume, pretreated by oven-drying at 80°C show an upward deviation commencing at $P/P_0 = 0.20$ at which the finest pores are filled [4.72]. Upward deviations in the *V–t* plots for hardened cement pastes have been interpreted as indicating that the pore system consisted mainly of mesopores of limited size and number [4.74].

The specific surface area values calculated from *t*-plots on cement pastes have been found to agree well with the specific surface values calculated from the BET-method (see also Section 4.7.3) [4.72,4.74].

The *t*-method for determining the surface area has been predominantly used with nitrogen. R. Mikhail and S. Selim, 1966 [4.75], have used a *t*-plot for cyclohexane. J. Hagymassy *et al.*, 1969, 1972, have used the technique with water vapor using the *t*-curve they proposed [4.20,4.42]. Their results show good agreement of the surface area determined by the *t*-plot and the surface area determined using the BET-method (see also Table 4.1).

4.5.6   The α_s-plot

The *t*-plot cannot be applied to a Type II isotherm without a well-defined B point or to a Type III isotherm. To avoid these difficulties, the *t*-plot method was modified by K. Sing, 1969, who proposed the use of a plot of adsorbed volume, V_a, vs. a quantity α_s that is called the α_s-plot [4.76]. The quantity α_s is defined as the ratio of the volume of gas adsorbed at any given relative pressure P/P_0 to the volume adsorbed at a standard relative pressure $P/P_0 = 0.40$, as is shown by Equation (4.25). The standard relative pressure $P/P_0 = 0.40$ was recommended by Sing as the appropriate value because both monolayer adsorption and micropore filling usually occur at $P/P_0 < 0.40$, whereas capillary condensation, in association with hysteresis, takes place at $P/P_0 > 0.40$ and might distort the isotherm even for non-porous solids [4.76]:

$$\alpha_s = \frac{V_a}{V_{a,0.40}} \tag{4.25}$$

where α_s is the normalized relative volume adsorbed (unitless), V_a is the volume of adsorbed gas (m^3 per gram of adsorbent), $V_{a,0.40}$ is the volume of adsorbed gas at the standard relative pressure $P/P_0 = 0.40$ (m^3 per gram of adsorbent).

For the construction of the α_s-plot, the adsorption isotherms of both a non-porous reference material and the adsorbent under test need to be obtained using the same adsorbate. For cement pastes, usually a non-porous silica is used as reference material (e.g. TK 800 with surface area approximately $154\, m^2/g$ as calculated by the BET-method, or Aerosil 200 with surface area $194\, m^2/g$ as calculated by the BET-method). From the isotherm of the reference material, the values α_s are calculated, and the α_s vs. P/P_0 plot is obtained, which is called the *reference (or standard) α_s-curve*. The shape of the reference α_s-curve is identical to the shape of the adsorption isotherm of the reference material. Using equal P/P_0 values, the α_s values of the reference material and the V_a values of the test material are obtained and a plot of V_a vs. α_s may be constructed, which is the α_s-*plot*. Since it is unlikely that the values of adsorbed gas V_a from the specimen under test and the α_s values of the reference material can be available at the exact same P/P_0 values, interpolation is often necessary.

From the α_s-plot, the surface area of the material under test can be determined from the slope of the linear region of the α_s-plot by using a proportionality (or normalizing) factor (see Equation (4.26)). The value 2.87 used in Equation (4.26), is the proportionality factor determined when nitrogen is used as adsorbate after calibration of the surface area obtained from the α_s-plot to the surface area of the reference material obtained using the BET-method [4.7,4.76]. From similar experiments, the proportionality factor was determined to be 1.92 when using carbon tetrachloride as adsorbate [4.76]:

$$S = 2.87\frac{V_a}{\alpha_s} \tag{4.26}$$

where S is the specific surface area (m²/g), V_a is the volume of adsorbed gas (m³ per gram of adsorbent), α_s is the normalized relative volume adsorbed (unitless).

Deviations of α_s-plots from linearity are explained in terms of capillary condensation, which may accompany multilayer formation (at $P/P_0 > 0.40$), and micropore filling, which only occurs at low P/P_0.

The adsorption isotherms and corresponding α_s-plots are shown in Figure 4.15. In the case of isotherm of type I, pronounced micropore filling is followed by multilayer adsorption (the linear branch) on a small external surface. For isotherm of type I, the α_s-plot can also be used to obtain the micropore volume by backward extrapolation of the linear branch to $\alpha_s = 0$. The linear α_s-plot in Types II and III is the result of unrestricted adsorption on a non-porous or macroporous solid; in this case, the isotherm may be of Type II, Type III, or Type VI, and is identical in shape to the isotherm of the reference material. The α_s-plot in Types IV and V, provides clear evidence of capillary condensation, taking place in mesopores.

A modified procedure of the α_s-plot is the *comparison plot*, which is obtained by comparing the shapes of two isotherms using the same adsorbate on two different adsorbents, the adsorbent under test, and a reference material. The amount adsorbed by one adsorbent is directly plotted against the amount adsorbed by the other adsorbent, by interpolation of the adsorption data at a series of identical pressures. Alternatively, the α_s values for the reference material can be plotted against the values of the material under test, resulting in a *reduced plot*. The comparison plot for an isotherm is linear if the test isotherm and reference isotherm have the same shape, which means that the same adsorption mechanisms are occurring. Differences in shape between the two isotherms will appear as departures from linearity in the resulting plot and its slope will reflect differences in adsorptive capacity over particular ranges P/P_0. Deviations above the diagonal at higher relative pressures indicate that capillary condensation is taking place during adsorption, while deviations below

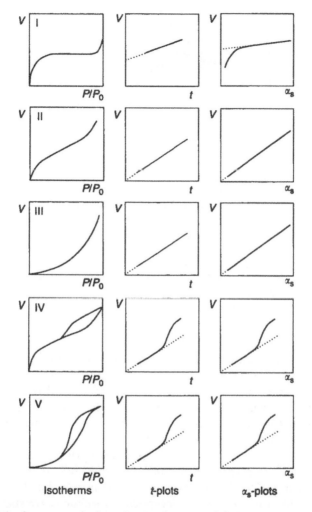

Figure 4.15 The five types of adsorption isotherms and their corresponding *t*-plots and α_s-plots.

the diagonal indicate either the presence of slit-shaped mesopores or the presence of micropores. The comparison plot can provide evidence of differences in the surface chemical structure of two or more adsorbents.

The comparison plot has been used by C. Lawrence and colleagues, 1980, 1989, using nitrogen and butane as adsorbates [4.61,4.62]. The shape of comparison plots obtained from cement pastes and non-porous silica are shown in Figure 4.16 indicating the effect of different pretreatment techniques of cement pastes [4.61]. Capillary condensation (upward deviation) occurs in D-dried cement pastes and in cement pastes soaked in methanol. A downward deviation of the plots occurs in the case of isotherms in specimens that are vacuum-dried and in specimens pretreated with ethanol. From Figure 4.16, it can be seen

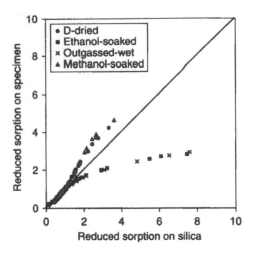

Figure 4.16 Comparison plot showing the effect of different pretreatment techniques on sorption on cement pastes and a reference material, silica TK 800 (nonporous silica with surface area 152.5 m²/g and unit size 10 g).

Source: Adapted from C. Lawrence, 1980 [4.61].

that until $P/P_0 = 0.40$, the shape of the isotherm is similar to non-porous silica, which probably means that it contains micropores and slit-shaped pores [4.61]. Similar results on cement pastes using butane as adsorbate showed the effect of degree of hydration on cement paste pore structure [4.62].

The α_s-plot method is an additional method to the BET-method and t-plot method for determining the specific surface and is distinguished from them because it does not assume that multilayer adsorption occurs. The α_s-plot provides a means of separating a complex experimental isotherm into its component parts [4.76]. In general, surface areas calculated from nitrogen α_s-plots on a wide variety of non-microporous specimens are shown to agree well with the corresponding BET surface areas. In addition, unlike the t-plot, which has limited use of adsorbates, the α_s-plot method can be used with any adsorbate. The α_s-plot method is the only method applicable to Types III and V isotherms for calculation of surface area.

4.6 Adsorption hysteresis

Experimental data from gas adsorption and MIP show similarities when the results are plotted in a comparable way. It was mentioned in Section 4.2.1 earlier that the hysteresis loop is a characteristic feature of the Type IV isotherm. It has been recognized that the pore structure of the adsorbent plays an important role in determining the size and shape of the hysteresis loop. Numerous authors have found that the relative pressure P/P_0 at which the hysteresis loop closes is usually a characteristic of the adsorbate rather than of the adsorbent. In addition, a comparison of the hysteresis loops of adsorption isotherms from various kinds of active charcoals supports the contention that there is a connection between pore dimensions and the size of the hysteresis loop [4.19]. The relative pressure at which the loop closes depends on the temperature and the adsorbate, being around $P/P_0 = 0.40$ for nitrogen [4.7,4.77,4.78]. Rarely have hysteresis loops been found that have not closed at the low end usually by relative pressures of 0.30 when nitrogen is used as adsorbate. According to the Kelvin equation, pore radii corresponding to relative pressures less than 0.30 are smaller than 1.5 nm, i.e. the pores are micropores. It is possible that adsorption and desorption in micropores proceed as a completely reverse procedure of each other, thereby precluding hysteresis [4.13]. An important experimental characteristic of adsorption hysteresis is that within the limits of observation, all adsorption and desorption curves are smooth and show no discontinuities.

Several types of hysteresis can be observed for the Type IV isotherm and can be classified into different types. A convenient classification of adsorption hysteresis may be based on the pressure range over which the main boundary loop extends. An alternative and somewhat more detailed classification has been proposed by J. de Boer, 1958, by taking into account the steepness of the adsorption and desorption branches [4.79]. At relative pressures above 0.30, de Boer distinguished five fundamentally different types of hysteresis loop, characterized by at least one vertical (or very steep) branch (see Figure 4.17a). Three of them, Types A, B, and E, are frequently encountered in practice, while Types C and D are very rare. Combinations of the different types of hysteresis may occur occasionally. De Boer described the characteristics on the types of loops observed and proposed the kind of pores that can generate these adsorption–desorption loops. He indicated that hysteresis phenomena during physical adsorption occur neither in macropores nor in micropores, but nearly exclusively in mesopores, and that experimental hysteresis loops may be the result of the combination of two or more types of pores.

In Type A hysteresis loop, both adsorption and desorption branches are steep at intermediate relative pressures. This hysteresis type is connected with

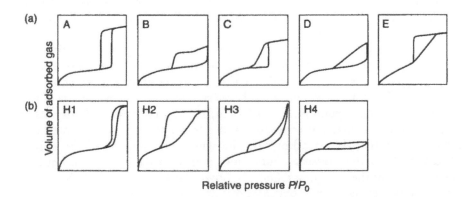

Figure 4.17 Types of adsorption hysteresis for isotherm Type IV (shown in Figure 4.3). (a) Types A, B, C, D, and E were proposed by J. de Boer, 1958 [4.79]. (b) Types H1, H2, H3, and H4 were proposed by IUPAC [4.80].

pores in the form of capillary tubes open at both ends, wide ink-bottle pores, and wedge-shaped capillaries. In Type B, the adsorption part is steep at the saturation vapor pressure, and the desorption part is steep at an intermediate relative pressure. Type B hysteresis loops are associated usually with the existence of open slit-shaped pores with parallel walls, and ink-bottle pores. In Type C, the adsorption branch is steep at an intermediate pressure and the desorption part is sloping. These types of hysteresis loops are typical of heterogeneous distribution of pores in the shape of capillary tubes with open ends. In Type D, the adsorption branch is steep at the saturation vapor pressure and the desorption branch is sloping. These hysteresis loops occur with heterogeneous pores in the form of capillary tubes having wide bodies and narrow necks, and with wedge-shaped capillary tubes open at both ends. In Type E, the adsorption branch has a sloping character and the desorption branch is steep at an intermediate relative pressure. Type E hysteresis loops have closer resemblance to Type A hysteresis loops and may be connected with ink-bottle pores or with interconnected capillaries [4.77]. Hysteresis loops of Types A and B are fairly well-described in terms of simple pore models, but the interpretation of Type E hysteresis loops is not straightforward. If the ink-bottle effects are more pronounced in a porous material than other pore shapes, then a Type E hysteresis loop is found rather than a Type B [4.77]. Theoretical evaluation of hysteresis loops for cylindrical, ink-bottle, and parallel plate pores is given elsewhere [4.7].

A new classification of hysteresis loops was recommended by IUPAC and consists of four types, as shown in Figure 4.17b [4.80]. The classification by IUPAC includes the hysteresis Types A, B, and E, that were proposed by de Boer, which correspond to the new hysteresis Types H1, H3, and H2 respectively. Progress has been made in linking the characteristic shapes of certain hysteresis loops with the nature of the pore structure, which are not significantly different from the shapes proposed by de Boer. H1 loops are often obtained from materials with cylindrical pores or spheroidal particles of fairly uniform size. Materials with ink-bottle shaped pores tend to give H2 loops, but in these cases the distribution of pore size and shape is not well defined. Types H3 and H4 have been obtained with adsorbents having slit-shaped pores or plate-like particles [4.8].

Several theories have been formulated in order to explain the difference between the state of the adsorbate during adsorption and during desorption. For example, it has been postulated that the hysteresis phenomenon was caused by a difference in contact angle between adsorbent and adsorbate during adsorption and desorption [4.1]. V. McBain, 1935, accounted for hysteresis by assuming that the material contains ink-bottle shape pores [4.81]. McBain's model asserts that during adsorption the wide inner portion of the pore is filled at high relative pressures, but cannot empty during desorption until the narrow neck of the pore first empties at low relative pressures. However, his model cannot explain how condensation into the wider inner portion of the pore can occur, once the narrow neck has been filled at low relative pressure. S. Brunauer *et al.*, 1967, argued that if this theory were the sole and correct explanation of hysteresis, then the adsorption branch would represent true capillary condensation equilibrium [4.39]. Brunauer *et al.* proposed that hysteresis arises because the processes of capillary condensation and evaporation inside the pores do not take place as exact reverses of each other: capillary condensation in pores of a certain size during adsorption does not occur exactly at the same relative pressure as capillary evaporation from the same pores during desorption.

Another theory states that the mechanism of adsorption by the porous material and the type and size of pores have an effect on adsorption hysteresis. Generally speaking, two cases of pore filling may be distinguished. The first case is pore filling as a result of adsorbent–adsorbate forces. Micropores are usually filled during adsorption in this manner, e.g. adsorption of organic vapors in active charcoals. In this case, the adsorbate exists under an increased hydrostatic pressure resulting from the presence of adsorption forces, and the filling of pores proceeds without adsorption hysteresis. In the second case, adsorbate–adsorbate forces are responsible for pore filling. Capillary condensation in mesopores and adsorption of water in micropores of active charcoals are examples of this case. The adsorbate can be under a negative hydrostatic pressure caused by adsorbate–adsorbate forces and hysteresis is possible. Depending on which forces play the decisive role in pore filling, pore filling may proceed by either the first or the second mechanism, and even when the pores are of the same dimensions [4.19]. Other theories have also been proposed by other researchers [4.30,4.82].

The presence of the hysteresis loop introduces a considerable complication: within the region of the hysteresis loop there are two relative pressures corresponding to a given quantity of gas adsorbed, with the lower value always residing on the desorption isotherm. One main question is whether to adopt the adsorption or desorption branch of the hysteresis loop for pore analysis. The desorption value of relative pressure corresponds to the more stable adsorbate condition, and it has been suggested that the desorption isotherm should, with certain exceptions, be used for pore size analysis [4.13,4.46]. Many investigators have used the desorption branches of nitrogen isotherms for pore structure analysis (see also Section 4.4), and offered as the main criterion of the correctness of their analysis the agreement between the cumulative surface area of the pore walls and the surface area calculated using the BET-method. Cranston and Inkley however used both the adsorption and desorption branches in their analysis, and found that for most of the adsorbents they tested, the adsorption branch gave surface areas in better agreement with the BET surface area than when the desorption branch was used [4.17].

The shape of hysteresis loop obtained from nitrogen adsorption experiments on cement pastes comes close to Types B and D as they were classified by de Boer, or to Type H3 as classified by IUPAC (see also Figure 4.5). The shape of hysteresis loop in cement pastes indicates capillaries with very wide bodies having a varying range of narrow short necks, or plate-like pores. Several models for the structure of the C–S–H gel, which is largely responsible for the pore structure of the cement paste within the range of applicability of gas adsorption, have suggested that the C–S–H gel consists of very thin sheets (see also Chapter 1). Such sheets may well produce plate-like or slit-shaped pores.

Similar adsorption experiments on cement pastes have been carried out by several researchers who provided nitrogen isotherms [4.61,4.73,4.54,4.57, 4.74,4.30,4.47,4.72,4.83]. An example of the effect of pore size on the size of hysteresis loop is shown in Figure 4.18 for nitrogen adsorption on cement pastes having different water–cement ratios. An increase in water–cement ratio, and consequently increase in pore size, results in larger hysteresis loops. Similar results have been provided by other studies [4.73,4.54,4.30,4.83]. Similar behavior has also been observed when using other gases as adsorbates, such as cyclohexane, isopropanol, and methanol [4.75].

Adsorption experiments on cement pastes using water vapor as an adsorbate shows an isotherm with a large hysteresis (see Figure 4.19) [4.84]. Similar results have been published by other researchers [4.29,4.60,4.85]. The large hysteresis loop exhibited by the sorption isotherm when using water vapor as adsorbate over the entire range of partial pressure indicates that irreversible processes other than adsorption are taking place. Water, and to a less extent other molecules, could cause some expansion of the C–S–H gel structure. Due to the swelling process, all isotherms exhibit a very large hysteresis loop, down to zero relative pressure. Water molecules must be highly oriented to the surface, and in narrow pores the repulsion between dipoles similarly oriented on the two opposite walls, might cause such a measurable diminution in the heat of adsorption [4.86]. J. Hagymassy *et al.*, 1972, published water isotherms with small loops [4.42]. They attributed their differences to short times other researchers allowed for equilibration: water vapor adsorption proceeds slowly so that equilibrium cannot be established within a reasonable time. In their work, they allowed four-and-a-half months for the equilibration of each desorption point, compared to only one day allowed by other researchers.

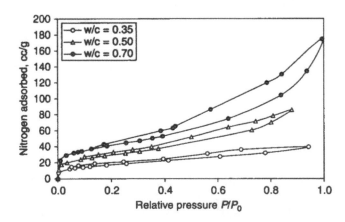

Figure 4.18 Isotherms of nitrogen adsorption near the boiling point of nitrogen on hardened Portland cement pastes having different water–cement ratios.

Source: Reprinted from the *Canadian Journal of Chemistry*, Vol. 42, No. 2, R.Sh. Mikhail, L.E. Copeland, S. Brunauer, "Pore structures and surface areas of hardened Portland cement pastes by nitrogen adsorption," pp. 426–438, copyright 1964, with permission from the National Research Council of Canada.

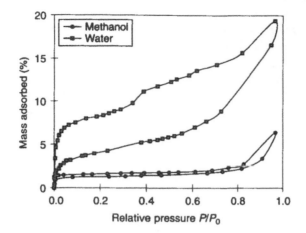

Figure 4.19 Adsorption isotherms on bottle-hydrated Portland cement using different adsorbates. The cement pastes had w/c = 5.0, and were degassed for 3 hours under vacuum and at 80 °C prior to the experiment.

Source: Adapted from R.F. Feldman, 1968 [4.84].

4.7   Factors affecting the results

The values of surface area and pore size distributions reported in the technical literature vary significantly, sometimes by an order of magnitude. Microstructural factors can affect the results such as water–cement ratio, type of cement used, temperature of hydration, mineral and chemical admixtures, degree of hydration etc., as well as type and duration of storage. Such effects were first investigated by R. Blaine and H. Valis, 1949 [4.6], and results from other studies have been partly demonstrated in Figures 4.9, 4.14, and 4.18. In addition to microstructural factors, parameters that relate to experimental and analysis procedures can affect the results; these parameters include specimen pretreatment, adsorbate used, and type of analysis. The effect of experimental and analysis parameters are discussed in the following sections.

4.7.1   Pretreatment method

Before a gas adsorption experiment is carried out, the specimen must first be dried and outgassed. Several studies in the technical literature demonstrate the effect of pretreatment on the pore structure results obtained [4.12,4.19,4.38,4.61]. An example of the effect of pretreatment on the results has already been shown in Figure 4.5. Figure 4.5 shows that the volume of nitrogen adsorbed in a cement paste that has been pretreated by methanol replacement is greater than the volume adsorbed by a cement paste specimen that has been oven-dried. Pretreatment with solvent replacement results also in a higher surface area to a value of 70 and 105 m²/g, compared to values obtained from oven-dried specimens that is about 38 m²/g (see Figure 4.20) [4.12]. G. Litvan, 1976, has demonstrated how specimen preparation affects the measured surface area using nitrogen as adsorbate (see Table 4.3) [4.58]. His experimental results show that P-drying and D-drying resulted in the smallest value of surface area for the cement pastes tested. It has been reported from experiments on identical specimens that different drying methods will give different results of pore analysis when using nitrogen as adsorbate, because the amount of water removed by different procedures is not identical [4.38,4.87].

It has also been shown that differences in specific surface areas determined using water vapor and nitrogen as adsorbates vanish when special techniques for drying are applied, such as vacuum-drying of saturated cement paste

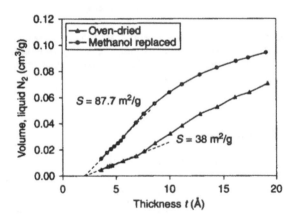

Figure 4.20 Effect of pretreatment on the surface area calculated from a nitrogen t-plot for hardened cement pastes having water–cement = 0.40.

Source: Reprinted from *Pore Structure and Permeability of Cementitious Materials*, L.R. Roberts, J.P. Skalny (eds), *MRS Symposium Proceedings Vol. 137*, Y. Abdel-Jawad, W. Hansen, "Pore structure of hydrated cement determined by mercury porosimetry and nitrogen sorption techniques," pp. 105–118, copyright 1989, with permission from the Materials Research Society, Warrendale, PA.

Table 4.3 Effect of pretreatment of cement pastes on the measured specific surface area using nitrogen adsorption [4.58]

Pretreatment method	Surface area (m²/g ignited paste)
P-drying	70
D-drying	105
Oven-drying	122
Replacement with methyl alcohol	194
Replacement with methyl alcohol and n-pentane	249

specimens [4.61] and drying by solvent replacement [4.58]. Drying without solvent replacement considerably decreases the pore volume accessible to nitrogen [4.73]. These decreases are slightly greater than the variations obtained from duplicate runs and are attributed to changes in the pore structure caused by the drying procedure. When dried rapidly, hardened cement pastes appear to contain slit-shaped pores whose widths lie between 2 and 4 nm for a range of paste densities. Slowly-dried pastes have much reduced surface areas and appear to contain wider more symmetrical pores, an effect that is also demonstrated in Figure 4.16 [4.62].

Alcohols have lower surface tension than water (see Table 2.2). For example, methanol has surface tension 22.07 mN/m at 25°C, which is lower than the surface tension of water which is 71.99 mN/m. Therefore, methanol is expected to cause less distortion to hydrated cement paste structures during evaporation, since hydrostatic tensions at curved menisci will be smaller, and this effect might explain the higher surface areas found in cement pastes that have been pretreated with methanol [4.61]. Drying by solvent replacement seems to increase the accessibility of nitrogen into the small pores in hardened cement pastes, such that nitrogen adsorption and water adsorption yield approximately similar surface areas [4.58]. Pretreatment with methanol replacement seems to increase the volume of micropores in cement pastes accessible to nitrogen, because cement pastes pretreated with methanol show increased downward deviations in their α_s-plots in comparison with those on corresponding untreated pastes [4.61]. Another possible mechanism proposed is the chemisorption of alcohol molecules onto hydrate surfaces, preventing in this way the bonding of adjoining surfaces [4.61].

4.7.2 Type of adsorbate used

A number of different adsorbates have been used for gas adsorption measurements on cement pastes, leading to variations in the measured values, because the pore structure parameters determined are often dependent on the adsorbate used during the experiment. Any non-corrosive gas can be used as an adsorptive without special calibration. Some typical adsorbates suitable for gas adsorption measurements and their parameters are shown in Table 4.4. Nitrogen is commonly used for routine gas adsorption measurements, because much experimental evidence has been obtained in the past using nitrogen making comparisons easy and it has been shown that surface areas in cement pastes can be calculated accurately using nitrogen.

Two adsorbates have been predominantly used to examine the pore structure parameters of hardened cement pastes: *nitrogen* at 77.3 K [4.6,4.30,4.47,4.51,4.54,4.57,4.72,4.73,4.74,4.85], and *water vapor* at room temperature [4.29,4.47,4.51,4.42,4.85,4.86]. Other adsorbates used in cement pastes have been argon [4.6], oxygen [4.6], and organic solvents, such as methanol [4.75,4.84,4.88], ethanol [4.37], cyclohexane [4.75], isopropanol [4.75], and butane [4.62]. Surface area and pore volume obtained on hardened cement pastes using different adsorbates are presented in Figures 4.21 and 4.22 [4.75].

Table 4.4 Parameters of some inorganic and organic vapors used for adsorption experiments [4.93]

Gas	σ (nm^2)	M_w (g/mol)	ρ (g/cm^3) (at °C)	$V_{mol} = M_w/\rho$ (cm^3/mol)
Argon	0.142	39.95	1.633×10^{-3} (25°C)	24.46×10^3
Benzene	0.430	78.11	0.8765 (20°C)	89.12
Carbon dioxide	0.195	44.01	1.799×10^{-3} (25°C)	24.46×10^3
Carbon monoxide	0.163	28.01	1.145×10^{-3} (25°C)	24.46×10^3
Cyclohexane	0.390	84.16	0.7785 (20°C)	108.11
Ethanol	—	46.07	0.7893 (20°C)	58.37
Isopropanol	0.277	60.10	0.7855 (20°C)	76.51
Krypton	0.208	83.80	3.425×10^{-3} (25°C)	24.46×10^3
Methanol	0.181	32.04	0.7914 (20°C)	40.49
Nitrogen	0.162	28.02	1.145×10^{-3} (25°C)	24.46×10^3
Oxygen	0.141	32.00	1.308×10^{-3} (25°C)	24.46×10^3
Water	0.108	18.02	0.9982 (20°C)	18.05
Xenon	0.025	131.30	5.366×10^{-3} (25°C)	24.46×10^3

Notes
The parameters include: the area σ occupied by one molecule, the molecular weight, M_w, density, ρ, and the molar volume, V_{mol}. These parameters can be used in Equation (4.3) to determine the thickness of a monomolecular layer.

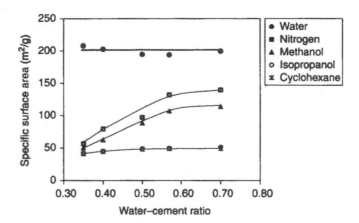

Figure 4.21 Variation of the specific surface area of hardened cement pastes with change in the water–cement ratio using different adsorbates.

Source: Reproduced with permission of the Transportation Research Board. In Special Report 90: *Symposium on Structure of Portland Cement Paste and Concrete*, Highway Research Board, National Research Council, Washington DC, 1966, R.Sh. Mikhail, S.A. Selim, "Adsorption of organic vapors in relation to the pore structure of hardened Portland cement pastes," pp. 123–134.

The surface areas and porosities determined using water vapor are substantially greater than the surfaces areas and porosities determined using nitrogen [4.58,4.30,4.73,4.42,4.84]. The values for specific surface area determined using nitrogen as adsorbate range between 10 and 200 m^2 per gram of dried paste, while the values for specific surface areas using water vapor as

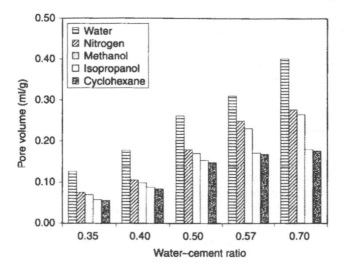

Figure 4.22 Pore volumes of hardened cement pastes determined using gas adsorption and different gases as adsorbates.

Source: Drawn from experimental results from R. Mikhail and S. Selim, 1966 [4.75].

adsorbate are typically around 140–200 m² per gram of dried paste [4.5,4.58, 4.30,4.89,4.90]. Many investigations carried out over the last 20 years have given conflicting reports regarding the mechanism of water adsorption. Water settles on the surface of a solid first as a monolayer and with increasing moisture as multilayers. The first layer is hydrogen-bonded to the surface of the solid and is immobile but in some cases water behaves like a liquid on flat surfaces and in wide pores. Additional layers can behave as a liquid. There has been a debate regarding the specific interactions of nitrogen and water vapor with cement paste and there is a sharp difference in opinion as to which gas yields the "correct" pore structure parameters [4.42,4.47,4.51,4.53, 4.54,4.85]. Different explanations have been given for the difference in results obtained by water and nitrogen adsorption, making use of the idea of accessible and inaccessible pores to nitrogen.

R. Feldman and P. Sereda, 1968, have claimed that the surface areas of cement pastes measured using nitrogen as an adsorbate are more correct than surface areas obtained when using water vapor [4.60]. According to the researchers, a fraction of the water molecules taken up by the specimen enters spaces between the layers of the C–S–H gel sheets. The consequence is an excessively high amount of water bound by the material, rather than water adsorbed on the pore walls' surface, which leads to erroneously high measured surface areas. Contrary to the water molecules, the nitrogen molecules do not enter the C–S–H gel interlayer spaces, but remain adsorbed only on the surface; in this way the surface areas obtained by the BET method using nitrogen reflect the true surface area of the material. Feldman and Sereda reported that the hydraulic radius of the pore system inaccessible to nitrogen is about 1.1 nm.

Their theory has been questioned by several researchers. Work by S. Brunauer *et al.*, 1958, has shown that water cannot penetrate between the layers of strongly dried C–S–H gel, even when the gel is soaked in water; therefore the difference between water and nitrogen adsorption is not a consequence of the layer structure of the C–S–H gel [4.91]. I. Odler *et al.*, 1972, and R. Mikhail and S. Abo-El-Enein, 1973, argue that a value of 1.1 nm for the hydraulic radius of the pores inaccessible to nitrogen corresponds to average separation between layers of 2.2 nm, if the molecular layers of C–S–H gel are assumed as parallel plates, and to a diameter of 4.4 nm, if the pores are assumed cylinders [4.87,4.90]. According to them, such separation would lead to a complete dispersion of the gel by water. They also conclude that a hydraulic radius of 1.1 nm does not correspond to distances between molecular layers, but to large pores with narrow necks, i.e. ink-bottle pores. I. Odler *et al.*, 1972, showed that the surface area measured using water vapor as adsorbate increases with increasing degree of hydration, and thus decreasing pore dimensions, while the surface area measured using nitrogen as adsorbate strongly decreases. They explained this observation by the view that nitrogen molecules are unable to penetrate a large fraction of the wide pores, and, in general, nitrogen molecules penetrate only fraction of the pore system that water vapor can access [4.90].

It is generally believed that the pore structure of cement paste is such that nitrogen molecules cannot penetrate into many large pores and cover the total surface area of the pores [4.30]. The difference between the penetration abilities of nitrogen and water molecules depends on several factors including the size of the molecules, the dipole nature of water molecules, and the diffusion process of the water molecules.

- The nitrogen molecules are larger than the water molecules: the diameter of the nitrogen molecule is about 3.5 Å, while the diameter of the water molecule is 2.8–3 Å. The small size of the water molecule enables it to penetrate into pores that are inaccessible to the nitrogen molecule [4.29,4.30,4.90].
- Another factor that may also play a role is the fact that the water molecule has a strong dipole, which is strongly attracted to the ionic surfaces of the compounds in hydrated cements [4.51]. The nitrogen molecule has only a quadrupole, which is attracted to the surface much more weakly than water [4.90].
- As it was mentioned earlier, there is indication that most of the pores in cement pastes are ink-bottle pores. As hydration proceeds, the necks of the ink-bottle pores become narrower. The occurrence of constrictions in the necks of the pores presents an energy barrier to the diffusion of the adsorbate molecules into the pores. The process by which molecules enter ink-bottle pores is an activated diffusion process and is called *activated-entry process*. The molecules are adsorbed at the entrance of the pore and eventually obtain enough energy from the vibrations of the surface atoms to pass over the energy barrier and enter into the pore. The energy barrier is the energy of activation of the process that has a strong temperature

dependence. If it is assumed that the energy barrier is the same for the nitrogen and water molecules (it is probably smaller for the water molecule because of its smaller size), one can calculate the relative rates for reaching adsorption equilibrium. The adsorption isotherms are measured at 25°C (298 K) for water vapor and at −196°C (77 K) for nitrogen, which means that the rate of equilibration of water vapor should be about 50 times as fast as that of nitrogen. The time needed for water vapor to reach adsorption equilibrium is two to three weeks; therefore, it would take approximately two to three years for nitrogen to reach equilibrium [4.90].

The difference between the specific surface area obtained by the BET-method using water and nitrogen adsorption isotherms on hydrated cement paste were found to vanish, if special techniques were employed for drying the specimens before the experiment. It has been found that pretreatment with solvent replacement, methanol and then pentane, and then vacuum-drying, results in specific surface areas approximately 200 m²/g using either nitrogen or water vapor as adsorbate [4.58].

Surface areas obtained from adsorption experiments using *oxygen* and *argon* have been determined to be 20.8 m²/g and approximate the value determined using nitrogen, which is 25.9 m²/g [4.6]. Surface areas obtained using cyclohexane and isopropanol as adsorbates are about 50 m²/g [4.75]. From Figure 4.22 it can be seen that the surface area is about 200 m²/g when water is used as adsorbate, around 50 m²/g when isopropanol and cyclohexane are used as adsorbates, and does not change as the water–cement ratio changes.

Isotherms using *methanol* as an adsorbate were first measured by S. Mikhail and S. Selim, 1966, and showed fairly large low-pressure hysteresis [4.75]. R. Feldman, 1968, confirmed this observation, and investigated the nature and reversibility of methanol and water isotherms to compare their sorption characteristics (see also Figure 4.19) [4.84]. The isotherm obtained using methanol as adsorbate was very flat and it was concluded that the alcohol molecule binds strongly to the pore wall surface, and orients vertically to the surface preventing multilayer formation at low pressures (autophobicity of alcohol adsorption). The isotherms exhibit strong monolayer binding with little tendency toward multilayer formation; multilayer formation occurred almost asymptotically at high vapor pressures. The result would approximate to a Langmuir isotherm, but for methanol, whose hydrocarbon chain is not so long, this approximation is not completely attained [4.84].

4.7.3 Analysis method used

The analysis theories reported in the early sections of this chapter apply for different ranges of pore sizes. Furthermore, none of the adsorption theories (Langmuir, BET, DR, Kaganer, HJ) is able to describe an experimental adsorption isotherm over the complete pressure range. For this reason, sometimes more than one analysis theory needs to be used. The methods of analysis commonly used that have been described in the previous sections and the parameters they determine are summarized in Table 4.5.

Table 4.5 Analysis methods used for gas adsorption and parameters determined

Equation/Plot	Adsorption mechanism	Parameter	Limitations
Kelvin	Monolayer	Pore size distribution	Macropores and mesopores
Barrett–Joyner–Halenda	Multilayer, condensation	Pore size distribution Surface area	Macropores
Dubinin–Radushkevich	Micropore filling	Pore size distribution Micropore volume	Micropores
Halsey	Monolayer	Thickness of adsorbed film	Macropores and mesopores
de Boer	Multilayer	Thickness of adsorbed film	Macropores and mesopores
Cranston–Inkley	Multilayer	Thickness of adsorbed film	Macropores and mesopores
Langmuir	Monolayer	Surface area	Types I, II
BET	Multilayer	Surface area	Types II, IV, mesopores and macropores
Dubinin–Kaganer	Micropore filling	Surface area	Micropores
Harkins–Jura	Monolayer	Thickness of adsorbed film	Macropores and mesopores
t-plot (Lippens–de Boer)	Multilayer	Micropore volume Surface area	Types I, II, IV
α_s-plot	Multilayer	Surface area	Types I, II, III, IV, mesopores and macropores
t-curve	Multilayer	Thickness of adsorbed film	Macropores and mesopores
α_s-curve	Multilayer	Surface area	—

Table 4.6 Comparison of the values of surface area of cement pastes with different SF contents at different ages determined using the BET-method, S_{BET}, and the t-plot method, S_t [4.72]

SF content (%)	Age (days)	C_{BET}	S_{BET} (m^2/g)	S_t (m^2/g)	V_p (ml/g)	r_h $(Å)$
0	28	8	35.02	32.8	0.168	47.97
0	90	10	32.05	31.5	0.158	49.29
10	28	6	42.3	40.8	0.176	41.61
10	90	6	34.4	34.2	0.167	48.54
20	28	8	46.1	45.6	0.184	39.91
20	90	14	39.1	38.8	0.177	45.26

Notes
The pore volume, V_p, the hydraulic radius, r_h, and the BET-constant, C_{BET}, are also reported. The cement pastes were dried at 80°C and then evacuated at room temperature for 8 hours. The total pore volume was taken at the saturation value of the isotherm.

Comparisons have usually been made in the past between the surface area determined using the BET-method and the t-plot method [4.30,4.42,4.54,4.72,4.74]; some of these results are presented in Table 4.1 and Table 4.6 [4.42,4.72].

Kh. Khalil, 1996, has shown that the surface area using the BET-method and the t-plot method agree very well for Portland cement pastes containing

Table 4.7 Specific surface area and pore volume of hydrated cement pastes from nitrogen isotherms using the BET-, the BJH-, and the DR-method [4.29]

Specimen	S_{BET} (m^2/g)	S_{BJH} (m^2/g)	$S_{BJHdes.}$ (m^2/g)	$V_{BJHads.}$ (mm^3/g)	$V_{BJHdes.}$ (mm^3/g)	V_{DR} (mm^3/g)
Portland cement	1.09	1.59	2.63	8.58	8.92	0.46
Blast furnace slag cement	2.20	2.56	3.41	13.23	13.50	0.92

Notes
The specimens were air dried for 24 hours prior to testing. For the pore volume using the BJH-method, both the adsorption and desorption isotherms have been used.

Table 4.8 Surface areas of hardened cement pastes obtained using the BET-method, the Langmuir method, and the DK-method

Analysis method	S (m^2/g) N_2 [4.47]	S (m^2/g) H_2O [4.47]	S (m^2/g) (H_2O) [4.84]
BET	14	116	120
(BET constant)	(49)	(725)	(119)
Langmuir	19	91	71
(Langmuir constant)	(23)	(3152)	(630)
DK	19	115	150
(constant D)	(-0.08)	(-0.01)	(-0.04)

Notes
Results of nitrogen adsorption measurements and water adsorption measurements on hardened cement paste. The cement paste was made from ordinary Portland cement that had a w/c = 0.40, and was dried over ice trap at $-40°C$, after 28 days curing.

various amounts of silica fume and at different ages [4.72]. As it was mentioned in Section 4.5.5 such close agreement should be expected because the way of calculating the surface areas by the BET-method and the t-plot method is not independent from each other.

A comparison of the results on pore volume and surface area determined using the BET-method and the BJH-method are shown in Table 4.7 when using the adsorption and desorption branch of the isotherm [4.29]. From the experimental data presented, it seems that analysis using the adsorption isotherm gives surface area results closer to the BET values, while analysis using the desorption branch of the isotherm gives higher values.

A comparison of surface areas obtained using the Langmuir- and DK-methods to the surface areas obtained using the BET-method are presented in Table 4.8 using nitrogen and water vapor as adsorbates, and compared to previously published results [4.47]. Even though there are limited results available to compare and draw conclusions, the values seem to be within a reasonable range.

4.8   Advantages and limitations

The sorption of nitrogen is a very well-established procedure for porous materials in general. The nitrogen adsorption experiment on cement pastes is a complex method that requires some experience and can be time-consuming: one measurement might last 6 hours or even longer. Experiments using water vapor adsorption might require longer times for equilibration to be achieved.

It has been reported that the volumetric method is not ideally suitable for measurements on materials with large pore volume, since dosing large quantities of vapor is tedious to perform, and might increase cumulative errors [4.35]. Gravimetric systems can be adapted for fully automatic operation, producing results of high precision. Adsorption experiments that use mixtures of the adsorbate and an inner carrier gas are generally considered unsuitable in the capillary condensation region [4.35].

Small specimens are used during testing and precautions are required to ensure that these specimens are representative of the material. The technique becomes experimentally difficult when a specimen has a small specific surface, because little gas is adsorbed.

Nitrogen adsorption is a method that is considered to provide a description of a specimen's microstructure. Correct interpretation, however, is complicated by the dependence of the results on the method of specimen preparation. Studies have shown that the shape of nitrogen isotherms on hardened cement pastes depends on the conditions during production and storage. In addition, precautions must be taken to ensure that the necessary preparation and drying do not produce large microstructural changes and give misleading results. Drying using a solvent replacement technique has been reported to increase the sensitivity of the nitrogen sorption test. Vacuum-drying at elevated temperatures is a customary procedure with capillary condensation methods.

Use of gases having low saturation vapor pressures at the required adsorption temperature permits more accurate measurement of the small adsorptions obtained with materials having small specific surface areas.

The interactions that occur during adsorption between nitrogen molecules and the cement paste are assumed to be purely physical in nature. However, this is questionable since detailed investigations on hardened cement pastes reveal that over the range of relative pressures $0 < P/P_0 < 0.4$ several different shapes of isotherms may be obtained quite reproducibly. Surfaces in hydrated cement pastes have some ionic character, so it is believed that changes in the nitrogen isotherm shapes may be attributed to alterations in the chemical nature of the internal surfaces. It is also possible that adsorption of water vapor leads to chemical interactions with the cement paste that may obscure the correct interpolation of the results in these cases.

The nitrogen adsorption technique is considered of limited value because there is evidence that nitrogen cannot penetrate into many large pores. Many of the ink-bottle pores have such narrow entrances that nitrogen molecules cannot penetrate into them [4.30]. In addition, it has been suggested that nitrogen may not be the most suitable adsorptive for characterizing microporous solids and that a number of carefully selected molecules should be used as probes for microstructural evaluation. The same argument may be applied to hardened cement pastes, especially those containing pores whose sizes are only marginally greater than micropores (about 2 nm diameter). In these cases, other sorptives should be studied in order to confirm structure evaluations based on nitrogen as a probe molecule. One such example is butane: its molecule contains no permanent electrical charge separation and its physical interactions with solid surfaces should simply involve nonspecific van der Waals forces.

It is difficult to draw conclusions comparing surface areas from the literature because of inconsistencies in the way results are reported. Surface area values have been calculated per gram of dried paste, per gram of ignited paste, or per gram of C–S–H gel, and it is often unclear which measurement unit was reported.

The area occupied per molecule of adsorbate must be known but is often uncertain. The exact value is mainly dependent on the net heat of adsorption: when the heat of adsorption is very large, the adsorbate is more like a solid than a liquid.

The relative pressure P/P_0 cannot be controlled with adequate accuracy above a value of about 0.98. Even before this stage is reached, the physical properties of the adsorbate (surface tension and density) may differ sufficiently from the bulk properties to introduce appreciable errors in calculation of pore size [4.92]. The adsorption of nitrogen at its boiling point is not recommended, even though it is commonly used in specific surface measurements. The reason is that, unless an elaborate cryostat system is used, fluctuations in the boiling point of the liquid nitrogen used as refrigerant, arising both from the solution of impurities and from variations of atmospheric pressure, may cause unacceptable fluctuations in the saturation vapor pressure within the system. This problem can be avoided by using other adsorbates such as butane at 0°C, or tetramethylsilane at 15 or 20°C. Because of its specific interactions with many materials of practical interest, water may be an interesting adsorbate especially when compared with other, more inert adsorbates. This possibility has been utilized, e.g. by Brunauer and co-workers in their comparative studies of water and nitrogen adsorption by cement pastes.

Dense hardened cement paste systems of limited porosity and limited pore connectivity may not respond correctly to the capillary condensation model. At the temperature of liquid nitrogen, under which the experiments are carried out, diffusion of nitrogen through very fine interconnections is extremely slow. As a result, capillary condensation may not occur in some pores simply because of local unavailability of the necessary nitrogen vapor. Nitrogen adsorption measurements could also be subject to misinterpretations due to the tortuosity of the pore structure [4.35].

The application of the BET-equation assumes that a monomolecular layer is covering the pores' surface. It is possible that a second layer of adsorbed gas molecules may be forming before the first adsorbed layer is complete. However, this is not taken into account and the monolayer capacity is determined from the isotherm equations [4.7]. The BET-theory assumes the same energy for all adsorption sites on the surface and neglects horizontal interactions between neighbor molecules in the same layer, considering only the forces between the adsorbing molecule and the surface. The BET-constant C is assumed constant, but this is not true; for C to be constant, the surface must be energetically homogenous. The first molecules to be adsorbed on the pore surface generate more energy than subsequent molecules, therefore the BET-theory is not applicable at relative pressures less than about 0.05 [4.7].

The method of nitrogen adsorption makes no assumptions about the shape of the pores, but assumes that the pores of interest all lie in the range that can be detected by capillary condensation, which is pores of diameter up to about 40 nm. This is a substantial limitation for the study of cement-based materials, which have significant amounts of pore volume with sizes ranging up to and well beyond 100 nm. For most cement pastes, other methods need to be used to study pores with size greater than 40 nm, which is often the major part of the pore system.

The mechanisms by which pores fill and empty are not completely understood for micropores. Micropores are thought to fill by a homogeneous mechanism. When the opposite walls of a pore are so close together that their adsorption fields overlap, then the whole pore may be regarded as forming part of the adsorption space and it will become filled with adsorbate at very low relative pressures. Thus, such pores will appear more as an extension of the surface of the solid than as part of the pore volume [4.35]. The applicability of the Kelvin equation is limited, and the capillary condensation model cannot be used in the case of micropores.

A likely source of error in the capillary condensation measurements involving condensates other than nitrogen, may lie in the use of *t*-curves. Any error in the *t*-curve will have relatively insignificant effects on the pore size distribution except in the smallest diameter region of pore sizes, where the effect may be considerable.

When hysteresis accompanies capillary condensation, it is necessary to establish the boundary curve of the hysteresis loop. It is therefore essential that adsorption measurements be carried up to the adsorption limit at saturation vapor pressure before desorption is begun, and that desorption also be completed without premature reversal.

References

4.1   Zsigmondy R.A. "Über die Struktur des Gels der Kieselsäure. Theorie der Entwässerung," *Zeitschrift für Anorganische Chemie*, Vol. 71, No. 4, 1911, pp. 356–377.

4.2   Langmuir I. "The constitution and fundamental properties of solids and liquids. Part I. Solids," *Journal of the American Chemical Society*, Vol. 38, No. 11, 1916, pp. 2221–2295.

4.3   Langmuir I. "The adsorption of gases on plane surfaces of glass, mica and platinum," *Journal of the American Chemical Society*, Vol. 40, No. 9, 1918, pp. 1361–1403.

4.4   Brunauer S., Emmett P.H., Teller E. "Adsorption of gases in multimolecular layers," *Journal of the American Chemical Society*, Vol. 60, No. 2, 1938, pp. 309–319.

4.5   Powers T.C., Brownyard T.L. "Studies of the physical properties of hardened Portland cement paste," *Research Bulletin 22*, Portland Cement Association, Chicago, IL, 1948, Reprinted from the *Journal of the American Concrete Institute*,

October 1946–April 1947, in *Proceedings* Vol. 43, pp. 101, 249, 469, 549, 669, 845, 933.

4.6 Blaine R.L., Valis H.J. "Surface available to nitrogen in hydrated Portland cements," *Journal of Research of the National Bureau of Standards*, Vol. 42, 1949, pp. 257–267.

4.7 Allen T. *Particle Size Measurement*, 5th Edition, Chapman and Hall, London, 1997.

4.8 Gregg S.J., Sing K.S.W. *Adsorption, Surface Area and Porosity*, 2nd Edition, Academic Press, London, 1982.

4.9 ASTM Designation D 4222 "Determination of nitrogen adsorption and desorption isotherms of catalysts by static volumetric measurements," in *Annual Book of ASTM Standards, Vol. 05.03 Petroleum Products and Lubricants*, American Society for Testing and Materials, West Conshohocken, PA.

4.10 Brunauer S., Deming L.S., Deming W.E., Teller E. "On a theory of the van der Waals adsorption of gases," *Journal of the American Chemical Society*, Vol. 62, No. 7, 1940, pp. 1723–1732.

4.11 Joyner L.G., Weinberger E.B., Montgomery C.W. "Surface area measurements of activated carbons, silica gel and other adsorbents," *Journal of the American Chemical Society*, Vol. 67, No. 12, 1945, pp. 2182–2188.

4.12 Abdel-Jawad Y., Hansen W. "Pore structure of hydrated cement determined by various porosimetry and nitrogen sorption techniques," in *Symposium Proceedings Vol. 137: Pore Structure and Permeability of Cementitious Materials*, L.R. Roberts, J.P. Skalny (eds), Materials Research Society, Warrendale, PA, 1989, pp. 105–118.

4.13 Lowell S., Shields J.E. *Powder Surface Area and Porosity*, 3rd Edition, Chapman and Hall, New York, 1991.

4.14 Linsen B.G., Van den Heuvel A. "Chapter 35: Pore structures," in *The Solid–Gas Interface*, Vol. 2, E.A. Flood (ed.), Marcel Dekker Inc., New York, 1967, pp. 1025–1053.

4.15 Emmett P.H., Brunauer S.J. "The use of low temperature van der Waals adsorption isotherms in determining the surface area of iron synthetic ammonia catalysts," *Journal of the American Chemical Society*, Vol. 59, No. 8, 1937, pp. 1553–1564.

4.16 Lippens B.C., Linsen B.G., de Boer J.H. "Studies on pore systems in catalysts I. The adsorption of nitrogen; apparatus and calculation," *Journal of Catalysis*, Vol. 3, No. 1, 1964, pp. 32–37.

4.17 Cranston R.W., Inkley F.A. "The determination of pore structures from nitrogen adsorption isotherms," *Advances in Catalysis, Vol. IX*, in *Proceedings of the International Congress on Catalysis*, Philadelphia, PA, 1956, A. Farkas (ed.), Academic Press Inc., New York, 1957, pp. 143–154.

4.18 De Boer J.H., Linsen B.G., Osinga Th.J. "Studies on pore systems in catalysts VI. The universal t curve," *Journal of Catalysis*, Vol. 4, No. 6, 1965, pp. 643–648.

4.19 Kadlec O., Dubinin M.M. "Comments on the limits of applicability of the mechanism of capillary condensation," *Journal of Colloid and Interface Science*, Vol. 31, No. 4, 1969, pp. 479–489.

4.20 Hagymassy J., Brunauer S., Mikhail R.Sh. "Pore structure analysis by water vapor adsorption I. *t*-curves for water vapor," *Journal of Colloid and Interface Science*, Vol. 20, No. 3, 1969, pp. 485–491.

4.21 Harkins W.D., Jura G. "Surfaces of solids. XII. An absolute method for the determination of the area of a finely divided crystalline solid," *Journal of the American Chemical Society*, Vol. 66, No. 8, 1944, pp. 1362–1366.

4.22 Harkins W.D., Jura G. "Surfaces of solids. XIII. A vapor adsorption method for the determination of the area of a solid without the assumption of a molecular area, and the areas occupied by nitrogen and other molecules on the surface of a solid," *Journal of the American Chemical Society*, Vol. 66, No. 8, 1944, pp. 1366–1373.

4.23 Halsey G. "Physical adsorption of non-uniform surfaces," *The Journal of Chemical Physics*, Vol. 16, No. 10, 1948, pp. 931–937.

4.24   Thomson W.T. "On the equilibrium of vapour at a curved surface of liquid," *Philosophical Magazine and Journal of Science*, Vol. 42, No. 282, 1871, pp. 448–452.

4.25   Diamond S. "A critical comparison of mercury porosimetry and capillary condensation pore size distributions of Portland cement pastes," *Cement and Concrete Research*, Vol. 1, No. 5, 1971, pp. 531–545.

4.26   Wheeler A. *Presentations at Catalysis Symposia*, Gibson Island, AAAS Conferences, June 1945/1946.

4.27   Dubinin M.M., Radushkevitch L.V. "Equation of the characteristic curve of activated charcoal," in *Proceedings of the Academy of Sciences, Physical Chemistry Section*, USSR, Vol. 55, 1947, pp. 331–333.

4.28   Dubinin M.M. "The potential theory of adsorption of gases and vapors for adsorbents with energetically nonuniform surfaces," *Chemical Reviews*, Vol. 60, No. 2, 1960, pp. 235–241.

4.29   Robens E., Benzler B., Büchel G., Reichert H., Schumacher K. "Investigation of characterizing methods for the microstructure of cement," *Cement and Concrete Research*, Vol. 32, No. 1, 2002, pp. 87–90.

4.30   Mikhail R.Sh., Copeland L.E., Brunauer S. "Pore structure and specific surface area of hardened Portland cement pastes as determined by nitrogen adsorption," *Canadian Journal of Chemistry*, Vol. 42, 1964, pp. 436–438.

4.31   Barrett E.P., Joyner L.G., Halenda P.P. "The determination of pore volume and area distributions in porous substances. I. Computations from nitrogen isotherms," *Journal of the American Chemical Society*, Vol. 73, No. 1, 1951, pp. 373–380.

4.32   Mikhail R.Sh., Brunauer S., Bodor E.E. "Investigations of a complete pore structure analysis I. Analysis of micropores," *Journal of Colloid and Interface Science*, Vol. 26, No. 1, 1968, pp. 45–53.

4.33   Brunauer S., Mikhail R.Sh., Bodor E.E. "Pore structure analysis without a pore shape model," *Journal of Colloid and Interface Science*, Vol. 24, No. 3, 1967, pp. 451–463.

4.34   Orr C., Dallavalle J.M. *Fine Particle Measurement. Size, Surface and Pore Volume*, The Macmillan Company, New York, 1959.

4.35   Haynes J.M. "Pore size analysis according to the Kelvin equation," *Matériaux et Constructions*, Vol. 6, No. 33, 1973, pp. 209–213.

4.36   Shull C.G. "The determination of pore size distribution from gas adsorption data," *Journal of the American Chemical Society*, Vol. 70, No. 4, 1948, pp. 1405–1410.

4.37   Wittmann F., Englert G. "Bestimmung der Mikroporenverteilung in Zementstein," *Materials Science and Engineering*, Vol. 2, 1967, pp. 14–20.

4.38   Juenger M.C.G., Jennings H.M. "The use of nitrogen adsorption to assess the microstructure of cement paste," *Cement and Concrete Research*, Vol. 31, No. 6, 2001, pp. 883–892.

4.39   Brunauer S., Mikhail R.Sh., Bodor E.E. "Some remarks about capillary condensation and pore structure analysis," *Journal of Colloid and Interface Science*, Vol. 25, No. 3, 1967, pp. 353–358.

4.40   Bodor E.E., Odler I., Skalny J. "An analytical method for pore structure analysis," *Journal of Colloid and Interface Science*, Vol. 32, No. 2, 1970, pp. 367–369.

4.41   Lippens B.C., de Boer J.H. "Studies on pore systems in catalysts V. The t method," *Journal of Catalysis*, Vol. 4, No. 3, 1965, pp. 319–323.

4.42   Hagymassy J., Odler I., Yudenfreund M., Skalny J., Brunauer S. "Pore structure analysis by water vapor adsorption III. Analysis of hydrated calcium silicates and Portland cements," *Journal of Colloid and Interface Science*, Vol. 38, No. 1, 1972, pp. 20–34.

4.43   Odler I., Skalny J. "Pore structure of hydrated calcium silicates II. Influence of calcium chloride on the pore structure of β-dicalcium silicate," *Journal of Colloid and Interface Science*, Vol. 36, No. 3, 1971, pp. 293–297.

4.44 Skalny J., Odler I. "Pore structure of hydrated calcium silicates III. Influence of temperature on the pore structure of hydrated tricalcium silicate," *Journal of Colloid and Interface Science*, Vol. 40, No. 2, 1972, pp. 199–205.

4.45 "Surface Area, Density, and Porosity of Metal Powders," *Metals Handbook, Vol. 7: Powder Metallurgy*, 9th Edition, American Society for Metals, Materials Park, OH, 1986, pp. 262–271.

4.46 Sing K.S.W. "Characterization of porous solids: an introductory survey," *Studies in Surface Science and Catalysis, Vol. 62: Characterization of Porous Solids II, Proceedings of the IUPAC Symposium*, Alicante, Spain, 6–9 May 1990, F. Rodriguez-Reinoso, J. Rouquerol, K.S. Sing, K.K. Unger (eds), Elsevier Science Publishers, Amsterdam, The Netherlands, 1991, pp. 1–9.

4.47 Adolphs J., Setzer M.J. "Description of gas adsorption isotherms on porous and dispersed systems with the excess surface work model," *Journal of Colloid and Interface Science*, Vol. 207, No. 2, 1998, pp. 349–354.

4.48 Brunauer S., Skalny J., Bodor E.E. "Adsorption on nonporous solids," *Journal of Colloid and Interface Science*, Vol. 30, No. 4, 1969, pp. 546–552.

4.49 Brunauer S. *The Adsorption of Gases and Vapors, Vol. 1: Physical Adsorption*, Princeton University Press, Princeton, NJ, 1943.

4.50 Fergusson R.R., Barrer R.M. "Derivation and development of Hüttig's multilayer sorption isotherm," *Transactions of the Faraday Society*, Vol. 46, 1950, pp. 400–407.

4.51 Odler I. "The BET-specific surface area of hydrated Portland cement and related materials," *Cement and Concrete Resarch*, Vol. 33, No. 12, 2003, pp. 2049–2056.

4.52 Tennis P.D., Jennings H.M. "A model for two types of calcium silicate hydrate in the microstructure of Portland cement pastes," *Cement and Concrete Research*, Vol. 30, No. 6, 2000, pp. 855–863.

4.53 Odler I., Köster H. "Investigation on the structure of fully hydrated Portland cement and tricalcium silicate pastes. III. Specific surface area and permeability," *Cement and Concrete Research*, Vol. 21, No. 6, 1991, pp. 975–982.

4.54 Bodor E.E., Skalny J., Brunauer S., Hagymassy J., Yudenfreund M. "Pore structure of hydrated calcium silicates and Portland cement determined by nitrogen adsorption," *Journal of Colloid and Interface Science*, Vol. 34, No. 4, 1970, pp. 560–568.

4.55 Rarick R.L., Thomas J.J., Christensen B.J., Jennings H.M. "Deterioration of the nitrogen BET surface area of dried cement paste with storage time," *Advanced Cement Based Materials*, Vol. 3, No. 2, 1996, pp. 72–75.

4.56 Thomas J.J., Hsieh J., Jennings H.M. "Effect of carbonation on the nitrogen BET surface area of hardened Portland cement paste," *Advanced Cement Based Materials*, Vol. 3, No. 2, 1996, pp. 76–80.

4.57 Weiß R., Schneider U. "N2-Sorptionsmessungen zur Bestimmung der Spezifischen Oberfläche und der Porenverteilung von Erhitztem Normalbeton," Vol. 6, No. 5, 1976, pp. 613–622.

4.58 Litvan G.G. "Variability of the nitrogen surface area of hydrated cement paste," *Cement and Concrete Research*, Vol. 6, No. 1, 1976, pp. 139–143.

4.59 Mikhail R.Sh., Abo-El-Enein S.A. "Studies on water and nitrogen adsorption on hardened cement pastes I. Development of surface in low porosity pastes," *Cement and Concrete Research*, Vol. 2, No. 4, 1972, pp. 401–414.

4.60 Feldman R.F., Sereda P.J. "A model for hydrated Portland cement as deduced from sorption-length changes and mechanical properties," *Matériaux et Constructions*, Vol. 1, No. 6, 1968, pp. 509–520.

4.61 Lawrence C.D. "The interpretation of nitrogen sorption isotherms on hydrated cements," *Cement and Concrete Association Technical Report 530*, 1980, p. 28.

4.62 Gimblett F.G.R., Lawrence C.D., Sing K.S.W. "Sorption studies of the microstructure of hydrated cement," *Langmuir*, Vol. 5, No. 5, 1989, pp. 1217–1222.

4.63 Powers T.C., Brunauer S., Copeland L.E. "Discussion of a paper by L.F. Gleysteen and G.L. Kalousek: simplified method for the determination of apparent surface area of concrete products," *Journal of the American Concrete Institute*, Vol. 27, No. 4, 1955, in *Proceedings Vol. 51*, pp. 448–1 to 448–5.

4.64   Kaganer M.G. "New method for determining the specific surface of adsorbents and other finely divided substances," *Russian Journal of Physical Chemistry*, Vol. 33, No. 10, 1959, pp. 352–356.

4.65   Emmett P.H. "Multilayer adsorption equations," *Journal of the American Chemical Society*, Vol. 68, No. 9, 1946, pp. 1784–1789.

4.66   De Boer J.H., Lippens B.C. "Studies on pore systems in catalysts II. The shapes of pores in aluminum oxide systems," *Journal of Catalysis*, Vol. 3, No. 1, 1964, pp. 38–43.

4.67   Lippens B.C., de Boer J.H. "Studies on pore systems in catalysts III. Pore-size distribution curves in aluminum oxide systems," *Journal of Catalysis*, Vol. 3, No. 1, 1964, pp. 44–49.

4.68   De Boer J.H., van den Heuvel A., Linsen B.G. "Studies on pore systems in catalysts IV. The two causes of reversible hysteresis," *Journal of Catalysis*, Vol. 3, No. 3, 1964, pp. 268–273.

4.69   De Boer J.H., Linsen B.G., van der Plas Th., Zondervan G.J. "Studies on pore systems in catalysts VII. Description of the pore dimensions of carbon blacks by the t method," *Journal of Catalysis*, Vol. 4, No. 6, 1965, pp. 649–653.

4.70   Sing K.S.W. "Empirical method for analysis of adsorption isotherms," *Chemistry and Industry*, 2 November 1968, pp. 1520–1521.

4.71   Haynes J.M. "Alternative adsorption methods for determination of the specific surface of porous solids," *Matériaux et Constructions*, Vol. 6, No. 33, 1973, pp. 247–250.

4.72   Khalil Kh.A. "Pore structure and surface area of hardened cement pastes containing silica fume," *Materials Letters*, Vol. 26, No. 4–5, 1996, pp. 259–264.

4.73   Hansen W., Almudaiheem J. "Pore structure of hydrated Portland cement measured by nitrogen sorption and mercury intrusion porosimetry," in *Symposium Proceedings, Vol. 85: Microstructural Development During Hydration of Cement*, L.J. Struble, P.W. Brown (eds), Materials Research Society, Warrendale, PA, 1987, pp. 105–114.

4.74   Al-Noaimi K.Kh., El-Hosiny F.I., Abo-El-Enein S.A. "Thermal and pore structural characteristics of polymer-impregnated and superplasticized cement pastes," *Journal of Thermal Analysis and Calorimetry*, Vol. 61, No. 1, 2000, pp. 173–180.

4.75   Mikhail R.Sh., Selim S.A. "Adsorption of organic vapors in relation to the pore structure of hardened Portland cement pastes," *Special Report 90: Symposium on Structure of Portland Cement Paste and Concrete*, Highway Research Board, National Research Council, Washington DC, 1966, pp. 123–134.

4.76   Sing K.S.W. "Utilization of adsorption data in the BET region," in *Proceedings of the International Symposium on Surface Area Determination*, held at Bristol UK, 16–18 July 1969, D.H. Everett, R.H. Ottewill (eds), Butterworth, London, 1970, pp. 25–42.

4.77   Everett D.H. "Chapter 36: Adsorption hysteresis," in *The Solid–Gas Interface*, E. Alison Flood (ed.), Vol. 2, Marcel Dekker Inc., New York, 1967, pp. 1055–1113.

4.78   Burgess C.G.V., Everett D.H. "The lower closure point in adsorption hysteresis of the capillary condensation type," *Journal of Colloid and Interface Science*, Vol. 33, No. 4, 1970, pp. 611–614.

4.79   De Boer J.H. "The shapes of capillaries," in *The Structure and Properties of Porous Materials*, in *Proceedings of the Tenth Symposium*, Colston Research Society University of Bristol, D.H. Everett, F.S. Stone (eds), Butterworths Science Publications, London, 1958, pp. 68–94.

4.80   Sing K.S.W., Everett D.H., Haul R.A.W., Moscou L., Pierotti R.A., Rouquérol J., Siemieniewska T. "Reporting physisorption data for gas/solid systems with special reference to the determination of surface area and porosity," *Pure and Applied Chemistry*, Vol. 57, No. 4, 1985, pp. 603–619.

4.81   McBain J.W. "An explanation of hysteresis in the hydration and dehydration of gels," *Journal of the American Chemical Society*, Vol. 57, No. 4, 1935, pp. 699–700.

4.82 Katz S.M. "Permanent hysteresis in physical adsorption. A theoretical discussion," *Journal of Physical Chemistry*, Vol. 53, No. 8, 1949, pp. 1166–1186.

4.83 Hunt C.M. "Nitrogen sorption measurements and surface areas of hardened cement pastes," *Special Report 90: Symposium on Structure of Portland Cement Paste and Concrete*, Highway Research Board, National Research Council, Washington DC, 1966, pp. 112–122.

4.84 Feldman R.F. "Supplementary paper III-23. Sorption and length-change scanning isotherms of methanol and water on hydrated Portland cement," in *Proceedings of the Fifth International Symposium on the Chemistry of Cement*, 7–11 October 1968, Tokyo, 1968, Part III Properties of Cement Paste and Concrete (Volume III), pp. 53–66.

4.85 Alford N.McN., Rahman A.A. "An assessment of porosity and pore sizes in hardened cement pastes," *Journal of Materials Science*, Vol. 16, No. 11, 1981, pp. 3105–3114.

4.86 Mikhail R.Sh., Abo-El-Enein S.A. "Studies on water and nitrogen adsorption on hardened cement pastes. II-Thermodynamics of adsorption," *Cement and Concrete Research*, Vol. 3, No. 1, 1973, pp. 93–106.

4.87 Dollimore D., Gamlen G.A., Mangabhai R.J. "Thermal analysis studies on cement pastes treated with organic solvents," in *Proceedings of the Second European Symposium on Thermal Analysis (ESTA 2)*, University of Aberdeen, UK, 1–4 September 1981, D. Dollimore (ed.), Heyden, London, pp. 485–488.

4.88 Parrott L.J. "Effect of drying history upon the exchange of pore water with methanol and upon subsequent methanol sorption behavior in hydrated alite paste," *Cement and Concrete Research*, Vol. 11, No. 6, 1981, pp. 651–658.

4.89 Gleysteen L.F., Kalousek G.L. "Simplified method for the determination of apparent surface area of concrete products," *Journal of the American Concrete Institute*, Vol. 26, No. 5, January 1955, in *Proceedings Vol. 51*, pp. 437–446.

4.90 Odler I., Hagymassy J.Jr., Bodor E.E., Yudenfreund M., Brunauer S. "Hardened Portland cement pastes of low porosity IV. Surface area and pore structure," *Cement and Concrete Research*, Vol. 2, No. 5, 1972, pp. 577–589.

4.91 Brunauer S., Kantro D.L., Copeland L.E. "The stoichiometry of the hydration of β-dicalcium silicate and tricalcium silicate at room temperature," *Journal of the American Chemical Society*, Vol. 80, No. 4, 1958, pp. 761–767.

4.92 Fagerlund G. "Determination of specific surface by the BET method," *Matériaux et Constructions*, Vol. 4, No. 33, 1973, pp. 239–245.

4.93 Lide D.R. *CRC Handbook of Chemistry and Physics*, 84th Edition, CRC Press, Boca Raton, FL, 2003–2004.

5 Pycnometry and thermoporometry

Two techniques that are simple and relatively fast to use are pycnometry and thermoporometry. Liquid and gas pycnometry provides information on total pore volume, but only limited or no information on other pore structure parameters, like pore size distribution and specific surface area. Therefore, the technique is not systematically used for characterization of cement-based materials. Thermoporometry was theoretically formulated almost 30 years ago, but is still under investigation and development for experimental applications. Thermoporometry, in combination with Nuclear Magnetic Resonance, is known as NMR cryoporometry and is discussed in Chapter 6. In this chapter, the principles and applications of pycnometry and thermoporometry are presented.

5.1 Pycnometry

Pycnometry determines the density of a material by measuring the difference between the real (specific) and total (bulk) volumes of a specimen. Archimedes determined the volume of an object using the displacement principle by weighing the object, immersing it in water, and measuring the volume of water displaced. Modern pycnometry represents a refinement of the displacement principle and uses either a liquid or a gas substance as the displaced material.

For a porous material of total volume V_{tot} containing pores with volume V_p (see also Figure 5.1), two different densities may be defined: the apparent density determined using the total volume that includes pores, and the true density determined using the real volume that excludes pores [5.1].

Apparent density, also mentioned as bulk or envelope density, ρ_{app}, is defined as the mass of material per unit of total volume, and is expressed by Equation (5.1).

True density, also mentioned as real or skeletal density, ρ_{tr}, is defined as the mass per unit of real (non-porous) volume of the material and is expressed by Equation (5.2). In certain applications these definitions can vary, e.g. when mention is made to a material consisting of porous particles that includes inter-particle and intra-particle spaces; as a result, density and porosity can be determined in several ways:

$$\rho_{app} = \frac{m_D}{V_{tot}} \tag{5.1}$$

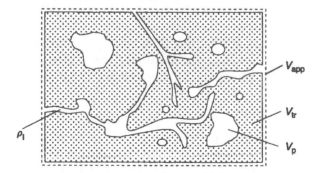

Figure 5.1 A schematic two-dimensional representation of the apparent volume, V_{app}, of a porous material, including the true volume, V_{tr} (shadowed area) and the volume of pores, V_p. The true and apparent density of the material can be calculated using the true and apparent volume, respectively. In the saturated state, the pores of the specimen are filled with a liquid that has density ρ_l.

$$\rho_{tr} = \frac{m_D}{V_{tr}} \qquad (5.2)$$

where ρ_{app} is the apparent density of the material (kg/m³), m_D is the dry mass of the material (kg), V_{tot} is the total volume of the material (m³), V_{tr} is the true volume of the material (m³), ρ_{tr} is the true density of the material (kg/m³).

Pore volume is the difference between the total volume and the real volume of a material; therefore, the porosity of a material may be determined from Equation (5.3) using the apparent and true density of the material, with the parameters defined as before:

$$\varepsilon = \frac{V_{tot} - V_{tr}}{V_{tot}} = 1 - \frac{V_{tr}}{V_{tot}} \quad \text{or} \quad \varepsilon = \left(1 - \frac{\rho_{app}}{\rho_{tr}}\right)100 \qquad (5.3)$$

where ε is the porosity of the material (%).

Pycnometry as a pore structure analysis technique can be used to determine the total pore volume of a material using properties that relate to density. However, pycnometry does not provide information on the material's pore size distribution. For the measurements, the fluid used must be capable of penetrating all the continuous pores of the material without being adsorbed at the pore surface [5.1]. Depending on the fluid used, pycnometry can be divided into liquid and gas pycnometry.

In *liquid pycnometry*, a liquid is used as the displacement medium and the displaced volume is measured directly during the experiment because liquids are incompressible. Inability of the liquid to penetrate some pores, chemical reaction between the liquid and the material, and evaporation of the liquid can contribute to errors during a density measurement.

In *gas pycnometry*, a high purity, dry, inert, non-adsorbing gas is used as the displacement medium, such as argon, neon, dry nitrogen, dry air, or helium.

The displaced volume is not measured directly during the experiment, but is determined from the pressure–volume relationship of a gas under controlled conditions using gas laws.

In the following sections the liquid and gas pycnometry techniques are described separately, including the theoretical principles, the experimental procedure, and their applicability on cement-based materials.

5.1.1   Liquid pycnometry

The bulk density and the apparent density of a material are determined by measuring the mass of a dry specimen, the mass of specimen saturated with a liquid, and the total volume of the specimen, which is evaluated using the immersion technique. The mass of the specimen saturated with a liquid can be expressed by Equation (5.4) (see also Figure 5.1), and the mass of the specimen while it is immersed in a liquid is expressed by Equation (5.5):

$$m_s = m_D + (V_{tot} - V_{tr})\rho_l \tag{5.4}$$

$$m_{sub} = m_D - V_{tr}\rho_l \tag{5.5}$$

where m_s is the mass of saturated, surface-dry sample (kg), m_D is the mass of dry sample (kg), V_{tot} is the total volume of the sample (m^3), V_{tr} is the true volume of the material (m^3), ρ_l is the density of the liquid in the pores of the sample (kg/m^3), ρ_{tr} is the true density of the material (kg/m^3), m_{sub} is the mass of the saturated sample when it is immersed in water (kg).

Solving Equation (5.4) for the total volume V_{tot} and after substituting the true volume V_{tr} from Equation (5.5), Equation (5.6) is obtained for the total volume. By substituting Equation (5.6) in Equation (5.1), Equation (5.7) is obtained for determining the apparent density of the material. By substituting the true volume V_{tr} from Equation (5.5) in Equation (5.2), Equation (5.8) is obtained for determining the true density of the material. Then the porosity of the material can be expressed by Equation (5.9) using only the masses of the specimen at different states:

$$V_{tot} = \frac{m_s - m_{sub}}{\rho_l} \tag{5.6}$$

$$\rho_{app} = \frac{m_D}{m_s - m_{sub}}\rho_l \tag{5.7}$$

$$\rho_{tr} = \frac{m_D}{m_D - m_{sub}}\rho_l \tag{5.8}$$

$$\varepsilon = \left(1 - \frac{m_D - m_{sub}}{m_s - m_{sub}}\right)100 \tag{5.9}$$

where V_{tot} is the total volume of the sample (m^3), m_s is the mass of saturated, surface-dry sample (kg), m_D is the mass of dry sample (kg), ρ_l is the density of the liquid in the pores of the sample (kg/m^3), ρ_{app} is the apparent density of the material (kg/m^3), m_{sub} is the mass of the saturated sample when it is immersed in water (kg), ρ_{tr} is the true density of the material (kg/m^3), ε is the total porosity (%).

It is necessary, to completely saturate the material with the liquid, otherwise the displaced volume of the liquid will not represent the real volume of the material, but will also include air that is trapped inside the pore network [5.1]. It should be noted here that the term *imbibition* is also used to describe the uptake of a liquid by a gel or porous substance that might or might not cause swelling. Even though the term is used often in Materials Science for a variety of materials, it is used infrequently in Concrete Science.

The liquid used for the measurements must be inert towards the material. The most commonly used immersion liquids for cement-based materials are water and alcohols, and are discussed in the following paragraphs.

5.1.1.1 *Water absorption*

The technique of determining the total porosity of a sample using water absorption is relatively fast and simple to carry out and does not require the use of expensive equipment. The method is used to determine the total pore volume and the apparent density of the porous material, and for this reason it is also referred to as *water displacement pycnometry*. Even though it does not provide any other information on the pore structure besides total porosity, it is used quite extensively for testing hardened concrete. For this procedure, it is assumed that volume changes of the specimen during drying or saturation with water are negligible [5.2]. It is also assumed that the volume of closed pores that are unreachable by water is negligible.

The experimental setup is very simple and is shown in Figure 5.2. The apparatus consists of a tightly sealed glass container that is connected to a vacuum pump and to an input that allows the addition of air-free deionized water in the container. The specimen is required to be free of cracks [5.3]. The mass or volume of the specimen to be tested is defined by the standard or

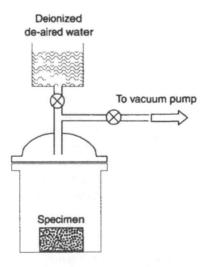

Figure 5.2 Schematic setup of the apparatus used for water absorption.

recommendation followed during testing. For example, RILEM CPC 11-3 recommends a specimen volume at least $1000\,cm^3$, while ASTM C 948 mentions a specimen volume not greater than $650\,cm^3$ or $1300\,g$. The specimen is first dried to constant mass in a carbon dioxide-free ventilated oven at a temperature of $105° ± 5°C$. Immediately after cooling to $20° ± 3°C$, the specimen's dry mass, m_D, is determined. The specimen is then placed in the container, where a vacuum is maintained with pressure below 100–250 Pa for a period of at least 24 hours. Deionized, de-aired water is released into the container until the specimen is fully submerged in water and there is an excess height from the surface of the specimen of approximately 20 mm water [5.3]. Then the pressure is slowly raised to atmospheric pressure, and water is being sucked into the pores saturating the specimen. The specimen must be kept under water at atmospheric pressure for 24 hours or until it has reached a constant saturated mass [5.1]. The surface of the specimen is dried and the saturated surface-dry mass, m_s, is determined. In addition, the mass of the specimen while submerged in water m_{sub}, is obtained. From the three masses obtained, m_D, m_s, m_{sub}, the total porosity of the specimen can be determined using Equation (5.9), and is usually referred to as "water porosity." The apparent and true densities of the specimen can also be determined using Equations (5.7) and (5.8), respectively [5.2,5.3]. The accuracy of the pycnometric method for determining the porosity depends on the accuracy of the three weighings for obtaining the masses m_s, m_D, and m_{sub} [5.1].

With the water absorption technique, changes to the cement paste structure can be followed as water is removed from the cement paste pores or as water re-enters the pores [5.4]. Porosity determined using water is quite different from the porosity determined using other liquids, mainly because of the interaction of water molecules with cement paste and the penetration of water molecules into interlayer spaces that have partially collapsed after drying. Even though this technique provides a quick estimate of porosity, there are concerns involving the state of water removed during drying. As it was mentioned in Chapter 2, oven-drying of cement paste specimens tends to collapse the delicate microstructure, and removes some chemically bound water, besides capillary and interlayer water [5.1,5.2]. An example of the effects of different drying techniques and consequently different water removal levels on the water porosity is shown in Figure 5.3 [5.5]. Oven drying at 105°C results in the highest value of porosity, which can be attributed to the greater water removal from the cement paste compared to the other methods, as it has already been discussed in Chapter 2. Similar studies have been carried out by other researchers [5.6]. In addition, resaturation of dried specimens with water to determine porosity can be inaccurate because some of the products may rehydrate. For these reasons, solvents are used instead of water as pycnometric liquids and this procedure is described in the following section.

5.1.1.2 *Water replacement using an alcohol*

The method of water replacement is also mentioned as solvent exchange or solvent replacement, and the procedure is similar to the procedure followed

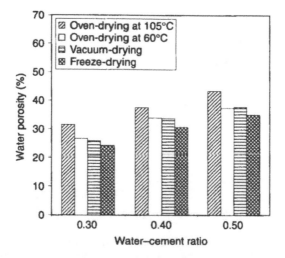

Figure 5.3 Total water porosity of ordinary Portland cement pastes having different water–cement ratios pretreated by different drying techniques.

Source: Reprinted from *Cement and Concrete Research*, Vol. 31, C. Gallé, "Effect of drying on cement-based materials pore structure as identified by mercury intrusion porosimetry: a comparative study between oven-, vacuum-, and freeze-drying," pp. 1467–1477, copyright 2001, with permission from Elsevier Science, Amsterdam, The Netherlands.

for pretreatment of specimens that was discussed in Chapter 2. The first research work using solvents for liquid pycnometry has been reported by T. Powers and T. Brownyard, 1948 [5.7]. The solvents most commonly used by researchers are the alcohols methanol and isopropanol, but any other solvent mentioned in Chapter 2 could also be used. Some points of attention when using alcohols in cement pastes have already been discussed in Section 2.1.2.

The procedure followed during solvent exchange consists of immersing a water-saturated surface-dry specimen, with thickness of approximately 3 mm into a large container with solvent at room temperature [5.2,5.8]. The solution–sample volume ratio is important and is taken to be 100:1, since the point of saturation of water in most organic solvents is quite low. The alcohol in the container should be renewed regularly, e.g. every hour during the first 24 hours of experiment and then less frequently. During the experiment, the mass of the specimen is measured at regular intervals in order to monitor the penetration of the alcohol within the specimen. The reduction of the specimen's mass with time is monitored until mass equilibrium is achieved (see Figure 5.4) [5.9]. The difference in densities between water and alcohol enables mass changes during the exchange of pore water with the alcohol to be interpreted quantitatively using Equations (5.7)–(5.9), and the appropriate alcohol density (see also Table 2.2). After the specimen has reached constant mass, i.e. water has been completely replaced by alcohol, the specimen is dried.

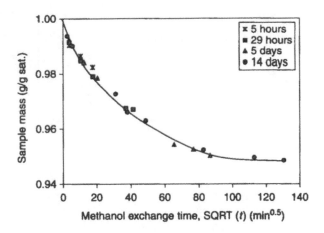

Figure 5.4 Reduction of the mass with time for water-saturated specimens of hydrated C₃S pastes as the pore water is replaced by methanol, for different exchange times. After 14 days the mass is reaching an asymptotic value.

Source: Reprinted from *Cement and Concrete Research*, Vol. 13, L.J. Parrott, "Thermogravimetric and sorption studies of methanol exchange in an alite paste," pp. 18–22, copyright 1983, with permission from Elsevier Science, Amsterdam, The Netherlands.

Two estimates of porosity can be obtained, one by using the original water-saturated mass and the other using the alcohol-saturated mass. The water porosity is determined from Equation (5.9) using the water-saturated mass of the specimen as it was described in Section 5.1.1.1. The first estimate of porosity based on water saturation corresponds to the porosity before any contact of the specimen with alcohol. The alcohol exchange porosity is determined from Equation (5.9) using the alcohol-saturated mass of the specimen. The corresponding densities of the specimen, i.e. the density using water and the density using alcohol, can also be determined.

Replacement of the pore water by an organic solvent is a simple physical process of counter diffusion. The rate of exchange of pore water by alcohol depends on the thickness of the specimen and its density [5.10]. It has been pointed out that very long immersion times (up to several weeks in some cases) are required to achieve a steady state condition and constant mass in dense cement pastes [5.2,5.9,5.11,5.12].

Several researchers have used the solvent-exchange technique to measure alcohol porosities of cement pastes and compare them to water porosities [5.2,5.10,5.13]. The use of alcohols instead of water to determine porosity has the advantage of avoiding the problem of rehydration in many of the poorly oriented layered materials. Even though theoretically the porosities of the material characterized using alcohols or water should be independent of the liquid used for testing, the porosities determined using methanol and isopropanol give different results. It has been reported by researchers that the porosities determined using methanol are higher than the porosities measured using propanol,

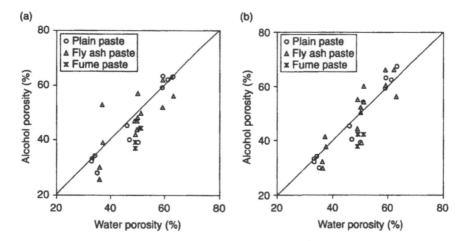

Figure 5.5 Alcohol exchange porosities vs. water porosity measured on plain and
blended cement pastes containing fly-ash or silica fume. The alcohol used
were (a) propanol and (b) methanol. The cement pastes had w/c = 0.47 and
were cured at 20°C.

Source: Reprinted from *Cement and Concrete Research*, Vol. 18, R.L. Day, B.K. Marsh,
"Measurement of porosity in blended cement pastes," pp. 63–73, copyright 1988, with permission
from Elsevier Science, Amsterdam, The Netherlands.

indicating that the pore volume of cement paste penetrated by methanol was
on average 0.7% greater than the pore volume occupied by the original pore
water [5.2,5.10]. The observed differences are due to a combination of several
factors such as the accessibility of alcohols inside the pore space and chemical
interaction with cement paste and they have been discussed in Section 2.1.2.

An example on how the water and alcohol porosities compare in cement
pastes is given in Figure 5.5, both for methanol and propanol [5.2]. The water
porosity compares well to the alcohol porosity. A different study however, has
shown that for D-dried cement pastes, the water porosity can be higher by
14–90%, the difference becoming higher as the water–cement ratio decreases
[5.14]. The porosities determined by water and alcohols depend on a number
of factors, such as the age of the cement pastes, their mineral composition, the
water–cement ratio, and the method of pretreatment.

5.1.2 *Gas (helium) pycnometry and helium flow*

In an alternate way to liquid displacement and liquid pycnometry, the total
volume of a specimen V_{tot} may be determined by gas pycnometry. Even though
several gases can be used for gas pycnometry, helium is recommended and
used predominantly. Helium is supposed to give the most reliable density
results, because it has the following characteristics [5.8,5.14,5.15]:

- Helium is a monomolecular gas with the smallest atomic diameter of
  about 0.22 nm (for comparison, the molecular diameter of water is about

0.28 nm, and the molecular diameter of mercury is about about 0.4 nm).
The small size of the helium atom enables it to penetrate pores as small as
its size.
- Helium does not interact with most materials and does not adsorb on the
surface of a material.
- Helium behaves as an ideal gas.

Early forms of the gas pycnometry apparatus were given by E. Washburn
and E. Bunting, 1922, and by G. Karns, 1926 [5.16,5.17]. Many modifications,
improvements, and adaptations have subsequently been made to the gas
pycnometry apparatus, and the currently available models offer extreme simplicity of operation along with great speed and accuracy. Helium pycnometry has
already been described and applied to hydrated Portland cement systems by
several researchers. T. Powers and T. Brownyard, 1948, first determined the
real volume of the solid phases and the volume of pores in hardened cement
pastes by the helium displacement method [5.7].

5.1.2.1   Theoretical aspects

Helium pycnometry provides a more rapid and more accurate measurement of
true densities than does the Archimedean technique using liquid pycnometry;
helium pycnometry has a density resolution of approximately $0.0001 \, g/cm^3$.
The helium pycnometry method uses the gas laws and the assumption of ideal
gas behavior to calculate the real volume of a specimen. Under precisely
known pressure, helium is used to fill the pores within a specimen. The pores
and spaces into which helium cannot instantaneously flow are assumed to be
negligible. The change of helium volume in a chamber of constant volume
allows determination of the true volume of the specimen from pressure measurements using Mariotte's law. The ratio of sample mass to the sample true
volume gives the true density [5.18].

The equipment consists of two vessels, one that holds the sample, and one
of known volume that acts as a reference. A schematic diagram of the helium
pycnometry setup is shown in Figure 5.6. The specimen to be tested must first
be dried, using one of the techniques mentioned in Chapter 2. Usually, water
is removed incrementally by evacuation alone and later by evacuation and
heating at increasing temperatures for different periods of time. This is usually
done in a separate vacuum vessel and the sample is then transferred to the pycno-
meter. There are no specific requirements for the dimensions of the specimen,
but small fragments of cement paste or mortar with a volume of about $1 \, cm^3$
are used for the experiment. The sample can also be in the form of a disk,
e.g. 32 mm diameter $\times$ 1 mm thick, and the total mass varying from 15 to 30 g
[5.19]. It takes about 48 hours for helium to flow into the pores of cement
pastes. All runs are performed in a temperature-controlled laboratory at 22°C.

The pressures applied to helium range between 40 and 55 kPa. The equipment is first evacuated until a vacuum level of $10^{-3} \, Pa$ is reached. The
reference cell having a known volume V_R is first filled with helium gas at
atmospheric pressure P_1. Then helium is allowed to fill the empty sample cell

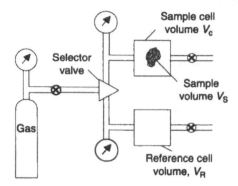

Figure 5.6 Schematic out-of-scale setup for helium pycnometry. The selector valve allows connection between two paths only, blocking flow to the third path.

by opening the selector valve (see Figure 5.6) and the two vessels are filled with helium at atmospheric pressure [5.1,5.19]. The state of the system can be defined by Equation (5.10) for the sample cell, and by Equation (5.11) for the calibrated reference cell [5.20]:

$$P_1 V_c = nRT \tag{5.10}$$

$$P_1 V_R = n_R RT \tag{5.11}$$

where P_1 is the atmospheric pressure (N/m^2), V_c is the volume of the sealed empty sample cell (m^3), V_R is the volume of a carefully calibrated reference cell (m^3), n is the moles of gas in the sample cell at pressure P_1, n_R is the moles of gas in the reference cell at pressure P_1, R is the gas constant, T is the ambient temperature (K).

After the prepared specimen of volume V_S is placed in the sample cell, Equation (5.10) can be modified to Equation (5.12) in order to take into account the presence of the sample:

$$P_1(V_c - V_S) = n_1 RT \tag{5.12}$$

where V_S is the volume of the porous sample (m^3), n_1 is the moles of gas occupying the remaining volume in the sample cell at pressure P_1.

Then the sample cell is pressurized to a pressure P_2, which is around 2 atm. The state of the sample cell including the specimen is defined by Equation (5.13):

$$P_2(V_c - V_S) = n_2 RT \tag{5.13}$$

where n_2 is the moles of gas occupying the remaining volume in the sample cell at P_2.

The selector valve is opened to connect the sample cell with the calibrated reference cell. Then the pressure in the connected system drops to an equilibrium value P_3, which has a value between the atmospheric pressure P_1 and the applied pressure P_2 $(P_1 < P_3 < P_2)$. The total amount of helium moles in the system is $n_2 + n_R$, and the state of the system is described by Equation (5.14):

$$P_3(V_c - V_S) + P_3 V_R = n_2 RT + n_R RT \qquad (5.14)$$

By substituting the number of moles n_2 and n_R in Equation (5.14) from expressions obtained from Equations (5.13) and (5.11), in order to eliminate them in the expression, Equation (5.14) becomes Equation (5.15), which can be rearranged to Equation (5.16):

$$P_3(V_c - V_S) + P_3 V_R = P_2(V_c - V_S) + P_1 V_R \qquad (5.15)$$

$$V_S = V_c - \frac{P_1 - P_3}{P_3 - P_2} V_R \qquad (5.16)$$

where V_S is the volume of the specimen (m³), V_R is the volume of the calibrated reference cell (m³), V_c is the volume of the sample cell (m³), P_2 is the applied pressure (N/m²), P_3 is the equilibrium pressure after pressurization (N/m²).

By measuring the pressures P_2 and P_3, and knowing the volume of the sample cell, V_c, and the volume of the calibrated reference cell, V_R, the volume of the sample, V_S can be determined. From the mass of the sample measured before the experiment and the true volume measured by helium pycnometry, the true density can be determined, which is frequently referred to simply as the helium density, being obvious that open pores are excluded in such cases.

In cement pastes, at 11% RH the adsorbed water is approximately equal to a monolayer and it has been recommended that samples be conditioned at 11% RH prior to volume displacement with helium; in this way, we can avoid dissociation of hydrates due to further removal of water on drying [5.21,5.22].

5.1.2.2 Helium flow

During the same experiments carried out to measure density, helium flow can also be measured as the volume of helium adsorbed with time [5.4,5.23]. Immediately after the compression period when the pressure in the sample cell is raised to 2 atm, helium flow readings are recorded for a period of approximately 40 hours. The helium flow technique follows volume changes that occur in the solid and pores resulting from the removal of interlayer and physically adsorbed water from cement pastes. R. Feldman, 1971–1978, in a series of papers used the helium flow to study the state of water in-between the layers composing the C–S–H gel and to define the Feldman–Sereda model (see also Figure 1.1) [5.18,5.21,5.22,5.24,5.25,5.26]. In this way, Feldman could study the hydration of Portland cement pastes, permitting observation of the changes that occur in the pore structure during removal and re-entry of water. For his experiments, he used cement pastes with different water–cement ratios and different water contents by changing the RH.

Helium flow is presented as a plot of helium volume, ΔV, at 2 atm per 100 g of sample vs. time (see Figure 5.7) [5.21]. All cement pastes with different

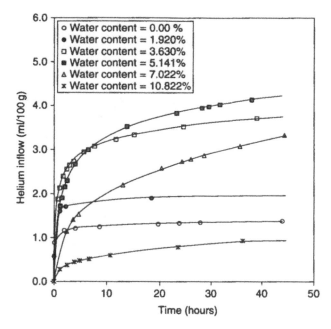

Figure 5.7 Helium inflow vs. time for hydrated Portland cement pastes with w/c = 0.40, conditioned to different water contents. Conditioning to water content of approximately 11% corresponds to a monolayer of water adsorbed on the pore walls.

Source: Reprinted from *Cement and Concrete Research*, Vol. 1, R.F. Feldman, "The flow of helium into the interlayer spaces of hydrated Portland cement paste," pp. 285–300, copyright 1971, with permission from Elsevier Science, Amsterdam, The Netherlands.

water–cement ratios and moisture contents show flow curves of an exponential type. However, the initial flow rate and total intake of helium varied with material and moisture content in the cement pastes.

The kinetic curves of helium inflow depend on the volume of vacated space and the size of the entrances of the pores. The solid volume of C–S–H gel includes interlayer spaces into which helium does not enter during the initial period of exposure to helium gas and its compression during the beginning of the experiments. When part of the vacated interlayer space has collapsed, resulting in a change in solid volume and helium inflow, an assessment of the space occupied by the water molecules can be made by combining these parameters. Collapse of the spaces or the entrances are indicated by a decrease in helium inflow rate [5.4,5.23]. As water is removed from the interlayer spaces more space is vacated and some collapse of the C–S–H sheets occurs. At first however, the collapse is not as great as the space created and helium inflow increases in rate. As entrances to interlayer spaces get significantly smaller, the rate of flow decreases, even though a larger volume of helium can ultimately flow in. The collapsing layers not only present "narrow necks" to the helium atoms, but also long narrow slits, which greatly restrict inflow. The collapse of the layers

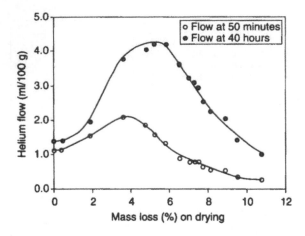

Figure 5.8 Helium flow at 50 minutes and at 40 hours plotted as a function of mass
loss for a cement paste with w/c = 0.40. The cement paste was conditioned
to 11% RH prior to testing.

Source: Reprinted from *Cement and Concrete Research*, Vol. 1, R.F. Feldman, "The flow of
helium into the interlayer spaces of hydrated Portland cement paste," pp. 285–300, copyright
1971, with permission from Elsevier Science, Amsterdam, The Netherlands.

traps the space vacated by water, and helium cannot enter this space even after
40 hours of exposure [5.27].

Feldman concluded from his studies that helium flows rapidly into all pores
except the interlayer spaces. By measuring helium flow for the same cement
paste specimen at different levels of water content, he concluded that the
"small pores" into which helium is flowing are interlayer spaces and not fixed-
dimension narrow-necked pores, and that water can be removed and added
between the layers composing the C–S–H gel [5.21,5.25].

In addition to the plot of helium flow vs. time, the samples are weighed after
the helium inflow run, and the change in flow characteristics can be plotted as a
function of change in specimen mass, which can be done in two ways [5.4,5.21].

1   One may plot "total inflow" vs. mass loss, as it is shown in Figure 5.8
    [5.21]. This plot was used by Feldman to illustrate how the helium flow
    changes with moisture content and represents the volume that flowed into
    the sample at 40 hours as a function of the moisture that was removed
    from the state of cement paste at 11% RH. Total inflow refers to the helium
    flow into the small pores into which nitrogen and methanol essentially
    cannot enter.

2   For the specimens from the same cement paste, i.e. having the same pore
    structure characteristics, as the RH changes, the volume of the specimen
    changes. Removal of moisture leads to a change in real volume of the
    specimen, ΔV, while the change in helium flow is ΔD. Adding these values,
    a parameter $\Delta V - \Delta D$ is obtained, where the decrease in specimen volume
    ΔV is negative and the increase in flow, ΔD, due to mass loss is positive. In

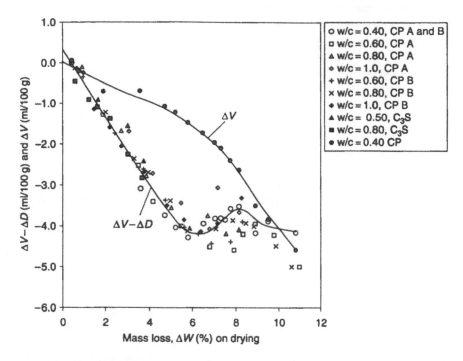

Figure 5.9 Plot of $\Delta V - \Delta D$ and ΔV as a function of mass loss during helium flow for cement pastes (CP) and C_3S pastes, having different water–cement ratios (w/c). The cement pastes in series A were D-dried, and the cement pastes in series B were dried well beyond the D-dried state. All samples were conditioned for 6 months to 11% RH prior to the experiments.

Source: Reprinted from *Cement and Concrete Research*, Vol. 1, R.F. Feldman, "The flow of helium into the interlayer spaces of hydrated Portland cement paste," pp. 285–300, copyright 1971, with permission from Elsevier Science, Amsterdam, The Netherlands.

this way, a plot of $(\Delta V - \Delta D)$ vs. mass loss can be obtained (see Figure 5.9) [5.21]. The slope of the plot $\Delta V - \Delta D$ vs. mass loss gives the average density of the water [5.28]. However, the correctness of the values determined and its association with a model has been questioned by S. Brunauer [5.29].

The change in volume ΔV due to water removal in the solid part of the cement paste includes interlayer spaces and measures the shrinkage of the cement paste. The change in volume ΔD is the helium taken at 40 hours into these interlayer spaces. The change in volume $\Delta V - \Delta D$ measures the space that adsorbed and interlayer water occupy in the cement paste. The change in volume $\Delta V - \Delta D$ plotted vs. mass loss gives a straight line up to about 5.5% mass loss.

It is assumed that at the beginning of the flow experiments, little helium enters into the small spaces vacated by water for the short period at 1 atm and during compression. Correction for this amount of helium can be made from the results of the sample conditioned at 11% RH, which are subtracted from the helium volumes at other RH conditions to obtain the changes in volume ΔV and ΔD. It is also assumed that helium does not interact with the surface or other body forces and is at 2 atm within all the small spaces that water has vacated.

Feldman found that for w/c = 0.40 the maximum helium inflow was 4.2 ml per 100 g of specimen, and for w/c = 0.60 the maximum helium inflow was 3.2 and 3.4 ml/100 g [5.21].

Based on helium flow experiments, Feldman identified as "missing porosity" the interlayer space normally occupied by approximately one layer of water when the material is in the wet condition. On drying and removal of the interlayer water, this space partially collapses. The effort and controversy related to the study of the missing porosity and its importance relating to mechanical properties has justified its inclusion in a discussion on porosity [5.4].

The helium inflow results during rewetting of the cement pastes are different from those obtained during removal of water, which is expected from the model of hydrated Portland cement for dehydration and rehydration of interlayer water [5.24].

5.1.2.3   Determination of surface area and hydraulic radius

R. Feldman suggested that the surface area and hydraulic radius of interlayer space of the C–S–H and capillary pores can be calculated from helium inflow data [5.27]. This can be done using length changes and determination of the solid volume change by helium pycnometry as the sample is exposed to different RH conditions. Solid volume includes the interlayer spaces into which helium does not enter during the initial period of exposure to helium gas and its compression during the beginning of the experiments.

When part of the vacated interlayer space has collapsed, resulting in a change in solid volume and helium inflow, an assessment of the space occupied by the water molecules can be made by combining these parameters [5.23]. The surface area may be calculated assuming that one water molecule covers an area of 10.8 Å². By measuring the solid volume at the dry state (0% RH) and at 11% RH, where the adsorbed water is approximately equal to a monolayer, one would obtain the volume of adsorbed water of a monolayer plus the increase in volume of the solid. The volume changes would be equal to the volume of the adsorbed water as a fraction of the total volume of the solid. This may be written as expressed by Equation (5.17) [5.4]. If the linear expansion of the porous body, $\Delta l/l$, is small, then the quantity $3\Delta l/l$ would be equivalent to the volumetric expansion:

$$\frac{\Delta V}{V} - 3\frac{\Delta l}{l} = \frac{v}{V} \tag{5.17}$$

where ΔV is the change in solid volume from a D-dried state to 11% RH, V is the volume of the cement paste specimen at the D-dried state, $\Delta l/l$ is the swelling of the specimen, v is the volume of the monolayer of adsorbed water on the internal surface of the pore space.

For example, for a cement paste with w/c = 0.40 that had been D-dried and conditioned at 11% RH, Feldman measured the following values:

$\Delta V/V = 2.827\%$, $\Delta l/l = 0.362\%$. The mass of water forming a monolayer per mass of D-dried sample was 91.83. The volume of the D-dried sample was $V = 42$, and from this, the surface area was calculated to be 34.8 m²/g. Based on these experimental studies and calculations, values of surface area were calculated to be 49.0 m²/g for w/c = 0.60, and 44.4 m²/g for w/c = 0.80, assuming that a water molecule covers an area of 10.8 Å² [5.27].

5.1.2.4 Effect of pretreatment

Since helium has the ability to enter different spaces in cement paste and in between the C–S–H layers, it is apparent that pretreatment will have an effect on the results. An example on the effect of pretreatment on porosities determined using helium pycnometry is shown in Figure 5.10 [5.30]. The results show that higher porosities are obtained for the blended cement pastes by direct oven-drying. For ordinary cement pastes the porosity values obtained by helium flow are fairly similar, when dried by isopropanol replacement and oven-drying. The differences observed in blended cement pastes could be due to collapse of the microstructure by oven-drying, which is not as significant in plain cement pastes.

It has been shown that a cement paste sample that has been D-dried is in a partially collapsed state and helium can enter only very slowly into the enclosed interlayer spaces [5.25]. Helium flow into interlayer space results in a pore volume significantly larger than the pore volume which can be accounted for by removal of water alone. The total helium inflow for specimens conditioned at 11% RH increases significantly when the water–cement ratio is below 0.40.

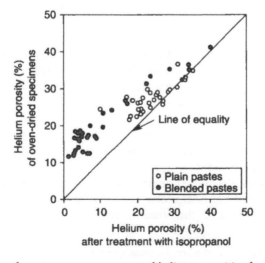

Figure 5.10 Effect of pretreatment on measured helium porosities for plain and blended cement pastes.

Source: Reprinted from *Very High Strength Cement-Based Materials*, J. Francis Young (ed.), *MRS Symposium Proceedings Vol. 42*, B.K. Marsh, R.L. Day, "Some difficulties in the assessment of pore-structure of high performance blended cement pastes," pp. 113–121, copyright 1985, with permission from the Materials Research Society, Warrendale, PA.

5.1.3   Advantages and limitations

The water absorption technique has the advantage that it can provide relatively fast an indication of the porosity of a cement-based material using simple and inexpensive equipment. The experiment can accommodate sizes of various shapes and dimensions. However, the porosities determined using the water absorption technique is used rather for comparative studies than for obtaining absolute values.

The solvent exchange technique is difficult to differentiate between the various types of water that are replaced. The solvent replacement method does not differentiate between normal pores and interlayer spaces, but is useful in detecting changes to the hydrated Portland cement paste structure due to various drying pretreatments. Complications due to possible interactions between certain alcohols and the cement pastes should be taken into account when selecting alcohols and when interpreting results.

Helium pycnometry is a rather simple technique and has been used extensively in the past to explore the microstructure of the C–S–H gel. The technique determines the pore volume without giving information on the pore size distribution. For this reason, it is not used for regular testing, but for comparison and modeling purposes. The helium inflow technique has been used to approximate the surface area of the interlayers in the C–S–H gel.

5.2   Thermoporometry

Thermoporometry (also mentioned as thermoporosimetry)* provides a simple method for determining the pore size distributions in a porous material that is fully or partially saturated with a liquid. The method is based on the thermodynamic conditions of the liquid–solid transformation (melting–solidification) of a capillary condensate inside a porous body and was initially applied as a relative method using calibration curves drawn from well-known samples. However, it has developed into an absolute method, enabling one to deduce the pore size distribution from the solidification thermogram purely by means of theoretical relationships. Low-temperature calorimetry or differential scanning calorimetry is used to monitor heat changes associated with freezing of the pore water or melting of the ice. Thermoporometry makes use of the triple point depression of a liquid in contact with its solid phase within the porous matrix of a material; this effect also depends on the curvature of the interface, and hence on the pore radius.

The method was developed mathematically by M. Brun *et al.*, 1977, using thermodynamic considerations and was further extended by his group to calculate the pore size distribution and the surface area of a porous

---

* The term "porometry" is used when the technique measures the minimum pore diameter in a porous system, and applies to capillary flow processes. The term "porosimetry" is used when the technique measures the total pore volume in a material. Therefore, the term thermoporosimetry would be more appropriate to identify the technique, rather than the term thermoporometry; however the technique is predominantly referred to in the technical literature as thermoporometry, and this is the term used here.

material [5.31,5.32,5.33]. Thermoporometry is useful for the study of porous materials because it does not require drying of the material, which can cause distortion of the pore structure. The sample is immersed in water or benzene and the amount of frozen liquid is measured as a function of decreasing temperature at ambient pressure by means of a highly sensitive calorimeter. As the temperature increases, the amount of solid melting is measured. From the freezing and melting curves obtained, the pore size distribution, the pore shape and the specific surface area can be calculated. The cyclic freezing and melting of the pore water that occurs during the experiments does not seem to cause any significant change to the pore structure [5.34]. The first study of the freezing relation to pore structure on cement pastes was made by C. Le Sage de Fontenay and E. Sellevold, 1980 [5.34]. However, thermoporometry has not become a very popular technique in cement pastes with the exception of NMR cryoporometry, which is discussed in Section 6.4.2.

5.2.1 Theoretical considerations

Thermoporometry analyzes the conditions of the solid–liquid phase transformation of a liquid inside the pores of a material with change in temperature. The method is based on the observation that the conditions of equilibrium of the solid, liquid, and gaseous phases of a pure substance inside a pore are determined by the curvature of the interfaces. M. Brun *et al.*, 1977, have given a detailed derivation of the solid–liquid transformation phenomena using thermodynamic equilibria [5.31].

Several studies have shown that when the water wetting a solid surface (i.e. a pore wall) is solidified, a liquid-like layer of the substrate remains at the surface. The solid–liquid interface determines the layer structure of water near the solid surface and results in a deformation of the hydrogen bonds between the water molecules of the different layers. The deformation of the hydrogen bonds at the boundary layer in comparison with the bonds in free water indicates the formation of heterogeneities in the liquid and results in the decrease of the temperature of crystallization in the boundary water layer. Water does not exist in a super-cooled metastable state, but it is thermodynamically at equilibrium and can coexist with ice at a temperature lower than the temperature of free water. The temperature of the water–ice phase transitions in porous solids is not affected by the nature of the pore wall material [5.35]. Therefore, a model is proposed in which the solid water (ice) is separated from the pore walls by a liquid water layer of thickness *t*, which does not change state as the temperature varies. For a porous material saturated by water or some organic solvents, during freezing, progressive penetration of the solid phase occurs in smaller pores as it is shown schematically for a capillary of varying size in Figure 5.11. Solidification in a capillary takes place either by a classical mechanism of nucleation or by a progressive penetration of the liquid–solid meniscus formed previously at the inlet of the pore. In both cases at the very moment of solidification the meniscus is almost spherical [5.35]. Studies have shown that the freezable pore water in pores larger than the size of the nucleus of bulk ice is supercooled, but the freezable pore water in the smaller pores is not significantly supercooled [5.36].

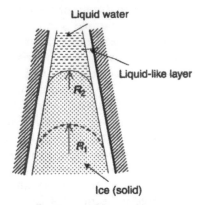

Liquid water

Liquid-like layer

Ice (solid)

Figure 5.11 Schematic representation of the solidification process of water inside a cylindrical pore. The curvature of the ice is almost spherical and equal to the radius of the pore (R_1). As the temperature decreases, the ice front progresses to smaller pores with smaller curvature (R_2). A liquid-like thin layer remains at the sides of the pore and does not freeze.

When a sample saturated with a liquid, such as water or some organic solvents, is cooled, solidification can occur progressively in the smaller pores when the size of the critical nuclei at a given temperature is the same as the size of the pores, and thereby supercooling cannot occur in the liquid confined in the pores [5.31].

A crystalline solid contained within a small pore necessarily consists of minute crystals and therefore has a large surface to volume ratio and a low melting point. The thermodynamic stability of small crystals has been treated by Gibbs, who demonstrated that the melting transition temperature of the liquid confined in a pore depends on the liquid surface curvature. The theory connecting surface curvature of liquid droplets and vapor pressure is due to Thomson (Kelvin) and correctly describes capillary condensation. M. Brun *et al.*, 1977, combined the Gibbs and Thompson equations to relate the decrease in melting temperature to crystal dimensions [5.31]. They obtained mathematically an equation showing that the curvature of the solid–liquid interface is related to the change in the transition temperature ΔT, as it is expressed by Equation (5.18). The full derivation is given analytically in the original paper and is not included here:

$$r = \frac{A}{\Delta T} \quad \text{with } A = \frac{2\gamma_{sl}v_m T_0 \cos\theta}{\Delta H_m} \tag{5.18}$$

where r is the curvature of the liquid–solid interface (m), T_0 is the temperature of triple point of water (K), ΔT is the depression of the freezing or melting temperature T of water $= T - T_0$ (K), v_m is the molar volume of water (m³/mol), θ is the contact angle between liquid, ice, and pore wall (degrees), ΔH_m is the change in the molar heat of the ice–water transition (J/mol), γ_{sl} is the interfacial tension between ice and water (N/m).

Considering that pore water consists of both freezable and nonfreezable water, the pore radius, R, should be equal to the sum of the curvature r of the solid–liquid interface and a parameter B that relates to the layer of nonfreezable pore water, and is expressed by Equation (5.19):

$$r_p = \frac{A}{\Delta T} + B \qquad (5.19)$$

where r_p is the pore radius, A, B are constants that can be derived theoretically.

The constant A is characteristic of the liquid and depends on the properties of the liquid and liquid/solid interface. Almost any substance can be used as the adsorbate. However, water is one of the most attractive substances to fill the pores, because its melting point lies in a temperature range that can easily be measured. For water, the molar volume of ice is $v_m = 19.6 \times 10^{-6}\, m^3/mol$, the temperature of the triple point is $T_0 = 273.14\,K$, and the change in molar heat is $\Delta H_m = 6.02\,kJ/mol$. The interfacial tension between water and ice can range from 0.01 to 0.04 N/m and can be measured experimentally in porous materials with controlled porosity. The variations of the interfacial tension between water and ice have been presented in the technical literature [5.36].

In principle, both the freezing point depression and the melting point depression can be used to determine the pore radius. However, delayed nucleation can have a large influence on the pore radii calculated from the freezing point depression, and for this reason, the melting point depression is more reproducible and most often used in experiments.

The pore size can be calculated for spherical pores from the solidification and melting process, and in the case of cylindrical pores, only from the melting process.

In the case of the *spherical* shape of pores, the curvature of the solid phase surface does not change upon melting and crystallization. The ratio of liquid volume to surface area is half the pore radius, $r_p/2$. The temperatures of melting and crystallization are equal and depend on the pore radius [5.35].

If the pores are *cylindrical*, the nuclei of the solid phase grow spontaneously and take the shape of the pores. In this case, on the basis of the geometrical consideration, the ratio of the liquid volume to the pore surface area is equal to the pore radius, r_p, upon freezing and to half the pore radius, $r_p/2$, upon melting. Therefore the freezing temperature depression should be twice as large as the melting temperature depression. This fact accounts for the hysteresis between melting and crystallization that is discussed in Section 5.2.2 [5.35].

5.2.2 Experimental procedure

The experimental setup is shown in Figure 5.12. The specimen is first degassed in vacuum and is subsequently water-saturated in a desiccator. Then, the specimen is sealed in an aluminum ampule and placed into a calorimeter. The temperature is reduced at a constant cooling rate, that can vary from 0.03 K/minute to 32 K/minute [5.35,5.37]. The crystallization thermogram is recorded, which is a plot of heat flow vs. temperature. After being cooled, the specimen is subjected to heating at the same rate, and the ice melting thermogram is obtained.

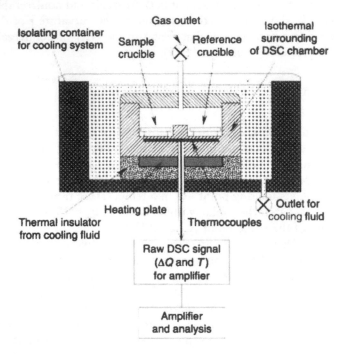

Figure 5.12 Schematic setup for the equipment used for the thermoporometry technique.

The crystallization and melting curves obtained during lowering and increasing the temperature in a saturated porous material will be slightly different, the difference being related to the pore geometry. The melting thermogram is used to determine the pore size distribution [5.38]. By comparing the melting and solidification thermograms, the pore shape can be determined from the existence and shape of hysteresis between the two thermograms (see Figure 5.13). It is possible then to present a model for the material made up of cylindrical and spherical pores [5.31,5.33,5.35]. In the case of spherical pores, the melting and crystallization temperatures are equal for a pore, and therefore, no hysteresis is observed between the freezing and melting curves (curves I and II in Figure 5.13). However, in the case of cylindrical pores, the freezing temperature depression is different than the melting temperature depression and a hysteresis is observed between the freezing and melting curves (curves I and III in Figure 5.13). The hysteresis is believed to be indicative of pore continuity, with ice front propagation during cooling being controlled by the narrowest necks in the pore system [5.34].

The value of constant A in Equation (5.18) is independent of the pore size and should be expected to be around 48 nm K on melting and 64 nm K on freezing of water [5.31,5.39]. It is also assumed that the liquid phase at the interface between ice and the pore wall consists of 2.5 molecular layers for the case of water, resulting in a liquid film with thickness of approximately 0.8 nm [5.31,5.40]. In a similar way, for benzene the liquid phase absorbed at the

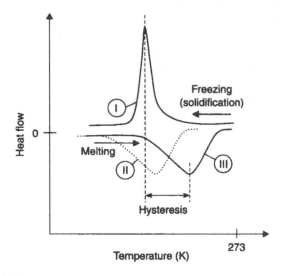

Figure 5.13 Schematic thermogram for water freezing (curve I) and water melting (curves II and III) in a porous material. The existence of hysteresis (between curves I and III) indicates the presence of cylindrical pores, while the absence of it (curves I and II) indicates almost spherical pores.

interior surface of the pore consists of 3.5 molecular layers, resulting in a thickness of 1.33 nm [5.31].

For water in spherical pores, the pore size is related to the temperature difference by Equation (5.20) that is valid for both freezing and melting. For cylindrical pores containing water, the pore size can be calculated from the melting process using Equation (5.21) for a temperature range −40–0 K [5.35]:

$$r_p = -\frac{64.67}{\Delta T} + 0.57 \tag{5.20}$$

$$r_p = -\frac{32.33}{\Delta T} + 0.68 \tag{5.21}$$

where r_p is the pore radius, ΔT is the depression of the melting temperature of water.

Pore size distributions in porous materials have been investigated by applying dynamic and thermodynamic relationships for the freezing and melting curves of water. The thickness B of the layer of nonfreezable pore water and the values of the coefficient A have been determined by simultaneous iterative optimization of both the peak radius of the pore size distribution and the entire pore volume. Assuming a cylindrical pore, the values for coefficient A during melting and freezing were estimated to be given by the relationship $33.30 + 0.32\Delta T$ and $56.36 + 0.90\Delta T$ (in nmK) respectively. The ice–water interfacial tension was also estimated to be given by the relationship $\gamma_{sl} = 20.4 \times 10^{-3} (1 - 15.9 \times 10^{-3}\Delta T)$ expressed in N/m determined from the temperature dependence of the coefficient A during melting [5.36].

Simulation based on spherical and cylindrical pores only enables one to determine at the same time the radius and the volume of those two extreme types of pores. The development of a model, by means of successive approximations, leads to a unique result whose agreement with the experimental results seems to be promising for the future of the method. More recent studies have used the shape factor in the analysis in the effort to include sizes of different shapes besides spherical and cylindrical [5.41]. The shape factor makes it possible to show to what extent the pore shape differs with the cylindrical model often propounded in the literature (see also Section 1.4.1.6).

5.2.3   Pore size distribution

Thermoporometry is applicable to pores with radius ranging between 2 nm and 200 nm. Pores with a radius less than 2 nm cannot be studied, using water as the probe molecule, because these pores contain only adsorbed water that does not freeze even below $-90°C$. Also pores of sizes less than 2 nm are comparable to the thickness of the adsorbed layer [5.31,5.32,5.42].

For a liquid contained inside a porous material, the solid–liquid interface curvature depends on the size of the pore and, therefore, the solidification temperature is different in each pore of the material. Since Equation (5.19) relates the change in temperature ΔT to the pore radius, if a distribution of ΔT is obtained, then a distribution of pore radii can be determined [5.31,5.33]. The thermogram gives information on the volume of pores through the measurement of the energy involved in the phase transformation of the pore liquid. The number of pores of each size can be determined from the total heat, because the quantity of solid material (ice) melting over any temperature range is a direct indication of the number of pores in that size range [5.37].

The abscissa of the DSC curve, temperature T, can be transformed into the pore radius, r_p, of freezable pore water using Equation (5.19). On the other hand, the ordinate displacement of the DSC curve, heat flow dq/dt, is transformed into a derivative volume change in the pores, as it is expressed by Equation (5.22) [5.36]:

$$\frac{dV_{fp}}{dR} = \frac{dq}{dt} \cdot \frac{dt}{dR} \cdot \frac{1}{m \cdot \Delta H_m(T)\rho(T)} \qquad (5.22)$$

where V_{fp} is the cumulative volume occupied by freezable pore water (m³), r is the pore radius (m), t is the time (s); dq/dt is the heat flow, m is the mass of the specimen (g), $\rho(T)$ is the density of water as a function of temperature T (g/m³), $\Delta H_m(T)$ is the change in the molar heat of the ice–water transition (J/g).

The parameters $\Delta H_m(T)$ and $\rho(T)$ in Equation (5.22) that are temperature dependent have been described from experimental studies by functions of temperature as it is described in the following paragraphs.

During cooling, the volume of freezable pore water will increase before and after freezing because the density of ice is smaller than the density of water; the volume of water must be calculated by dividing the mass by the density of freezable pore water. Similarly, during heating, the pore volume should be equal to the volume occupied by freezable pore ice, and accordingly, it can be calculated by dividing the mass by the density of freezable pore ice. By assuming that the density of freezable pore ice is the same as the density of bulk water, and that the density of freezable pore ice is the same as the density of bulk ice, Equations (5.23) and (5.24) have been reported for $\rho_{ice}(T)$ and $\rho_{water}(T)$ [5.36]:

$$\rho_{ice}(T) = 0.946 - 1.073 \times 10^{-4}T \tag{5.23}$$

$$\rho_{water}(T) = -7.1114 + 0.0882 \cdot T - 3.1959 \times 10^{-4} \cdot T^2 \\ + 3.8649 \times 10^{-7} \cdot T^3 \tag{5.24}$$

where $\rho_{ice}(T)$ is the density of frozen water (g/cm^3), $\rho_{water}(T)$ is the density of freezable pore water (g/cm^3).

$\Delta H_m(T)$ represents the heat of the ice–water transition concerning freezable pore water, which was also reasonably assumed to be the same as that of bulk water that can be approximately calculated using Equation (5.25) [5.36]:

$$\Delta H_m(T) = 334.1 + 2.119 \cdot (T - T_0) - 0.00783(T - T_0)^2 \tag{5.25}$$

The heat change indicates the quantity of water, and the temperature indicates the pore diameter, the smaller the pore the lower being the freezing temperature [5.31,5.34,5.43].

5.2.4 Determination of the surface area and the average radius

The surface areas of pores can be calculated from the pore size distribution using Equation (5.26) [5.41].

$$S = \int \frac{z}{R}\left(\frac{dV}{dR}\right)dR \tag{5.26}$$

The peak radii, the pore volumes, the surface area, and the shapes of the pore size distribution curves of pores determined in silica specimens obtained from the melting and freezing of pore water via thermoporosimetry were in good agreement with those obtained via nitrogen gas adsorption–desorption and via mercury porosimetry. Therefore, thermoporometry, if further developed by researchers, could be reasonably useful for the determination of pore analysis of porous materials [5.41].

5.2.5 Applications on cement paste

Even though calorimetry studies on cement pastes have been carried out by several researchers, thermoporometry as a pore structure characterization technique has not been studied extensively in cement-based materials. The technique has been used to study the hydration process and microstructure formation in cement pastes [5.44,5.45]. E. Robens *et al.*, 2000, carried out experiments in hardening cement pastes and found that swelling of C–S–H gel is a continuous process during which no stable pore structure is established and that by swelling of the gel, the space available for water is continuously extended without development of a stable pore system [5.45]. The researchers believe though that thermoporometry may be useful to characterize cement pastes after an extended hardening process.

Researchers have also investigated freezing phenomena in hardened cement pastes [5.34,5.46]. N. Stockhausen *et al.*, 1979, found that during the cooling process at least two phase transitions are observed in the temperature range between −10°C and −25°C and near −43°C respectively [5.46]. The researchers correlated the phase transitions to the radius of the water-filled pores and so to relative humidity and water content. A hysteresis obtained between cooling and heating indicates the existence of capillary-like pores. Comparable results were obtained in different kinds of cement pastes; these findings could be explained by the specific properties of water in porous systems induced by surface interaction. No thorough studies have been published yet to obtain pore size distributions or surface area calculations using thermoporometry in cement pastes.

5.2.6   Advantages and limitations

Thermoporometry is simple to handle, it gives the real size of the pores, and it only requires a hypothesis about the pore shape. The degree of hysteresis between the solidification and fusion thermograms gives information on the pore shape. The method provides information on pore size distributions and is continuously developing to determine useful pore structure parameters.

In the literature, theoretical derivations have only been developed for water and benzene. This limitation can be overcome by characterizing inert materials, such as porous glass beads, with water as the probe molecule and then determining the freezing point depression for a new probe molecule and fitting these data to the appropriate equation.

Thermoporometry is not difficult to implement and requires less time than the determination of a normal adsorption–desorption isotherm. It has been found that the surface areas determined in porous silica are in good agreement with surface areas calculated using the BET-method.

The method is limited to be used for pore radii in the range 2–200 nm and cannot be sensitive to pores less than 2 nm because their sizes are comparable to the thickness of the liquid-like layer (0.8 nm) on the pore walls.

References

5.1   Van Keulen J. "Density of porous solids," *Matériaux et Constructions*, Vol. 6, No. 33, 1973, pp. 181–183.

5.2   Day R.L., Marsh B.K. "Measurement of porosity in blended cement pastes," *Cement and Concrete Research*, Vol. 18, No. 1, 1988, pp. 63–73.

5.3   RILEM Code of Practice for Concrete, CPC 11-3. "Water adsorption of concrete specimens," *Reunion International des Laboratoires et d' Essais des Matériaux et Constructions*.

5.4   Sereda P.J., Feldman R.F., Ramachandran V.S. "Structure formation and development in hardened cement pastes," Sub-Theme VI-1 in *Proceedings of seventh International Congress on Chemistry of Cement*, 1980, pp. VI-1-3 to VI-1-44.

5.5 Gallé C. "Effect of drying on cement-based materials pore structure as identified by mercury intrusion porosimetry: a comparative study between oven-, vacuum-, and freeze-drying," *Cement and Concrete Research*, Vol. 31, No. 10, 2001, pp. 1467–1477.

5.6 Hughes D.C., Crossley N.L. "Pore structure characterisation of GGBS/OPC grouts using solvent techniques," *Cement and Concrete Research*, Vol. 24, No. 7, 1994, pp. 1255–1266.

5.7 Powers T.C., Brownyard T.L. "Studies of the physical properties of hardened Portland cement paste," *Research Bulletin 22*, Portland Cement Association, Chicago, IL, 1948, Reprint from *Journal of the American Concrete Institute*, October 1946–April 1947, in *Proceedings Vol. 43*, 1947.

5.8 Krus M., Hansen K.K., Künzel H.M. "Porosity and liquid absorption of cement paste," *Materials and Structures*, Vol. 30, No. 201, 1997, pp. 394–398.

5.9 Parrott L.J. "Thermogravimetric and sorption studies of methanol exchange in an alite paste," *Cement and Concrete Research*, Vol. 13, No. 1, 1983, pp. 18–22.

5.10 Parrott L.J. "Effect of drying history upon the exchange of pore water with methanol and upon subsequent methanol sorption behaviour in hydrated alite paste," *Cement and Concrete Research*, Vol. 11, No. 5–6, 1981, pp. 651–658.

5.11 Feldman R.F. "Diffusion measurements in cement paste by water replacement using Propan-2-ol," *Cement and Concrete Research*, Vol. 17, No. 4, 1987, pp. 602–612.

5.12 Mikhail R.Sh., Selim S.A. "Adsorption of organic vapors in relation to the pore structure of hardened Portland cement pastes," *Special Report 90: Symposium on Structure of Portland Cement Paste and Concrete*, Highway Research Board, National Research Council, Washington DC, 1966, pp. 123–134.

5.13 Parrott L.J. "An examination of two methods for studying diffusion kinetics in hydrasted systems," *Materials and Structures*, Vol. 17, No. 98, 1984, pp. 131–137.

5.14 Feldman R.F. "Density and porosity studies of hydrated Portland cement," *Cement Technology*, January–February 1972, pp. 5–14.

5.15 Lowell S., Shields J.E. *Powder Surface Area and Porosity*, 3rd Edition, Chapman and Hall, New York, 1991.

5.16 Washburn E.W., Bunting E.N. "Porosity. VI. The determination of porosity by the method of gas expansion," *Journal of the American Ceramic Society*, Vol. 5, 1922, pp. 112–129.

5.17 Karns G.M. "A modified type of gas volume-meter for the determination of the densities of solids," *Journal of the American Chemical Society*, Vol. 48, No. 5, 1926, pp. 1176–1178.

5.18 Beaudoin J.J., Feldman R.F. "The significance of helium diffusion measurements in studying the removal of structural water in inorganic hydrated systems," *Cement and Concrete Research*, Vol. 8, No. 2, 1978, pp. 223–231.

5.19 Ayral A., Phalippou J., Woignier T. "Skeletal density of silica aerogels determined by helium pycnometry," *Journal of Materials Science*, Vol. 27, 1992, pp. 1166–1170.

5.20 "Surface area, density, and porosity of metal powders," in *Metals Handbook, Vol. 7: Powder Metallurgy*, 9th Edition, American Society for Metals, Materials Park, OH, 1986, pp. 262–271.

5.21 Feldman R.F. "The flow of helium into the interlayer spaces of hydrated Portland cement paste," *Cement and Concrete Research*, Vol. 1, No. 3, 1971, pp. 285–300.

5.22 Feldman R.F. "Helium flow and density measurement of the hydrated tricalcium silicate-water system," *Cement and Concrete Research*, Vol. 2, No. 1, 1972, pp. 123–136.

5.23 Feldman R.F. "The porosity and pore structure of hydrated Portland cement paste," in *Symposium Proceedings Vol. 137: Pore Structure and Permeability of Cementitious Materials*, L.R. Roberts, J.P. Skalny (eds), Materials Research Society, Warrendale, PA, 1989, pp. 59–73.

5.24 Feldman R.F. "Helium flow characteristics of rewetted specimens of dried hydrated Portland cement paste," *Cement and Concrete Research*, Vol. 3, No. 6, 1973, pp. 777–790.

5.25　Feldman R.F. "Changes to structure of hydrated Portland cement on drying and rewetting observed by helium flow techniques," *Cement and Concrete Research*, Vol. 4, No. 1, 1974, pp. 1–11.

5.26　Feldman R.F., Beaudoin J.J. "Microstructure and strength of hydrated cement," *Cement and Concrete Research*, Vol. 6, No. 3, 1976, pp. 389–400.

5.27　Feldman R.F. "Application of the helium inflow technique for measuring surface area and hydraulic radius of hydrated Portland cement," *Cement and Concrete Research*, Vol. 10, No. 5, 1980, pp. 657–664.

5.28　Feldman R.F. "A reply to discussion by S. Brunauer of 1. 'The flow of helium into the interlayer spaces of hydrated Portland cement paste', *Cement and Concrete Research*, Vol. 1, 1971, pp. 285–300, and 2. 'Helium flow and density measurement of the hydrated tricalcium silicate water system,' *Cement and Concrete Research*, Vol. 2, 1972, pp. 123–136," *Cement and Concrete Research*, Vol. 2, No. 4, 1972, pp. 493–498.

5.29　Brunauer S. "A discussion of the helium flow results by R.F. Feldman," *Cement and Concrete Research*, Vol. 2, No. 4, 1972, pp. 489–492.

5.30　Marsh B.K., Day R.L. "Some difficulties in the assessment of pore-structure of high performance blended cement pastes," in *Symposium Proceedings Vol. 42: Very High Strength Cement-Based Materials*, J.F. Young (ed.), Materials Research Society, Warrendale, PA, 1985, pp. 113–121.

5.31　Brun M., Lallemand A., Quinson J.F., Eyraud C. "A new method for the simultaneous determination of the size and the shape of pores – the thermoporometry," *Thermochimica Acta*, Vol. 21, No. 1, 1977, pp. 59–88.

5.32　Brun M., Quinson J.F., Benoist L. "Determination of pore-size distribution by SETARAM DSC-101 (Thermoporometry)," *Thermochimica Acta*, Vol. 49, No. 1, 1981, pp. 49–52.

5.33　Quinson J.F., Astier M., Brun M. "Determination of surface areas by thermoporometry," *Applied Catalysis*, Vol. 30, No. 1, 1987, pp. 123–130.

5.34　Le Sage de Fontenay C., Sellevold E.J. "Ice formation in hardened cement paste-I. Mature water-saturated pastes," in *Durability of Building Materials and Components*, ASTM STP 691, P.J. Sereda, G.G. Litvan (eds), American Society for Testing and Materials, West Conshohocken, PA, 1980, pp. 425–438.

5.35　Usherov-Marshak A.V., Sopov V.P. "Thermoporosimetry of cement stone," *Colloid Journal*, Vol. 56, No. 4, 1994, pp. 527–530.

5.36　Ishikiriyama K., Todoki M., Motomura K. "Pore size distribution (PSD) measurements of silica gels by means of differential scanning calorimetry I. Optimization for determination of PSD," *Journal of Colloid and Interface Science*, Vol. 171, No. 1, 1995, pp. 92–102.

5.37　Ferguson H.F., Frurip D.J., Pastor A.J., Peerey L.M., Whiting L.F. "A review of analytical applications of calorimetry," *Thermochimica Acta*, Vol. 363, No. 1–2, 2000, pp. 1–21.

5.38　Quinson J.-F., Brun M. "Porous material characterisation by paranitrophenol thermoporometry," *High Temperatures–High Pressures*, Vol. 30, No. 6, 1998, pp. 677–682.

5.39　Janssen A.H., Talsma H., van Steenbergen M.J., de Jong K.P. "Homogeneous nucleation of water in mesoporous zeolite cavities," *Langmuir*, Vol. 20, 2004, pp. 41–45.

5.40　Usherov-Marshak A.V., Zlatkovskii O.A. "Relationship between the structure of cement stone and the parameters of ice formation during stone freezing," *Colloid Journal*, Vol. 64, No. 2, 2002, pp. 217–223.

5.41　Ishikiriyama K., Todoki M. "Pore size distribution measurements of silica gels by means of differential scanning calorimetry II. Thermoporosimetry," *Journal of Colloid and Interface Science*, Vol. 171, No. 1, 1995, pp. 103–111.

5.42　Titulaer M.K., den Exter M.J., Talsma H., Jansen J.B.H., Geus J.W. "Control of the porous structure of silica gel by the preparation pH and drying," *Journal of Non-Crystalline Solids*, Vol. 170, 1994, pp. 113–127.

5.43 Homshaw L.G. "High resolution heat flow DSC: application to study of phase transitions, and pore size distribution in saturated porous materials," *Journal of Thermal Analysis*, Vol. 19, 1980, pp. 215–234.

5.44 Usherov-Marshak A.V., Sopov V.P. "Microstructure of cement stone," *Colloid Journal*, Vol. 59, No. 6, 1997, pp. 785–789.

5.45 Robens E., Benzler B., Reichert H., Unger K.K. "Gravimetric, volumetric and calorimetric studies of the surface structure of Portland cement," *Journal of Thermal Analysis and Calorimetry*, Vol. 62, No. 2, 2000, pp. 435–441.

5.46 Stockhausen N., Dorner H., Zech B., Setzer M.J. "Untersuchung von Gefriervorgängen in Zementstein mit Hilfe der DTA," *Cement and Concrete Research*, Vol. 9, No. 6, 1979, pp. 783–794.

6 Nuclear magnetic resonance

The phenomenon of Nuclear Magnetic Resonance (NMR) in materials was first observed in 1946 independently by F. Bloch at Stanford University and E. Purcell at Harvard University, for which they were both awarded the 1952 Nobel prize in Physics [6.1,6.2,6.3]. In the period between 1950 and 1970, NMR was developed further by the introduction of the Fourier transformation by R. Ernst, 1966 [6.4], and by the extension of the NMR spectrum to more than one frequency coordinate, called multidimensional NMR [6.5]. In 1973, P. Mansfeld and P. Grannell and independently P. Lauterbur, obtained the first magnetic resonance images using linear gradient fields (Lauterbur and Mansfeld were awarded the 2003 Nobel Prize in Medicine) [6.6,6.7]. The first commercial pulse NMR spectrometers became available in the early 1970s. Advances in laboratory instrumentation, radiofrequency electronics, digital electronics, and computer capabilities, coupled with pulse techniques and Fourier transformation, have greatly enhanced the sensitivity of the NMR technique. It is now a routine procedure to obtain very high quality NMR spectra in few minutes using only a few hundred micrograms of sample. Physicists are still developing magnetic resonance to exploit a range of new applications on materials.

NMR spectroscopy is a physical method used for direct investigation of nuclear energy levels. Most organic and inorganic compounds exhibit the phenomenon of *nuclear magnetism*: in the absence of an external magnetic field, the compounds are not magnetic; however, when they are placed in a strong magnetic field, they develop within a short period of time a macroscopic magnetic moment. The nuclear magnetism arises from the magnetic moments of the nuclei in the compound. In a simplified form, NMR is the study of the properties of molecules containing magnetic nuclei by applying a magnetic field and observing the frequency at which they come into resonance with an electromagnetic field [6.8]. NMR branches include Magnetic Resonance Spectroscopy (MRS) and Magnetic Resonance Imaging (MRI). MRS is the study of chemistry of matter using the NMR absorption spectrum. MRI as a discovery in imaging is arguably as important as the discovery of X-rays, and forms images of the water in a specimen by virtue of the fact that the hydrogen nuclei of water are spinning protons that behave as tiny bar magnets.

Applications of NMR can be found in different fields of materials science. Latest developments in NMR combined with imaging techniques can be used for obtaining pore structure information as a function of position in a sample saturated with a pore fluid. The first application of NMR on cement pastes has been reported in Japan by Kawachi *et al.*, 1955, to study cement hydration; this led P. Seligmann in 1968 to use NMR to study the state of evaporable and bound water in hardened cement pastes [6.9]. NMR has been used systematically in cement-based materials since the early 1980s for several applications and can provide information on various aspects of interest, such as:

- Identify the various mineral phases that constitute Portland cement and special cements.
- Study the structure of the hydrated phases of plain cements, and cements containing mineral or chemical admixtures, and the kinetics of the hydration process.
- Distinguish the different states of water in cement pastes, i.e. chemically bound, physically bound, and free water.
- Determine the pore size distribution and its contribution to the transportation properties of structural materials.

Of these different applications, only the determination of pore structure is within the scope of the book and is discussed in this chapter.

6.1 Theoretical aspects/fundamentals

In order to understand the principles of the technique, some basic theoretical aspects of the structure of nuclei are reviewed briefly in the following sections. The reader can obtain more detailed information in other technical publications (e.g. [6.8,6.10,6.11,6.12,6.13,6.14]).

6.1.1 Single nucleus properties

NMR is based on the fact that matter is made up of moving electrically charged particles, electrons, protons, and neutrons, which have magnetic fields associated with their motions. Most of these particles also have additional magnetic fields associated with their spinning motions. Each individual unpaired electron, proton, and neutron possesses a spin of $+1/2$ or $-1/2$. Two or more particles with spins having opposite signs can pair up to eliminate the observable manifestations of spin; an example is helium that has two protons, two neutrons, and zero spin number. For NMR, the unpaired nuclear spins are of importance. When the spins of the protons and neutrons comprising the nucleus of an element are not paired, the overall spin of the charged nucleus is not zero and generates a magnetic dipole along the spin axis. The intrinsic magnitude of this magnetic dipole is a fundamental nuclear property called the *nuclear magnetic moment*, μ, depending upon the nucleus' *spin (quantum) number*, I, also called nuclear spin [6.8]. The spin number of the nucleus may have values of zero, half integers, and whole integers. A nucleus with a high value of spin number implies a more complex shape of magnetic field than a nucleus with a small spin number. Nuclei that have even atomic mass and even atomic number have a spin number equal to zero, like ^{12}C and ^{16}O. Nuclei with even atomic mass and odd atomic number have a spin number that is a whole integer, like ^2H. Nuclei with odd atomic mass, and even or odd atomic number have a spin number that is a half integer, like ^1H.

The nucleus' magnetic field strength is described by the *magnetogyric ratio*, γ (most commonly mentioned as gyromagnetic ratio). A nucleus with a large magnetogyric ratio has a stronger magnetic field than a nucleus with a small magnetogyric ratio. The strongest magnetic field is conveniently possessed by the proton, ^1H, and for this reason hydrogen NMR (also called proton NMR) is the most popular form of NMR today. The component μ_z of the nuclear magnetic moment on the z-axis is given by Equation (6.1) and depends on the orientation of the spin of a specific nucleus [6.8]. Different nuclei have different magnetic moments:

$$\vec{\mu_z} = \frac{\gamma I h}{2\pi} \quad \text{or} \quad \vec{\mu_z} = \gamma h m_I \quad \text{where } m_I = \frac{I}{2\pi} \tag{6.1}$$

where $\vec{\mu_z}$ is the component of the nuclear magnetic moment on the z-axis (T), γ is the magnetogyric ratio (MHz/T), I is the spin number (unitless), h is Planck's constant ($=6.626 \times 10^{-34}$ J s), m_I is the magnetic quantum number (unitless).

Only nuclei with a spin number I that is not zero can absorb or emit electromagnetic radiation. Nuclei with a spin number I equal to zero, have no magnetic moment, do not exhibit the NMR phenomenon, and hence, are invisible in NMR. Most nuclei have spins between 0 and 9/2 and experience the NMR phenomenon when they are subjected to an external magnetic field. A list of the nuclei routinely used in NMR studies is presented in Table 6.1.

A nucleus with a spin number I and a nuclear magnetic moment μ, behaves like a little spinning atomic magnet (see Figure 6.1). If a constant magnetic field B_0 is applied to the nucleus, the nuclear magnetic moment can align with the magnetic field B_0 in only $2I + 1$ ways [i.e. $I, I-1, I-2, \ldots, -I$], either reinforcing or opposing B_0. Each of the $2I + 1$ orientations of the nucleus has a different energy that is proportional to the magnetic moment of the nucleus and to the external magnetic field, and is given by Equation (6.2):

$$E_{m_I} = -\mu_z B_0 = -\gamma h B_0 m_I \tag{6.2}$$

Table 6.1 Spin properties of some commonly used nuclei in NMR [6.8]

Nuclei	Unpaired protons	Unpaired neutrons	Net spin	Magnetogyric ratio γ (MHz/T)	NMR frequency ω at 1T (MHz)
^1H	1	0	½	5.586	42.576
^2H	1	1	1	0.857	6.536
^{31}P	1	0	½	2.2634	17.238
^{14}N	1	1	1	0.4036	3.076
^{13}C	0	1	½	1.4046	10.705

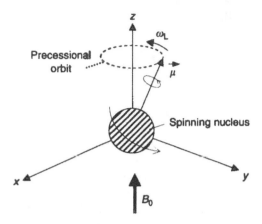

Figure 6.1 A spinning nucleus undergoing precession after being placed in an external magnetic field of strength B_0. The parameter ω_L (or ω_0) is the precession or Larmor frequency.

where E_{m_i} is the energy of a level (kcal/mol), μ_z is the component of the nuclear magnetic moment in the z-axis (T), B_0 is the externally applied magnetic field (T), γ is the magnetogyric ratio (MHz/T), $\hbar$ is Planck's constant ($=6.626 \times 10^{-34}$ J s), m_1 is the magnetic quantum number (unitless).

A nucleus with spin I has $(2I + 1)$ energy levels that are equally spaced, and the separation of the energy levels, ΔE, is given by Equation (6.3), with the symbols defined as previously:

$$\Delta E = \mu \frac{B_0}{I} \qquad (6.3)$$

The simplest case of a nucleus is the proton, ^1H, which has a spin number $I = 1/2$. When the proton is placed in an external magnetic field B_0, the magnetic moment vector of the particle aligns with the external magnetic field just like a magnet would. The proton may have two orientations and hence two energy states: a low energy state α, in which the nucleus' north pole is aligned to the south pole of the external magnetic field, and a high energy state β, in which the nucleus' north pole is aligned to the north pole of the external magnetic field. Since the nucleus is spinning on its axis, the external magnetic field causes the nucleus to undergo an orbital motion like a gyroscope, called *precession*. The vector representing a nuclear spin precesses about the direction of the field, moving on the surface of a cone at a frequency ω_0 (also noted ω_L), which is called the precession or the Larmor frequency (see also Figure 6.1). For protons, the precess angle is about 54°. The precession speed, which is much slower than the spin speed, depends on the properties of the nucleus (size and shape, spin speed etc.), and the strength of the magnetic field. The nuclear properties never change, so if one knows the strength of the applied magnetic field, one can accurately predict the precession frequency.

For the proton, the energy difference ΔE between the two energy levels is given by Equation (6.4):

$$\Delta E = \frac{1}{2}\gamma \hbar B_0 - \left(-\frac{1}{2}\gamma \hbar B_0\right) = \gamma \hbar B_0 \qquad (6.4)$$

where ΔE is the separation between two energy levels (J), γ is the magnetogyric ratio (MHz/T), $\hbar$ is Planck's constant ($= 6.626 \times 10^{-34}$ Js), B_0 is the externally applied magnetic field (T).

Electromagnetic radiation (a photon) of the correct frequency can cause a proton to undergo a transition between the two energy levels. The frequency of the electromagnetic radiation required to induce transitions from one nuclear spin state to the other is exactly equal to the precessional frequency ω_0 of the nucleus; any other frequency is quite ineffective. When the value of the electromagnetic radiation reaches the precessional frequency of the nucleus, absorption of energy occurs, which results in transition of the nucleus from the low energy state to the high energy state. By emission of energy, downward transition occurs and the nucleus will relax to the low energy state.

When a nucleus has $I = 1$, there will be three energy levels, but only transitions between adjacent energy levels are allowed, i.e. from a magnetic quantum number $m = 1$ to $m = 0$, and then to $m = -1$; transitions from $m = 1$ to $m = -1$ are not allowed.

A photon of frequency ν has energy E given by Equation (6.5):

$$E = h\nu \quad \text{or} \quad E = h\frac{\omega}{2\pi} = \hbar\omega \qquad (6.5)$$

where E is the energy of a photon (J), h is Plank's constant ($=6.626 \times 10^{-34}$ Js), ν is the frequency of the photon (Hz), ω is the angular frequency (Hz).

Therefore, when the nucleus is radiated with emitted photons of frequency ν, the energy separations come into resonance with the radiation when the frequency satisfies the resonance condition expressed by Equation (6.6) or (6.7) [6.8]:

$$h\nu_0 = \gamma \hbar B_0 \qquad (6.6)$$

$$\nu_0 = \frac{\gamma}{2\pi}B_0 \quad \text{or} \quad \omega_0 = \gamma B_0 \qquad (6.7)$$

where ν_0 is the Larmor frequency of a nucleus (cycles/s), B_0 is the externally applied magnetic field (T), γ is the magnetogyric ratio (MHz/T), ω_0 is the Larmor frequency (cycles/s).

Equation (6.7) is the Larmor equation and is the basic mathematical equation for NMR [6.10]. In NMR and MRI, the frequency ν_0 is called the resonance frequency or Larmor frequency. In an NMR experiment, the frequency of the photon is in the radio frequency range and is between 60 and 800 MHz for hydrogen nuclei. In clinical MRI, the frequency ν is typically between 15 and

80 MHz for proton imaging [6.14]. The Larmor frequency is very different for individual nuclei: e.g. in a magnetic field of 1.8 T, the Larmor frequencies are 79.54 MHz for the proton (^{1}H), 20 MHz for carbon-13 (^{13}C), and 8.06 MHz for nitrogen (^{15}N).

If two nuclei are of the same type, such as two protons, their two magnetogyric ratios are equal, both Larmor frequencies are approximately equal, and such a two spin system would be described as *homonuclear*. The opposite case is when the two nuclei are of different types, such as proton, ^{1}H, and carbon-13, ^{13}C, and such a system is described as *heteronuclear*. For a two nuclei heteronuclear system, the Larmor frequencies differ significantly, and the four energy levels have markedly different values.

6.1.2 Magnetization of a group of nuclei (bulk magnetization)

In the following, for simplicity in presentation the case is considered of a sample composed of many identical nuclei, with each nucleus having a spin quantum number $I = \frac{1}{2}$ such as the proton, ^{1}H. However, some of the equations and the majority of the concepts described apply equally well to nuclei with spin numbers greater that $\frac{1}{2}$. All the individual nuclear magnetic moments, μ_i, of the group of nuclei add up to a nuclear magnetization vector M. The magnetization vector M of the sample is manipulated during the NMR experiment.

In the *absence* of a magnetic field, the sample consists of equal numbers of low energy α and high energy β nuclear spins, both energy levels are equally populated, and the net magnetization of the sample is M = 0 (see Figure 6.2a). When the two spin states of a set of protons are equally populated, the total number of protons undergoing transitions from the lower spin state to the upper spin state is equal to the number of protons undergoing transitions in the opposite direction. In this case, no absorption or emission of energy can be observed, therefore there is no NMR signal [6.10].

If we apply a strong external *static magnetic field* B_0, the nuclear magnets will align with the direction of B_0 (see Figure 6.2b). Nuclei in a magnetic field can align themselves along the applied field direction (parallel) or opposed to it (antiparallel). The parallel orientations are energetically more favorable (low energy level) than the antiparallel ones (high energy level) [6.8]. Each energy level has a different population of spins and the population difference between the higher and lower energy levels is given by the Boltzman distribution, expressed by Equation (6.8). The energy difference between the two levels is given by Equation (6.4) that was mentioned in Section 6.1.1:

$$\frac{N_\beta}{N_\alpha} = \exp\left(-\frac{\Delta E}{kT}\right) \tag{6.8}$$

where N_α is the population of protons in the low energy state, N_β is the population of protons in the high energy state, ΔE is the energy difference between the two levels (cal/mol), k is the Boltzman constant, T is the temperature (K).

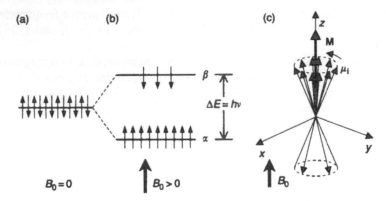

Figure 6.2 Schematic representation of the energy levels of a group of identical nuclei with spin $I = 1/2$, such as the proton, (a) in the absence, and (b) in the presence of an external magnetic field B_0. In the presence of a magnetic field there is an excess of spins that align parallel to the applied external field (low energy state α) than antiparallel (high energy state β). (c) The magnetic moments μ_i of the individual nuclei form a cone and contribute to the magnetization moment M of the sample.

As an example, for the case of the proton ^{1}H at 400 MHz ($B_0 = 9.5$ T), $\Delta E = 3.8 \times 10^{-5}$ kcal/mol and the population difference is given by Equation (6.9):

$$\frac{N_\alpha}{N_\beta} = 1.000064 \qquad (6.9)$$

The magnetic moments of all the individual nuclei in the sample, called the spin system, can be viewed as forming a cone about the direction of B_0 yielding a net magnetization M parallel to B_0. For simplicity in presentation, we will assume here that the direction of the magnetic field B_0 is aligned with the z-axis of an x–y–z coordinate system (see Figure 6.2c and 6.3a). The net magnetization M is proportional to the population difference between the α and β spins [6.8].

We can consider the case that on top of the static magnetic field B_0, we apply another *oscillating magnetic field* B_1 (e.g. RF electromagnetic pulse) with its magnetic component directed perpendicular to the static magnetic field B_0, e.g. aligned with the x-axis (see Figure 6.3b). The oscillating magnetic field will reorient nuclei with a non-zero magnetic moment and after a while, the nuclei are tipped so that they are rotating in the plane that is perpendicular to the permanent magnetic field, B_0. The external alternating magnetic field B_1 and the nucleus magnetic moment M interact, generating a torque on M. The system absorbs energy and the equilibrium of the system is altered by modifying the populations of the spins of the α and β energy levels. In particular, the magnetization of the nuclei M will be tilted to lie between the static and the oscillating magnetic fields. In addition, the magnetic moments of the nuclei

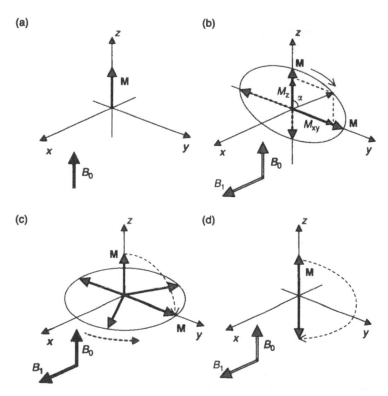

Figure 6.3 Schematic direction of the magnetic moment M of a sample with the application of a steady magnetic field (B_0) and an oscillating magnetic field (B_1). (a) Application of a constant magnetic field only. (b) Application of a constant magnetic field and an oscillating magnetic field B_1 at the Larmor frequency for an indefinite time. The magnetization moment shifts from z to y to $-z$ to $-y$ to z again. (c) Application of the oscillating magnetic field B_1 as a 90° pulse. M shifts from $+z$ to y and then precesses in the x–y plane. M lies in the xy-plane and begins to precess about the B_0 axis at the Larmor frequency. (d) B_1 is applied as a 180° pulse.

will no longer be distributed evenly about B_0. The angle through which M rotates away from the z-axis is known as the *flip angle* and is proportional to the frequency and duration of B_1 as it is described below.

If the frequency of the oscillating magnetic field B_1 is equal to the Larmor frequency of the nuclear spins, then the nuclei experience a steady magnetic field, B_1, because the rotating magnetic field is in step with the precessing spins and a *resonant condition* is achieved. Under the influence of the steady field B_1, the magnetization vector M of the sample begins to precess around the direction of B_1 and M rotates in a plane perpendicular to the direction of the magnetic field B_1 [6.8]. When the radio frequency is switched off, the nuclei fall back from their excited states to their original states, radiating their surplus energy to the surroundings at the same frequency as their precession frequency. The signal in NMR spectroscopy results from the difference between the energy absorbed by the spins that make a transition from the low energy state to the high energy state and the energy emitted by the spins that simultaneously make a transition from the high energy state to the low energy state [6.14].

If the B_1 field at the Larmor frequency is applied for an *indefinite* time, the magnetization moment M tilts away from the z-axis through the x–y plane towards the −z direction and then back towards the x–y plane and the z-axis, and the process begins again (see Figure 6.3b).

If the B_1 field at the Larmor frequency is applied as a *pulse of 90°*, the magnetization moment M lies in the x–y plane and precesses around the B_0 field (see Figure 6.3c). In the absence of the external magnetic field B_1 after the application of the 90° pulse, the magnetization vector of the sample, M, falls back (relaxes) to equilibrium into the resting alignment with B_0 by restoring the same N_α/N_β distribution. This phenomenon is called relaxation and is discussed in more detail in Section 6.3.

If the B_1 field is applied as a *pulse of 180°*, the magnetization moment M tilts to the −z direction and slowly relaxes to the +z direction again (see Figure 6.3d).

Interactions between nuclei   Molecules contain many nuclei in different electronic environments. Chemical bonding modifies the shape and density of the outer electron cloud, slightly changing the magnetic field at the nucleus; as a result, not all protons resonate at the same frequency. Instead of a single Larmor frequency, there is a characteristic spectrum of discrete frequencies. The corresponding change in the nuclear precession frequency ν_0 is called the *chemical shift* and allows chemists to distinguish the different types of hydrogen atoms in a molecule. Chemical shifts arise because the field actually experienced by a nucleus in an atom or molecule differs slightly from the externally applied magnetic field B_0 that would be experienced by a bare nucleus, stripped of its electrons. In an atom, the actual magnetic field B is slightly smaller than the applied magnetic field B_0, because the external field causes the electrons to circulate within their atomic orbitals. The induced motion generates a small magnetic field B' in the opposite direction to B_0. The higher the electron density about a nucleus, the greater the secondary magnetic field B' and, therefore, the greater the applied magnetic field must be to bring that nucleus into resonance. The induced magnetic field B' is proportional to B_0 and typically 10^4–10^5 times smaller [6.13]. In addition, since each nucleus behaves like a tiny bar magnet, it creates a weak additional magnetic field at the site of its neighbor, positive or negative depending on whether the nucleus is aligned spin-up or spin-down. This rather weak effect due to magnetic interactions between nuclei is called *spin–spin coupling*, is carried through the electrons that form the chemical bonds, and consequently only operates over relatively short distances.

A sample may contain many different magnetization components, each with its own Larmor frequency. Based on the number of allowed absorptions due to chemical shifts and spin–spin couplings of the different nuclei in a molecule, a NMR spectrum obtained after a magnetic excitation may contain many frequency lines. The sensitivity of the spectrum to the local environment in the sample enables NMR to provide detailed information about the structure for many classes of materials. For molecules randomly oriented in the rigid lattice of a solid, the variety of possible orientations often leads to a complicated spectrum. In liquids, the rapid and isotropic molecular motion averages the anisotropic interactions and thus narrows the spectral lines [6.15]. Development of new instrumentation and analytical techniques that produce better resolved and more detailed spectra continues to expand the application of NMR in various fields such as chemistry, biology, medicine, and materials science.

6.2 NMR experiment

The NMR experiment exploits the interaction of nuclear magnetic moments with externally applied electromagnetic waves in the radio frequency region. During the experiment, a specimen is placed in a strong magnetic field and irradiated with intense radio frequency pulses typically of duration 1–10 μs over a frequency range required to excite specific atomic nuclei from a low energy state to a high energy state. As soon as the RF signal stops, the nuclei relax, i.e. return to their natural and more comfortable state. As the nuclei relax, the NMR probe receives a very weak RF resonance response back from the sample and transmits it to the NMR console for amplification. The signal is amplified over 1 000 000 times, and is further analyzed to provide information about the composition and structure of the material tested.

6.2.1 Instrumentation

The NMR equipment is rather costly and its price can range from US$200 000 to 1 200 000, depending on the field strength desired. An NMR spectrometer consists of several parts and components, each of which is critical to the spectrometer operation. Even though there is a wide variety of equipment, there are several components common to all types of high-resolution NMR spectrometers. These include: a magnet, a magnetic field stabilization system, a sample probe, a radiofrequency oscillator, a transmitter, a detector, and a recorder for display of the spectrum (see Figure 6.4).

The NMR *magnet* is one of the most expensive components of the NMR spectrometer system. Different types of magnets can be used such as permanent magnet, electromagnet, or super-conducting solenoid. The magnet, regardless of its type, should be capable of providing a highly uniform, stable, and homogeneous magnetic field with a precision better than 10^{-9}–10^{-10} over all the sample volume [6.16]. Permanent magnets provide a fixed magnetic field and cannot be varied over a wide range. They are also subject to field variations induced by the placement of magnetic materials in the vicinity of the magnet. An electromagnet has the ability to vary the magnetic field strength by varying the current passing through the coils. In early equipment, the two magnetic fields applied during the NMR experiment were caused by different

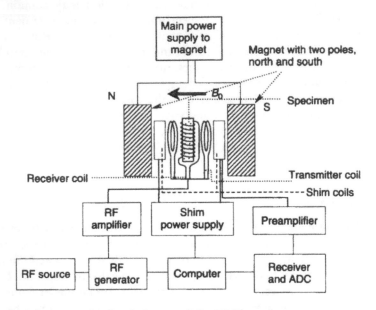

Figure 6.4 Schematic out-of-scale layout of the NMR equipment.

ways: the strong static magnetic field B_0 was caused by a permanent magnet and the oscillating magnetic field was caused by an electromagnet. Most modern magnets are of the superconducting type, which is an electromagnet made of superconducting wire. Superconducting wire has a resistance approximately equal to zero when it is cooled to a temperature close to absolute zero (0 K). Cooling is achieved by immersing the superconducting coil in a liquid helium bath, usually surrounded by a dewar vessel of liquid nitrogen. Extreme care is required to keep the helium cryogenic liquid high and to exclude any large tool made of iron or steel from a magnetic resonance laboratory because that could provoke a catastrophic quench of the magnetic field; this would lead to heat increase and damage the coil as it makes a sudden transition from superconducting to resistive wire [6.17]. The length of superconducting wire in the magnet is typically several miles. The superconducting magnet provides a strong, homogeneous magnetic field of at least 14 000 Gauss and can be as high as 100 000 Gauss (as a comparison, the earth's magnetic field is approximately 0.5 Gauss). In clinical MRI the magnetic field strength used is 0.1–0.2 T (1000 to 2000 Gauss).

As all magnets are not of high quality, it is often necessary to apply corrections, which is done by the *shim coils* located immediately within the bore of the magnet. A shim coil is designed to create a small magnetic field, which will oppose and cancel out an inhomogeneity in the B_0 magnetic field. Because the magnetic field variations may exist in a variety of functional forms (linear, parabolic etc.), shim coils are used that can create a variety of opposing fields. By passing the appropriate amount of current through each coil, a homogeneous B_0 magnetic field can be achieved. On most spectrometers, the shim coils are controllable by the computer, which via an algorithm can find the best shim value by maximizing the lock signal [6.16].

The *sample probe* is the part of the spectrometer that accepts the sample, sends RF energy into the sample and detects the signal emanating from the sample. The sample probe contains the RF coil, sample spinner, temperature controlling circuitry, and gradient coils [6.14]. The probe fits into the magnet and holds the sample, and the transmitter can be a frequency synthesizer or use a single frequency for manipulating the various nuclear spins. The probe is mounted between the pole caps of the magnet in such a way that its position may be adjusted to place the sample in the region of optimum field homogeneity. Molecules in different parts of the sample will experience slightly different magnetic fields due to residual field gradients in the region occupied by the sample [6.18]. A sample spinner can rotate the NMR sample tube about its axis, so that each spin in the sample at a given position will experience the average magnetic field for the specific position [6.14].

The *radio frequency (RF) coils* are also known as RF resonators and RF probes and serve two purposes [6.11,6.18,6.17]:

- They generate RF pulses at the Larmor frequency to excite the nuclei in the specimen tested. The *transmitter coil* for producing the oscillating magnetic field B_1 to excite the nuclear spins is wound in two sections such that it surrounds the receiver coil and has its magnetic axis parallel to the x-axis. The range of frequencies covered is 4–360 MHz for the NMR absorption of all magnetic nuclei with today's spectrometers. The RF coil must be able to produce a homogeneous B_1 field in the volume of interest at the Larmor frequency so that the nuclei can be excited uniformly; otherwise, the spins will rotate by a distribution of rotation angles, which will result to strange spectra.
- They receive the response RF signals emitted by the nuclei in the sample at the same frequency. The *receiver coil* is wound on a glass insert at a right angle to the transmitter coil so that its magnetic axis is also parallel to the x-axis. An RF coil used for reception must have a high signal-to-noise ratio and at the same time must be able to pick up RF signals with the same gain at any point in the volume of interest. The receiver amplifies, detects and filters the NMR signal, and is capable of producing an electrical signal suitable for display on an oscilloscope, or for input to a computer, and a recorder that displays the spectrum.

Usually but not always, the transmitter coil and the receiver coil are one and the same. Copper is employed for most radio frequency coils because it is an excellent conductor.

Some probes also contain a set of *gradient coils*. These coils produce a gradient in B_0 along the x, y, or z axis. Gradient coils are used to produce the gradients in the B_0 magnetic field needed for performing gradient enhanced spectroscopy, diffusion measurements, and NMR imaging.

The heart of the spectrometer is the *computer*, which controls all the components of the spectrometer: the RF frequency source and pulse programmer.

The source produces a sine wave of the desired frequency. The pulse programmer sets the width and in some cases the shape of the RF pulses. The RF amplifier increases the pulses' power from milli Watts to tens or hundreds of Watts. The computer also controls the gradient pulse programmer, which sets the shape and amplitude of gradient fields. The gradient amplifier increases the power of the gradient pulses to a level sufficient to drive the gradient coils.

6.2.2   NMR excitation and response

As it was discussed earlier, the application of the static magnetic field results in the generation of two distinct energy levels for protons. An oscillating magnetic field of a specified frequency, applied at right angles to the static magnetic field, induces transitions between the two energy levels. A quick burst of RF waves of frequency 88–108 MHz as the FM radio (88 million cycles/second), lasting about 10 μs can affect the nuclei for as much as several seconds.

There are different ways to apply a RF magnetic field. Early NMR experiments were performed by applying a *continuous wave* (CW) by slightly sweeping the RF applied to a sample in a fixed magnetic field, or by slowly sweeping the magnetic field with a fixed RF [6.14,6.19]. Such an experiment requires several minutes to record, since each transition is induced in succession, and at any given moment only one frequency is being observed. If many scans of the spectrum are required to obtain a useable signal, the total time required to complete the experiment can be extremely long (24 or 48 hours) [6.18].

Present day experiments are performed in a distinctly different mode using *pulse methods* or free precession methods applying RF pulses at a discrete frequency. Pulse methods were introduced almost as early as the CW methods but attracted the attention from chemists much later and have since been improved in sophistication and versatility [6.4]. A short RF pulse, which contains a bandwidth of frequencies sufficiently large to cover the entire resonance range of the nucleus of interest, is applied to the sample. Individual frequency components are absorbed by the sample and then reemitted when the pulse is turned off. The signal received from the sample in the receiver coil is a complex waveform that is a superposition of the resonances of all the nuclei in the sample resonating within the frequency range of the exciting device. Each set of distinct protons will produce a sine or cosine wave whose frequency matches their precession frequency and the intensity of which is related to the phase of the sine wave and to the magnitude of the magnetization M.

6.2.2.1   Free induction decay

Resonance radiation from a sample continues for a brief but electronically measurable time after the incident radiation is removed. The magnetic resonance response to the RF pulse is amplified electronically and detected. A plot of the signal induced in the receiver coil after the pulse is turned off vs. time as the nuclei return to equilibrium after the pulse is called the *free induction decay* (FID) (see Figure 6.5). The FID is the result of the resonating material's

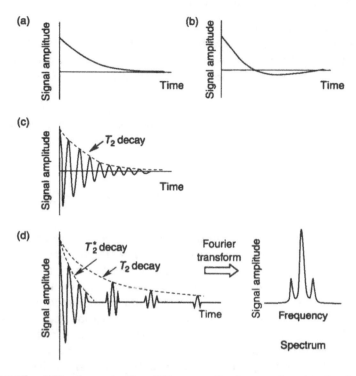

Figure 6.5 The FID of a spin for different radio frequencies. (a) At the Larmor frequency; (b) close to the Larmor frequency; (c) away from the Larmor frequency (off-resonance). It is worth noticing that far from resonance the decay exponential function appears as the envelope of the signal. (d) FID from a group of spins, showing the T_2 decay and the T_2^* decay. After Fourier transform, the spectrum is obtained.

"memory" of the radiation field and is usually obtained within a few seconds or less after the pulse is turned off [6.8,6.18].

The FID consists of one exponentially decaying sine wave for each single frequency absorbed from the pulse. If only a single frequency is present, then the time response signal following the RF pulse is a simple cosine wave that decays with time constant T_2 (see Figures 6.5a,b,c). The FID signal received from a sample is a sum of oscillating functions, and results from the super-position of many different frequencies, all jumbled together. The problem is to analyze the signal to its harmonic components by disentangling the many different frequencies present in the free induction decay. The conversion from the time-based FID of a sinusoidal form to the conventional frequency-based NMR spectrum made up of all the various resonance responses from the nuclear spins is carried out mathematically by *Fourier transformation* [6.10]. The time domain signal is converted into a convenient frequency-domain spectrum; in this way, the NMR spectrum $F(\omega)$ is obtained from the FID signal $f(t)$ (see Figure 6.5d).

6.2.2.2   *Fourier transformation and spectrum*

Fourier showed that any complex pattern can be represented by a superposition of sine waves of the appropriate amplitudes and relative phases. Any general, repetitive function $f(t)$ may be written as a Fourier series expansion, i.e. an infinite series of sines and cosines, as it is expressed by Equation (6.10). Expression (6.10) comprises a zero frequency component (A_0) and a series of harmonics of the basic frequency (ω_1) with their appropriate Fourier coefficients A_n and B_n that determine amplitude and phase. This expansion is valid over the region $-T \leq t \leq T$ and formulas are derived for determining the coefficients A_n and B_n:

$$f(t) = A_0 + \sum_{n=1}^{\infty} [A_n \cos(\omega_n t) + B_n \sin(\omega_n t)] \quad \text{where } \omega_n = \frac{n\pi}{T} \tag{6.10}$$

where $f(t)$ is the repetitive function with time t, A_0, A_n, B_n are coefficients.

However, when we actually carry out the mathematical operations of Fourier analysis, we do not deal with Fourier series, but with related integrals that can cope with functions in which the variable t is not restricted to the range $-T \leq t \leq T$, but can go to infinity. Sine and cosine functions can conveniently be combined to form a complex exponential [$\cos \omega t + i \sin \omega t = \exp(i\omega t)$]. The complex amplitudes of these exponentials constitute the spectrum $F(\omega)$ of the signal $f(t)$, and we can define the Fourier transform of $f(t)$ by Equation (6.11), with the parameters defined as previously [6.20]:

$$F(\omega) = \int_{-\infty}^{\infty} f(t) \exp(-i\omega t)\, dt = \int_{-\infty}^{+\infty} f(t)[\cos(\omega t) - i \sin(\omega t)]\, dt$$

$$\text{where } i = \sqrt{-1} \tag{6.11}$$

It can also be shown mathematically that the inverse relationship that converts a signal $F(\omega)$ from the frequency domain into a signal in the time domain $f(t)$ is given by Equation (6.12):

$$f(t) = \frac{1}{2\pi} \int_{-\infty}^{\infty} F(\omega) \exp(-i\omega t)\, d\omega \tag{6.12}$$

where ω is the angular frequency $= 2\pi\nu$.

The main advantages of the pulse method compared with the spectral sweep method (CW technique) are the much shorter time required to record a spectrum and the higher inherent sensitivity in detecting a signal. Using the Fourier transform pulse NMR, a spectrum can be acquired in a few seconds; e.g. with a repetition time of 1 second, 300 spectra can be obtained in the 5 minutes time required to record a single spectrum using CW techniques. In addition, it has been shown theoretically and experimentally that by using the pulse technique it is possible to enhance the sensitivity of high resolution proton MRS in a restricted time by a factor of ten than with sweep methods.

The FID can be converted using a Fourier transformation into either a spectrum (for high resolution NMR) or an image (for MRI). The spectrum is a plot of absorbed energy vs. frequency [6.8]. The position, width, and shape of each of the spectral peaks correspond to specific atoms and their relative concentrations, and is a sensitive indicator of structural and chemical bonding properties. For complicated spin systems in solution, a spectrum contains information about the system in a more explicit form than does the free induction decay. The three most important types of information obtained in the high resolution NMR spectrum are the frequency, area, and multiplicity of the resonance line. The exact resonance frequency of a nucleus is dependent on its chemical environment in the molecule and is measured against a reference line. The area under a resonance line can be integrated to give quantitative information about the relative number of nuclei giving rise to each peak. Multiplicity of a peak resulting from spin coupling provides structural information about near neighboring nuclei.

6.2.3 Pulse sequences

Many magnetic resonance experiments involve repeated excitation of the nuclear spins and manipulation of the effective spin interactions in different ways. A variety of pulse sequences can be formed by adding several pulses, varying the pulse angle, and inserting time delays. By suitable spacing of pulses, a variety of free induction decay signals can be obtained leading to a range of sophisticated experiments. The time constants for these FID signals can be related to fundamental properties of the material under test [6.9]. Two variables of interest in a pulse sequence are the echo time (t_E), and the repetition time (t_R). Echo time is the time from the application of an RF pulse to the measurement of the magnetic resonance signal. Repetition time is the time between two consecutive RF pulses usually measured in milliseconds. Some of the most common pulse sequences applied to a sample to produce a specific form of NMR signal and/or record an NMR spectrum are:

- The Hahn spin echo pulse sequence.
- The inversion recovery pulse sequence.
- The Carr–Purcell–Meiboom–Gill (CPMG) pulse sequence.

Since relaxation times can be as short as milliseconds (or even less), most pulse techniques developed employ pulses in the range of 1–100 μs.

6.2.3.1 Hahn spin echo pulse sequence

The Hahn spin echo pulse sequence starts with a 90° pulse, which rotates the magnetization vector of the nuclei in the sample from the z-axis into the xy plane (following the presentation in Figure 6.3) and has equal number of α and β spins [6.8]. After a selected time τ, when the FID has decayed significantly through spin–lattice relaxation, a second pulse of radiation with twice the duration of the first pulse (a 180° pulse) is applied in the y direction (see Figure 6.6).

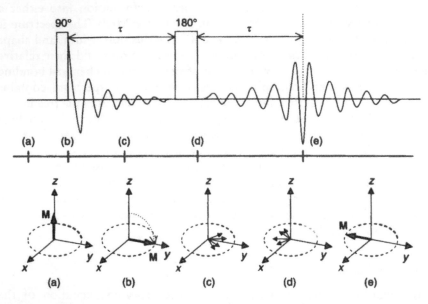

Figure 6.6 Hahn spin-echo pulse sequence for determining the spin–spin relaxation time. The top part shows the pulse sequence and the decay of the signal, and the bottom part indicates the modification of the magnetization vector M during the pulses: (a) at equilibrium, (b) right after the application of a 90° pulse, (c) dephasing after some time from the application of the 90° pulse, (d) right after application of a 180° pulse, and (e) refocusing after delay τ from the application of the 180° pulse.

This pulse will flip the phase of the individual spins and the initial decay of the signal during τ is reversed after the second pulse. A 180° pulse can also disturb the spin populations of the energy levels. A small excess of spins in the lower level are transferred to the upper levels and in this way there has been a population inversion. This is an unstable situation and the excess spins in the upper energy level fall back to the ground state at a rate determined by the spin–lattice relaxation time. At time $t_E = 2\tau$, the net phase shift is zero. The transverse components of all individual spins are aligned and give a maximum signal intensity, which is called *spin-echo*. If the translational motion of the nuclei of interest can be neglected, the height of the spin-echo is only influenced by the transverse relaxation mechanism, which will give an exponential decay of $[\exp(-t_E/T_2)]$ and is not affected by magnetic field inhomogeneities [6.9]. Spin echo is used to eliminate the effects of inhomogeneous broadening for the spin–spin relaxation time T_2 measurement (see Section 6.3.2) [6.8].

6.2.3.2   *Inversion recovery pulse sequence*

During the inversion recovery pulse sequence, the first step is to apply a 180° pulse to the sample. The 180° pulse is achieved by applying the B_1 field so that the magnetization vector rotates through 180° and points in the $-z$ direction. No signal can be seen at this stage, because there is no component of

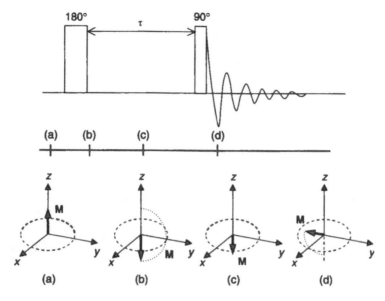

Figure 6.7 Inversion recovery pulse sequence for determining the spin–lattice relaxation time. The top part shows the pulse sequence and the decay of the signal and the bottom part indicates the modification of the magnetization vector M during the pulses: (a) at equilibrium, (b) right after the application of a 180° pulse, (c) after delay from application of the 180° pulse, and (d) right after the application of a 90° pulse.

magnetization in the xy plane where the detection coils are sensitive. The β spins begin to relax back into α spins, and the magnetization vector shrinks exponentially towards its thermal equilibrium value M_z [6.8]. After an interval τ, a 90° pulse is applied, which rotates the magnetization vector into the xy plane, where it starts to generate a free-induction decay signal (see Figure 6.7) [6.8].

The intensity of the spectrum obtained in this way depends on the length of the magnetization vector that is rotated into the xy plane. The length of the vector decreases exponentially as the interval between the two pulses is increased, and so the intensity of the spectrum also decreases exponentially with increasing τ. We can therefore measure T_1 by fitting an exponential curve to the series of spectra obtained after different values of τ [6.8]. The signal recovery is an exponential curve with a time constant T_1.

6.2.3.3 Carr–Purcell echo pulse sequence

H. Carr and E. Purcell, 1954, showed that a simple modification of Hahn's spin-echo method reduces drastically the effect of diffusion on the determination of the relaxation time T_2 [6.20]. The method may be described as a 90°, τ, 180°, 2τ, 180°, 2τ, 180°, 2τ, etc. sequence (see Figure 6.8). The dephasing of the spins after the spin-echo is not different from the dephasing of the spins

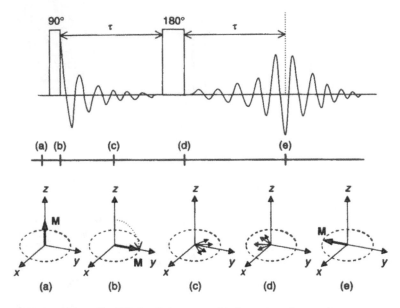

Figure 6.8 Carr–Purcell–Gill–Meiboom multiple spin-echo pulse sequence for determining the spin–spin relaxation time. The top part shows the pulse sequence and the decay of the signal, and the bottom part indicates the modification of the magnetization vector M during the pulses: (a) at equilibrium, (b) right after the application of a 90° pulse, (c) dephasing after delay from the application of the 90° pulse, (d) right after application of a 180° pulse, and (e) refocusing after delay τ from the application of the 180° pulse.

directly after the 90° pulse, except that the magnetization has decreased by an amount of $\exp(-t_E/T_2)$. So the same rephasing action, as done with the first 180° pulse, can be done again. If a second 180° pulse is applied at a time 3τ, a second spin-echo will appear at the time 4τ. One can extend this sequence to an arbitrary number of 180° pulses, which will give spin-echoes. The Carr–Purcell method of measuring T_2 is much faster than the repetitive Hahn spin-echo with increasing interpulse time. In addition, the effect of diffusion due to inhomogeneities may be virtually eliminated by making τ short (e.g. 0.3 ms), since it is only during a period of 2τ that diffusion is effective in reducing the amplitude of an echo [6.19].

6.3   Spin relaxation

After the application of a RF pulse, the complex free precession signal does not last forever, but is gradually attenuated becoming weaker and weaker with time until it eventually disappears. The signal decay with time implies that the component of the magnetization vector in the *xy*-plane must be shrinking. FID happens due to relaxation and dephasing. The nuclear spins transfer some magnetic energy to their surroundings, and in this way, the magnetization moment M decays in the *xy*-plane and returns to its equilibrium state (direction of the *z*-axis); this process is called *relaxation* [6.21]. In addition, phase differences occur at various points across the sample due to nuclei precessing at different frequencies, and this phenomenon is called *dephasing*. The phase differences increase with time and the magnetization moment in the xy-plane M_{xy} reduces with time.

The motion of the magnetic moment vector can be described by the Bloch differential equations, which in general describe the evolution of the spin system with time under the effects of magnetic fields, as well as relaxation of the spins. The generalized forms of the Bloch equations are given in detail elsewhere [6.22].

As an example, after the electromagnetic radiation has been removed $(B_1 = 0)$, the Bloch equations simplify to Equations (6.13), and (6.14), with solutions given by Equations (6.15), and (6.16) respectively:

$$\frac{dM_z(t)}{dt} = -\frac{M_z(t) - M_0}{T_1} \tag{6.13}$$

$$\frac{dM_{xy}(t)}{dt} = \frac{-M_{xy}(t)}{T_2} \tag{6.14}$$

$$M_z(t) = M_0 + [M_z(0) - M_0]\exp\left(-\frac{t}{T_1}\right) \tag{6.15}$$

$$M_{xy}(t) = M_{xy}(0)\exp\left(-\frac{t}{T_2}\right) \tag{6.16}$$

where $M_z(t)$ is the longitudinal magnetization moment at time t, M_0 is the equilibrium longitudinal magnetization moment, T_1 is the spin–lattice relaxation time, $M_{xy}(t)$ is the transverse magnetization at time t, T_2 is the spin–spin relaxation time.

Equations (6.15) and (6.16) take a different form depending on the initial conditions for the components of the magnetization vector. For example, after a 90° pulse (state (b) shown in Figure 6.6), when the magnetization vector is in the xy-plane and $M_{xy}(0) = M_0$, and $M_z(0) = 0$, the equations simplify to Equations (6.17) and (6.18) and are shown in Figure 6.9, with the parameters defined in the same way as stated earlier:

$$M_z(t) = M_0\left[1 - \exp\left(-\frac{t}{T_1}\right)\right] \tag{6.17}$$

$$M_{xy}(t) = M_0\left[1 - \exp\left(-\frac{t}{T_2}\right)\right] \tag{6.18}$$

For convenience during the experiment, the normalized magnetization is used, which is usually identified by $A(t)$ defined by Equation (6.19), and can be used accordingly for both longitudinal and transverse magnetization:

$$A(t) = \frac{M(t)}{M(0)} \tag{6.19}$$

where $A(t)$ is the normalized magnetization as a function of time t, $M(t)$ is the magnetization moment as a function of time t, $M(0)$ is the magnetization moment at time $t = 0$.

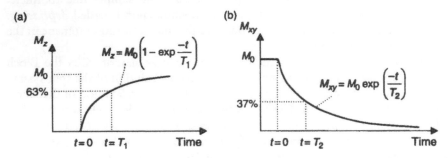

Figure 6.9 The decay process of the magnetization moment. (a) T_1 relaxation due to spin–lattice interactions. After withdrawal of the oscillating magnetic field B_1, the spins release their energy leading to regrowth of M_z. This corresponds to the transition of spins in their lower state shown in Figure 6.7(c). (b) T_2 relaxation due to spin–spin interactions. The decay of the magnetization component corresponds to the dephasing of the spins shown in Figure 6.6(c).

Relaxation theory is concerned with the description of the processes that lead to relaxation. In most cases, relaxation is due to rotational and translational molecular motions, which lead to fluctuating local fields at the sites of the nuclei. The origin of these fields is in one or more of spin interactions, the most important are the dipole–dipole interaction, the anisotropy of the chemical shift, and the quadrupole interaction [6.23]. The analysis of relaxation times can provide important information about the structure of the material.

The relaxation rate of the nucleus shows how fast the magnetic vector of the nucleus aligns with the main magnetic field initially or after being tilted by an RF pulse. The relaxation times of interest are the spin–lattice relaxation time T_1, the spin–spin relaxation time T_2, and the effective spin–spin relaxation time T_2^*. These relaxation times are described in the following sections. The relaxation times T_1 and T_2 can provide a lot of information about the environment of the nucleus, e.g. molecular motion, and properties of the fluid inside the pore system.

6.3.1   Spin–lattice relaxation

At equilibrium, the spins have a Boltzman distribution with more α spins than β spins. However, a magnetization vector in the xy-plane immediately after a 90° pulse has equal numbers of α and β spins. The populations revert to their equilibrium values exponentially so as to reestablish a Boltzmann population distribution (with excess of nuclei in the lower energy state) after the application of a RF pulse by a process that is called spin–lattice relaxation. Spin–lattice relaxation depends not only on the type and rapidity of the molecular motions of the molecule in the lattice, but also on the magnetogyric ratio of the nucleus. The time constant that describes how the longitudinal magnetization M_z returns to its equilibrium value, is called *spin–lattice relaxation time T_1*. T_1 is also referred to as longitudinal relaxation time because it is parallel to the applied magnetic field [6.10]. Several common mechanisms of spin–lattice relaxation include interactions of the nuclei with neighboring magnetic nuclei, free rotation of a functional group or module, the presence of significant anisotropy in the shielding, and interactions through the spin–spin coupling with another rapidly relaxing nucleus.

The magnetization decay follows an exponential curve, e.g. as it is shown in Figure 6.9a. The parameter T_1 is normally measured in milliseconds or seconds and varies significantly from one material to another. The spin–lattice relaxation time T_1 for most common nuclei is 0.1–10 seconds, but may be as long as several hundred seconds for some nuclei. The relaxation time T_1 may have very low values, around 10^{-11}–10^{-10} seconds, in which case no NMR signal can be observed since the nucleus returns to its equilibrium before the RF field will have reached the spin. On the other extreme, relaxation times as large as several days can be observed at very low temperatures [6.16]. Sometimes, a definition of the relaxation time T_1 is taken as the time for an initially saturated signal to recover a certain percentage of its full equilibrium intensity, e.g. 63% (see Figure 6.9a).

6.3.2   Spin–spin relaxation

The magnetization vector is large when all the spins are bunched together immediately after a pulse. However, at this state, the spins are not in equilibrium and are spread out so that they are uniformly distributed with all possible angles around the z-axis until the component of magnetization vector in the plane is zero. Two factors contribute to the decay of transverse magnetization T_2: molecular interactions and variations (inhomogeneities) in the applied constant magnetic field B_0 [6.14]. The time constant that describes the return of the transverse magnetization to equilibrium is called the *spin–spin relaxation time*, T_2. T_2 is also called transverse relaxation time because it is observed in the plane transverse to the applied magnetic field.

Molecular interactions occur because the tiny fluctuating local magnetic fields from neighboring nuclear spins slightly perturb the Larmor frequency of the nuclear spins, and thus lead to loss of synchronization. Dephasing occurs by the exchange of energy between nuclei, which means that some spins acquire some energy and precess more rapidly, and the spins that have lost energy precess at a lower frequency [6.16]. In this way, the spins have now all possible orientations in the xy plane, which gives a vectorial sum equal to zero and therefore no macroscopic transverse magnetization [6.16]. The magnetization decay follows an exponential curve characterized by a time constant T_2 (see Figure 6.9b). The relaxation time T_2 can sometimes be defined as the time required for the transverse magnetization to decay to a certain percentage, e.g. 37% of its value immediately following an excitation pulse.

6.3.3   Inhomogeneous broadening

Because the magnetic field can never be perfectly homogeneous, not all spins in the sample will be manipulated at the correct resonance frequency. In this way, an extra dephasing mechanism adds up to the dephasing caused by the transverse relaxation. At long times, the initially aligned transverse components of the individual spins will have accumulated largely different phases and the summation of these contributions is zero. This extra mechanism of dephasing is called inhomogeneous broadening and is absent in the longitudinal magnetization. To include broadening in the Bloch equation, the relaxation time T_2 has to be replaced by the effective relaxation time T_2^*, according to Equation (6.20) (see also Figure 6.5d) [6.17]. The effective spin–spin relaxation time T_2^* describes the effective disappearance of transverse magnetization and incorporates the effects of an inhomogeneous main magnetic field B_0. The combination of these two factors results in the decay of transverse magnetization having a combined time constant T_2^*:

$$\frac{1}{T_2^*} = \frac{1}{T_2} + \frac{\gamma}{2\pi}\Delta B \qquad (6.20)$$

where T_2^* is the effective transverse relaxation time (s), T_2 is the spin–spin relaxation time (s), γ is the magnetogyric ratio (MHz/T), ΔB is the variation in the main magnetic field over the region of interest (T).

An example of T_1 and T_2 values for different materials is shown in Table 6.2 [6.24,6.91]. The proton spin–lattice and spin–spin relaxation time measurements are used in studying the hardening of Portland cement and tricalcium silicate pastes [6.25].

The spin–lattice relaxation time T_1 is highly dependent on the type of nucleus and on factors such as the physical state of the sample and the temperature, the fluid, proton frequency, pore surface chemistry, and pore size [6.26]. Molecular motion strongly affects the relaxation times. In general, spin–lattice relaxation times T_1 in liquids range from 0.01 to 100 seconds, but in some cases may be as short as 10^{-4} seconds; in solids, T_1 may be much longer, sometimes days. The spin–lattice relaxation times T_2 in liquids tend to be in the range of 0.01–10 seconds, and may be as short as 10^{-6} seconds in solids [6.27].

Hydrogen atoms in water are of major importance when the hydration process of cement is studied, and several authors have utilized the differences in the spin–lattice and spin–spin relaxation rate of protons in different states in water i.e. free, bound etc. in order to characterize the cement pastes. The proton spin–lattice relaxation time T_1 of the solid component increases with time as hydration proceeds, while the spin–spin relaxation time, T_2, decreases

Table 6.2 Spin–lattice (T_1) and spin–spin (T_2) relaxation times for various porous building materials

Material	T_1 (ms)	T_2 (ms)	T_2^* (ms)	Source
Gypsum	50	4100	—	[6.24]
Sand-lime brick	45	850	—	[6.24]
Mortar	35	1000	—	[6.24]
Mortar (w/c = 0.30)	2.8	—	0.157	[6.91]
Mortar (w/c = 0.60)	4.8	—	0.309	[6.91]

with time as hydration proceeds. Values of T_2 in the same range as the calcium silicate hydrates indicate that the evaporable water in cement paste could be water of crystallization [6.9]. T_2 values do not lead to a unique model for the free (evaporable) water in the cement paste: free water has long T_1 and T_2 relaxation times, but on becoming chemically bound during hydration, the relaxation rate increases, i.e. the relaxation time decreases. Relaxation times can also be used to monitor evolution of the pore structure during hydration of cement pastes. Experimental results have clearly shown that there are several types of bound water [6.28,6.29,6.25].

Both spin–lattice and spin–spin relaxation processes occur simultaneously. The spin–lattice relaxation time T_1 is affected only by very fast dynamic processes (at about the Larmor frequency), while the transverse relaxation time T_2 is affected by both fast and slow molecular processes; therefore, T_2 is always equal to or less than T_1 [6.11]. In pure water, the spin–lattice and spin–spin relaxation times, T_1 and T_2 are approximately the same, about 2–3 seconds. In cement pastes $T_1 < 10$ milliseconds, $T_2 < 0.5$ milliseconds [6.30].

Paramagnetic effects on spin–spin relaxation times have been studied theoretically [6.31,6.32,6.30,6.33] and experimentally [6.31]. In common porous building materials, usually large amounts of paramagnetic and ferro-magnetic impurities (e.g. Fe(III)) are present. Portland cement contains Fe_2O_3 in an amount approximately 3% by mass, and this complicates NMR measurements by two effects [6.24]:

- The spin–spin relaxation time T_2 is drastically decreased by paramagnetic centers at the surface of the pores to values less than $100\,\mu s$ and this effect is more notable in small pores. It has been found that the relaxation time T_2 tends to decrease as the iron content increases.
- Due to the difference of the magnetic susceptibility of water and the porous material, local field gradients are induced, which broaden the resonance line and thereby limit the spatial resolution.

6.4 Pore size determination

Analysis of the relaxation curves of fluids is of particular interest for the characterization of the pore structure in materials. Considerable effort has been dedicated by researchers to deriving the proper mathematical formulations to extract pore sizes and pore size distributions from measurements of NMR magnetization. The amount of energy absorbed by a specimen after a RF pulse is applied, depends on the number of protons in the specimen; therefore, it can serve as a measure of the moisture content, since hydrogen nuclei in building materials in general occur in the form of water. Since the relaxation times are very different in solids ($T_2 < 50\,\mu s$) and in liquids ($T_2 > 50$ milliseconds), the two phases can be easily distinguished [6.34]. Pore structure parameters can be determined using the results of an NMR experiment by a variety of analysis methods. The most commonly employed methods are the Magnetic Resonance Relaxation Analysis (MRRA), the Cryoporometry, and the MRI, and these techniques are described in the following sections.

6.4.1   *Magnetic-resonance relaxation analysis*

MRRA, also called NMR relaxometry, provides insight into pore sizes, surface areas, and molecular dynamics in porous materials. Relaxation times are very sensitive to changes in the molecular environment and in the molecular mobility. Because longitudinal and transverse relaxation rates ($1/T_1$ and $1/T_2$) can be significantly increased in the vicinity of solid–liquid interfaces, the relaxation behavior of a proton and other nuclei of fluids confined in porous materials can provide important information about porosity, pore size distribution, and pore connectivity [6.32,6.35,6.36]. The suitability of NMR relaxation experiments in studying porous materials is due to an observed enhanced relaxation rate of liquids when confined in the pores of the material. Although performing a NMR spin–lattice relaxation experiment is straightforward, the extraction of the desired pore size distribution from the observed magnetization decay can be a complicated procedure due to interactions involving the probing molecules at the liquid–solid interface.

Several approaches have been used to extract pore structure information from NMR data. A common approach has been to determine the spin–lattice relaxation time from the FID. More modern approaches still under development include using algorithms to extract pore size distributions from relaxation data. The general methodology of the pore size analysis includes applying an appropriate pulse sequence and recording the magnetization decay, which represents the laboratory measured curves by an exponential function, and from the fitting parameters determining useful pore structure parameters. The theory on how the magnetization decay relates to pore structure parameters is discussed briefly in the following paragraphs.

It was mentioned in Section 6.3 that the decay of the magnetization moment is given by Equations (6.15), and (6.16). The decay of the magnetization of a porous material is in general complex, because most porous materials of interest consist of interconnected pores with a wide distribution of sizes. The decay of the nuclear magnetization moment can be characterized by many relaxation times and is described as the sum of exponential curves [6.37]. The most common exponential curve fits used by researchers to represent NMR experimental data and which are mentioned in this section are the single-exponential, the multi-exponential (two- and three-), and the stretched exponential, which are given by Equations (6.21), (6.22), and (6.23), respectively [6.37,6.38]. The stretched exponential is also referred to as the compressed exponential when the exponent has a value greater than 1. The modified stretched exponential, which is mentioned to be suitable in some occasions, is given by Equation (6.24) [6.39]:

$$M(t) = M_0 \exp\left(\frac{-t}{T_1}\right) \tag{6.21}$$

$$M(t) = \sum_{i=1}^{N} M_{0i} \exp\left(\frac{-t}{T_{1i}}\right) \tag{6.22}$$

$$M(t) = M_0 \exp\left(\frac{-t}{T_{1a}}\right)^a \quad \text{with } 0 < a < 1 \tag{6.23}$$

$$M(t) = M_0 \exp\left[\frac{-t}{\tau_0}\left(1 + \frac{t}{\tau_c}\right)^{\beta-1}\right] \tag{6.24}$$

Equation (6.22) is the double-exponential for $N = 2$, triple-exponential for $N = 3$ etc. Mathematically, the stretched exponential representation has the advantage of having fewer parameters than the two- and three-exponential representation, which is really important in order to correlate a few parameters to the properties of interest [6.40]. The stretched exponential curve is also more convenient to integrate than performing a more general Laplace inversion of raw data [6.41]. The stretched exponential curve is the single exponential when the exponent a is equal to 1.

MRRA can be performed on a liquid that is absorbed inside the pores of a material; consequently, the analysis is only applicable to materials with an interconnected pore structure. Many liquids can be used to saturate the porous material, provided that they are inert with the material tested so as not to modify the pore structure, and that they have a suitable NMR active nucleus such as hydrogen or fluorine, a condition that is satisfied for virtually all liquids [6.41]. In addition, the nuclear spin interactions between the liquid and the solid in the porous material should be much stronger than the spin interactions in the bulk liquid. In cement pastes, proton relaxation analysis is the common method used.

For identical spins with $I = \frac{1}{2}$ in a *bulk fluid*, the nuclear magnetization is characterized by a single-exponential process given by Equation (6.25) [6.42,6.43,6.41]. The equation is applicable both for longitudinal (with $i = 1$) and for transverse magnetization (with $i = 2$):

$$M_i(t) = M_0 \exp\left(\frac{-t}{T_i}\right) \tag{6.25}$$

where $M_i(t)$ is the magnetization moment at time t (T); M_0 is the equilibrium longitudinal magnetization moment (T); T_i is the relaxation time (s).

When the liquid is confined *inside a pore*, collision/interaction of water with the pore walls controls the relaxation behavior. Fluid near a surface will undergo spin–lattice relaxation and spin–spin relaxation at a faster rate than the bulk fluid. The pore size analysis is based on the difference between the relaxation times of protons in bulk water in the middle of the pore and the more constrained protons in the near-surface water at the pore boundary. In addition, it is well known from experiments, that relaxation times decrease as the pore size decreases and a variety of theories have been developed to explain this observation. In large pores, the liquid molecules have more room to move around without bumping into the pore walls than in small pores, so these collisions are less frequent. Therefore, in large pores, the NMR relaxation time is longer than in small pores [6.44,6.45]. Provided the liquid molecules diffuse sufficiently fast that they explore all parts of the pore volume during the experiment, we can relate the relaxation time T_1 to the pore structure.

An effort to develop models that relate magnetization data to pore size distributions has been attempted by several researchers [6.46,6.47,6.36,6.48, 6.49,6.50,6.51,6.40]. A porous material contains a range of pore sizes that have different relaxation rates in different pores. Two different approaches are used to extract pore size distributions from relaxation data: the *discrete pore models* and the *diffusion cell models* [6.41]. Discrete pore models treat the magnetization individually from each pore, assuming that pores are interconnected with throats that do not allow for diffusion from pore to pore. Discrete pore models are the models proposed by K. Brownstein and C. Tarr, 1979 [6.52], M. Cohen and K. Mendelson, 1982 [6.49,6.50] and S. Davies and K. Packer, 1990 [6.42,6.53]. Diffusion between pores was first treated by K. Mendelson, 1986 [6.51] and the diffusion cell model has been proposed by W. Halperin *et al.*, 1989, to account for diffusion from one pore to another [6.41]. Researchers developing theoretical models try to provide equations that give the magnetization decay $M(t)$ and the relaxation times T_1 and T_2 as a function of pore size. Some of the developed models are briefly discussed here and further information is given in the original papers.

J. Robinson and colleagues, 1970, made the first attempt to relate magnetization data to pore size distribution by developing a diffusion model in spherical and cylindrical pores [6.46,6.47]. They introduced the idea that water molecules at the pore surface have different relaxation times than bulk water. For a single pore they found that the spin–lattice relaxation follows a single exponential curve with a decay constant being a sum of two terms: (a) the relaxation rate of the bulk water inside the pore, and (b) the rate at which molecules diffuse to the pore surface and subsequently undergo surface relaxation. They considered water molecules at the pore surface to have increased relaxation rate, and therefore, a reduced relaxation time.

In later studies, K. Brownstein and C. Tarr, 1979, and M. Cohen and K. Mendelson, 1982, indicated that the spin–lattice relaxation times T_1 relate to the surface area-pore volume ratio [6.52,6.49]. K. Brownstein and C. Tarr, 1977, introduced a simple theoretical model, with its physical and mathematical formulation, the "*two-fraction fast-exchange*" model, sometimes also referred to as the BT model [6.48]. The model assumes that within a single pore with surface area S and volume V there are two magnetically distinct phases of the pore fluid: a bulk phase remote from the pore surface with relaxation rate that is characteristic of the bulk fluid, and a surface phase of thickness Δ with a much faster relaxation rate and therefore shorter decay time (see Figure 6.10). The relaxation time within the bulk phase $T_{1,b}$ is well-defined and uniform and the first molecular layer has a well-defined and uniform surface relaxation time $T_{1,s}$. The researchers did not consider the physical nature of the relaxation sources, but considered the pore surface as sinks of nuclear magnetization. They studied relaxation in pores of spherical and cylindrical shape, and in parallel plates [6.52]. Several researchers have used the principles of this model and fast-exchange to provide analysis of the magnetization in porous materials.

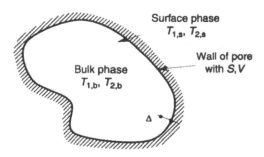

Figure 6.10 Two-fraction fast-exchange model in a pore of surface area S and volume V filled with liquid. The bulk phase of the liquid has relaxation times $T_{1,b}$ and $T_{2,b}$ for the longitudinal and transverse relaxation, respectively. The surface phase of the liquid has a thickness Δ and relaxation times $T_{1,s}$ and $T_{2,s}$ for the longitudinal and transverse relaxations, respectively.

Brownstein and Tarr considered that in both water subregions (bulk and surface), the nuclear spins inside the pore are experiencing two mechanisms: diffusion and decay (flip). Water is exchanged between the surface and bulk region via a diffusion mechanism. For liquid confined in a pore, diffusion and relaxation at the surface act simultaneously and contribute to the magnetization decay. If the diffusion is rapid (fast-exchange), then all water molecules interact with the pore surface during the decay time, magnetization across the pore is uniform, and the surface relaxation time is proportional to pore size.

Depending upon the mechanism of relaxation on the pore surface, Brownstein and Tarr identify three different behavioral regimes, the fast-, intermediate-, and slow-diffusion. The fast-diffusion regime relates to small pores, and the slow diffusion regime relates to large pores [6.52].

The *fast-diffusion regime*, also referred to as "surface-limited" relaxation, is characterized by the fact that the time taken for the molecules to diffuse to the pore surface is short compared to the surface relaxation strength. The relaxation behavior is uniform over the entire pore, and all the spins show a single relaxation time constant T_1. As a consequence, a single-exponential relaxation decay is observed for a given pore, and the rate of magnetization decay does not depend on the pore shape but only on the surface-to-volume ratio. Fast-diffusion is characteristic of small pores, rapidly diffusing fluids, and surfaces with low relaxation strength. The majority of natural rocks conforms with the fast-diffusion regime.

The *slow-diffusion regime*, also referred to as "diffusion-limited" relaxation, is characterized by the fact that the time taken for the molecules to diffuse a distance is long compared to the rate of relaxation at the pore surface. The magnetization decay is controlled by the transport of molecules to the surface. Thus, nuclear spins close to the relaxing surface are readily distinguished from spins further away from the surface. Slow-diffusion is characteristic of relatively large pores and/or a strong surface relaxation. In this regime, even in a single pore, the magnetization decay has a multi-exponential character and depends on the shape of the pore. In the centers of large pores the relaxation times approach the relaxation times of the bulk liquid, while the relaxation times of molecules near the surface of the pores are much shorter, often by several orders of magnitude. In the diffusion-limited case, the strength of the surface relaxation does not affect the observed NMR decay.

6.4.1.1   Relaxation inside a single pore

Following the formulation of the BT model, the relaxation time for the fluid within a single pore is given by Equation (6.26), which is valid both for longitudinal and transverse relaxation (with $i = 1$ and $i = 2$, respectively) [6.54,6.41,6.55,6.49,6.48]. For simplicity, the equation is shown for $i = 1$:

$$\frac{1}{T_1} = \frac{f_b}{T_{1,b}} + \frac{f_s}{T_{1,s}} \quad \text{where } f_b = \frac{V_b}{V}, \ f_s = \frac{V_s}{V}, \ \text{and } f_b + f_s = 1 \quad (6.26)$$

where T_1 is the longitudinal relaxation time (s), $T_{1,b}$ is the longitudinal relaxation time for water molecules in bulk (s), $T_{1,s}$ is the surface affected longitudinal relaxation time (s), f_b is the volume fraction of pore fluid with bulk properties (or proportion of fluid in the bulk phase) (unitless), f_s is the volume fraction of pore fluid with surface-affected properties (or proportion of fluid in the surface phase) (unitless), V_b, and V_s are the rapidly exchanging volumes of bulk and surface liquid respectively (m^3).

In order to use Equation (6.26) for pore structure analysis, the volume fractions of the two phases, f_b and f_s must be related to pore size. The volume of the surface phase can be approximated by the product of the liquid surface area S and the thickness Δ of the surface phase; therefore the volume fractions of the two phases are given by Equation (6.27) [6.54]:

$$f_s = \frac{\Delta \cdot S}{V} \quad f_b = 1 - \frac{\Delta \cdot S}{V} \quad (6.27)$$

where f_b is the volume fraction of pore fluid with bulk properties (or proportion of fluid in the bulk phase) (unitless), f_s is the volume fraction of pore fluid with surface-affected properties (or proportion of fluid in the surface phase), (unitless), S is the specific surface area of a pore (m^2), V is the total volume of a pore (m^3), Δ is the thickness of the surface phase (m).

If the volume of the surface phase is taken to be much smaller than the volume of the bulk phase ($f_s << f_b$), then $f_b \cong 1$, and by using Equation (6.27), Equation (6.26) may be re-written as Equation (6.28) [6.54,6.49]:

$$\frac{1}{T_1} = \frac{1}{T_{1,b}} + \frac{\Delta \cdot S}{V T_{1,s}} \quad \text{or} \quad \frac{1}{T_1} = \frac{1}{T_{1,b}} + \rho \frac{S}{V} \quad \text{with } \rho = \frac{\Delta}{T_{1,s}} \quad (6.28)$$

where T_1 is the longitudinal relaxation time (s), $T_{1,b}$ is the longitudinal relaxation time for water molecules in bulk (s), Δ is the thickness of the surface phase, S is the specific surface area of a pore (m^2), V is the total volume of a pore (m^3),

$T_{1,s}$ is the surface affected longitudinal relaxation time (s), ρ is the surface relaxivity parameter (m/s).

In order to relate T_1 to the desired pore size of a material containing water (or another liquid), the parameters $1/T_{1,b}$, and surface relaxivity parameter, ρ, must be determined. This can be accomplished by using porous materials that have a known surface area–pore volume ratio (S/V) containing the same liquid (water) inside the pores. By using Equation (6.28) with known values for $1/T_{1,b}$, the S/V, and the measured T_1, the value of the relaxivity parameter ρ can be obtained [6.54]. Then, by assuming a pore shape, the surface area–volume ratio can be related to the hydraulic radius of the pore. The constant $1/T_{1,b}$ depends upon the pore fluid, the temperature, and the magnetic field strength.

For the derivation of Equation (6.28) it is assumed that the volume of the surface phase is much smaller than the volume of the bulk phase, therefore, the applicability of Equation (6.28) is limited to pore sizes greater than about 3–5 nm [6.54]. For geometries characterized by pore sizes that are comparable to Δ, a correction is required.

Brownstein and Tarr, 1979, further developed their model into a more general one, invoking only the properties of bulk fluid [6.52]. They assumed that relaxation within the bulk fluid is negligible compared to the rate of relaxation at the pore surface. According to the fast exchange theory, all nuclei relax at the same rate, and neglecting bulk liquid relaxation, the relaxation time T_2 can be written as expressed in Equation (6.29):

$$\frac{1}{T_2} = \frac{\Delta}{T_{2,s}} \frac{S}{V} \tag{6.29}$$

where T_2 is the observed relaxation time (s), $T_{2,s}$ is the relaxation time for water molecules in the surface phase (s), Δ is the thickness of the water layer on the pore surface (m), S is the specific surface area of a pore (m^2), V is the total volume of a pore (m^3).

The parameter Δ is the thickness of water representing the extension of the surface interaction responsible for the enhanced relaxation. The thickness Δ has been determined both theoretically and experimentally, and is taken as the thickness of a water monolayer, approximately 0.3 nm [6.41,6.44,6.56].

The surface relaxivity parameter, ρ_1 (also called surface interaction parameter) includes the effect of material constants, such as the thickness of the surface layer within which relaxation can take place, the proportion of surface sites occupied by paramagnetic ions [6.32,6.31], temperature, magnetic field strength, and pore surface chemistry [6.54,6.26,6.44]. The paramagnetic effects of iron species on nuclear magnetic relaxation of fluid protons in porous materials have been tested experimentally [6.57]. The surface relaxivity parameter can increase linearly with increasing surface concentrations of iron. NMR relaxation measurements are so sensitive to paramagnetic impurities that most natural samples will probably have values for the surface relaxivity parameter above 1 μm/s [6.57]. Values of the surface relaxivity parameter for different materials are given in Table 6.3.

The two-fraction fast-exchange model and the fast-diffusion theory have been widely used by several research groups, to study a variety of porous

Table 6.3 Values for the surface relaxivity parameter
for different porous materials [6.57]

Material	Surface relaxivity parameter, ρ ($\mu m/s$)
Quartz	0.83
Silica gel	$(1.2 \text{ to } 7.7) \times 10^{-3}$
Porous silica glass	1.8×10^{-3}
Silica sand	2.89–3.06
Sandstone rock	9.0–46

Note
Some of the values have been obtained from the technical
literature, cited in the original reference.

materials including cement pastes [6.45,6.54,6.55,6.58,6.59,6.40]. The most obvious disadvantage of the two-fraction fast exchange model is that it represents a limiting behavior within a continuous system. The assumptions of two discrete phases and of fast-diffusion are unlikely to be realistic under many experimental conditions [6.42].

6.4.1.2   *Relaxation inside a distribution of pore sizes*

Porous materials have a distribution of pore sizes. Analysis of relaxation data becomes more complicated in the case of a distribution of pores rather than a single pore, and the discrete model and diffusion cell model have been proposed to describe the relaxation phenomena and facilitate analysis of magnetization data.

Discrete model   In a porous material there is usually a complex distribution of pore sizes and shapes, so the observed relaxation behavior will be a sum over all the contributing pores [6.42]. If the pores are identical in size, Equation (6.26) holds true, even if the pores are interconnected. However, for a distribution of pore sizes the observed relaxation is a sum of single-exponential decays [6.52,6.46]. Each pore contributes to the total signal intensity with its own transverse relaxation time, and the total magnetization is expressed by Equation (6.30):

$$M(t) = \sum_i M_i \exp\left(-\frac{t}{T_{2,i}}\right) \tag{6.30}$$

where $M(t)$ is the total magnetization at delay time t, M_i is the intensity of the ith pore, $T_{2,i}$ is the transverse relaxation time of the ith pore.

K. Brownstein and C. Tarr, 1979, explained theoretically the multiexponential decay of relaxation and provided an equation similar to Equation (6.30) for the magnetization $M(t)$, assuming a shape for the pores [6.52]. The general solution to the diffusion equation of the magnetization inside a pore system may be expressed in the form of Equation (6.31), where the intensity I_i and the relaxation time T_{1i} from each pore i are determined by the pore size and geometry, the rate of diffusion of the fluid inside the pores and the nature of the pore surface. Thus in general, the relaxation behavior of a fluid confined in a pore may be represented by a sum of exponential functions:

$$M_z(t) = M_0\left[1 - 2\sum_{i=0}^{\infty} I_i \exp\left(-\frac{t}{T_{1,i}}\right)\right] \qquad (6.31)$$

where $M_z(t)$ is the magnetization at time t, M_0 is the initial magnetization, I_i is the intensity from pore i, $T_{1,i}$ is the longitudinal relaxation time of the ith pore.

The intensities I_i and time constants $T_{1,i}$ for the contributing pores are calculated from the BT model using theoretical equations the researchers provided in their original papers and require the knowledge of the fluid self-diffusion coefficient D, and also the relaxation time of the bulk fluid $T_{1,b}$. Brownstein and Tarr quote analytical results for planar, cylindrical, and spherical pore geometries.

In a similar study, K. Mendelson, 1982, 1986, followed the work of Senturia and Robinson, and Brownstein and Tarr, and developed the theory for relaxation in spherical pores and interconnecrted pores [6.50,6.51]. They examined single spherical pores and separated interconnected pores using weak connections between the pores. They considered that the magnetization relaxation time is uniform within each pore, but different among pores.

S. Davies and colleagues, 1990, presented a discrete pore model assuming magnetically isolated spheres of radius a, that are connected by smaller channels [6.42,6.53]. Their model is an improved BT model and gives the magnetization moment as a function of time $M_z(t)$ including the pore size distribution and assuming a shape for the pores. They developed their model theoretically and tested it experimentally on rock samples by inversion of NMR experimental data to estimate the pore size distribution using a nonnegative least squares optimization. Their approach shows how the BT model can be used to describe spin–lattice relaxation as a function of the pore size distribution. Although it is ignored in the original BT model, relaxation of the bulk fluid often makes a significant contribution to the total relaxation and should be accounted for. Inclusion of the relaxation time of the bulk liquid $T_{1,b}$ has no effect upon the intensities I_i, and yields an additive term to the overall relaxation rates $T_{1,i}$ [6.42]. The pore surfaces are assumed to act as an efficient relaxation sink and provide a constant and uniform spin–lattice relaxation sink of strength M. For a pore of radius r, the contribution to the total NMR signal is proportional to the pore volume. In a distribution of pore sizes $P(r)$ the signal in the range r to $(r + dr)$ will, therefore, be proportional to the parameter $r^3 P(r) dr$. Integration over possible values of the pore radius r yields an expression for the magnetization as a function of the pore size distribution. Using the *spherical pore model*, the spin–lattice relaxation obeys Equation (6.32). The pore size distribution can be obtained at discrete pore radius values using a nonnegative least squares (NNLS) method:

$$M_z(t) = M_0\left\{1 - (1-x)\int_0^{\infty} \frac{r^3 P(r)\sum_{i=0}^{\infty} I_i(r)\exp(-t/T_{1,i}(r))}{N}\,dr\right\} \qquad (6.32)$$

where $M_z(t)$ is the magnetization after relaxation delay t, M_0 is the equilibrium longitudinal magnetization, x is the fractional inversion, a parameter to account for instrumental imperfections that prevent the observation of perfect inversion (full inversion corresponds to $x = 1$), $P(r)$ is the sample pore size distribution, r is the pore radius, N is a normalization constant, $I_i(r)$ is the intensity from pore i, $T_{1,i}$ is the longitudinal relaxation of the ith pore.

A variety of methods for the analysis of relaxation curves have been developed and most usually researchers try to fit a predetermined exponential function (some of which were mentioned in the beginning of Section 6.4.1) to the data, and relate the fitting parameters to pore structure parameters.

The basis of the developed methods, which determine a distribution of relaxation time T_1 values appears to be that a given T_1 value should be correlated uniquely with a particular pore size. Since a porous solid will have a range of pore sizes, a distribution of relaxation times T_1 will exist, which is related to the desired pore size distribution and must be extracted from the observed magnetization relaxation data. With some care it is possible to obtain accurate pore size distributions in saturated materials through NMR longitudinal relaxation times. The alternative approach is to determine a distribution of T_1 values (which may be continuous or discrete), $f(T_1)$, from the experimental data, where the total magnetization may be represented by an integral (or explicit summation) over the range of relaxation times T_1, as expressed by Equation (6.33) [6.43]:

$$M_z(t) = M_0\left[1 - 2\int_0^\infty f(T_1)\exp\left(\frac{-t}{T_1}\right)dT_1\right] \qquad (6.33)$$

where $M_z(t)$ is the measured magnetization at different delay times t, M_0 is the equilibrium magnetization, $f(T_1)$ is the distribution of longitudinal relaxation times.

The distributions of relaxation times can be converted to pore size distributions, usually on the basis that the relaxation behavior is confined to the fast-diffusion regime, where a particular T_1 may be correlated directly with a particular pore size, e.g. via the "two-fraction fast-exchange" model. It is possible that fluid in larger pores may show behavior characteristic of the slow-diffusion regime, i.e. multi-exponential for a single pore size; this may be corrected partially using an average T_1 value [6.42,6.53].

Diffusion cell model In interconnected pores, if there is no diffusion between pores, the relaxation is dominated by the largest pores, which have relaxation times similar to the relaxation times of bulk water. Magnetic resonance measurements on a liquid in many porous materials cannot be described by the isolated pores approach. Generally, the narrow throat condition is not fulfilled, and the fluid molecules may diffuse freely between different pores, before much of the magnetization relaxes. K. Mendelson, 1986, considered individual pores with connections between them and diffusion between pores [6.51]. Diffusion between pores allows the small pores to contribute to the relaxation at long times thus reducing the relaxation time.

Mendelson showed that when diffusion between pores is sufficiently fast, the nuclear magnetization is uniform, decays exponentially, and the relaxation rate is linearly related to surface area–volume ratio of the entire pore space. It is also believed that, if the small and large pores are mixed well within a diffusion cell, the magnetization recovery will be a single-exponential. If the small and large pores occupy regions of space which are separated by more than a diffusion length, the recovery will be non-exponential [6.60].

The diffusion cell model was developed by W. Halperin *et al.*, 1989, to account for diffusion in the pore network [6.41]. The sample is viewed as being divided into a number of non-overlapping cells. Each of these cells may be treated independently and may have its own characteristic relaxation time. For diffusion cells with different surface area–volume ratio in each one, the signal amplitude in time will be the superposition of various exponential relaxations, resulting in a non-exponential recovery. For this case, after treating each separate region as being locally in fast exchange, the magnetization recovery can be expressed by Equation (6.34). Numerical inversion of the measured magnetization recovery $M(t)$ similar to Laplace transform can yield the pore volume distribution function $P(r)$:

$$M(t) = M_0 \int P(r) \exp\left(-\frac{\Delta}{T_{2,s}} \frac{S}{V} t\right) dr \qquad (6.34)$$

where $M(t)$ is the magnetization at time t, M_0 is the equilibrium magnetization, $P(r)$ is the pore volume distribution function, r is the pore size, Δ is the thickness of the surface phase, $T_{2,s}$ is the transverse relaxation time, S/V is the ratio of surface area to pore volume.

6.4.1.3   *Pore size distribution in cement pastes*

MRRA has been used with considerable success to study many porous materials including glasses, natural sandstones, and ceramic materials. Magnetization decays with longitudinal relaxation T_1 have been found to be mono-exponential for synthetic porous ceramics and multi-exponential for natural materials, usually double- and triple-exponentials [6.39,6.32,6.40]. The stretched exponential can be applied for both longitudinal and transverse relaxation times, and has been used for exponential fits to data for a wide range of rocks, from carbonates to sandstones [6.39,6.40]. The stretch parameter a in the stretched exponential fit (see also Equation (6.23)) of water-saturated rocks has values within a small range of 0.5–0.7 [6.40]. The modified stretch exponential has also been found useful in describing the relaxation behavior of rocks [6.39].

MRRA is a relatively new and unconventional method for characterizing the pore structure of hardened cement pastes. The method has also been used to monitor cement hydration and separate signals from protons in H_2O and OH^- groups [6.25]. Relaxation time studies of pore structure can only be accurately performed on uniformly wet cement samples, as any drying will lead to spatial variations in relaxation properties [6.61]. Researchers have shown experimentally that transverse relaxation time distributions in cement pastes are bi- or tri-modal (exponential), depending on whether chemically bound water is included, and can be approximated by two or three-component fits [6.62]. In another study, through the characteristic relaxation of the water protons, four forms of water could be distinguished in cement paste by NMR: free water, adsorbed water, interlayer and pore water, and chemically bound water such as in $Ca(OH)_2$ [6.29]. A 5-exponential curve has also been used by researchers to fit to the decay, and explain the two largest porous modes to be macroporosity, and the three smallest porous modes to relate to C–S–H induced porosity [6.63].

As an example, the spin-echo signal is given as a function of the spin-echo time for a mortar sample in Figure 6.11a [6.58]. The signal intensity for concrete has been described as a two-exponential decay and could be attributed to two very distinct pore sizes, which are reflected in the relaxation: gel pores (with size 5–100 Å), and capillary pores (with size 0.05–10 μm). Assuming Gaussian pore size distributions, the pore size distribution could be calculated from the measured relaxation (see Figure 6.11b). Similar results have been reported by other researchers [6.63,6.25]. The double exponential behavior of the relaxation of the water has been described by Equation (6.35):

$$M(t) = M_{gel} \exp\left(-\frac{t}{T_{2,gel}}\right) + M_{cap} \exp\left(-\frac{t}{T_{2,cap}}\right) \tag{6.35}$$

where $M(t)$ is the magnetization at time t, M_{gel} is the magnetization due to the water in the gel pores, M_{cap} is the magnetization due to the water in the capillary pores, $T_{2,gel}$ is the transverse relaxation time of the gel pores, $T_{2,cap}$ is the transverse relaxation time of the capillary pores.

A fit to the reported experimental data with a double exponent, gives the following results: $T_{2,gel} \approx 600\,\mu s$ and $T_{2,cap} \approx 2700\,\mu s$ [6.64,6.65]. It has been previously reported that non-evaporable water has a transverse relaxation time $T_2 \cong 20\,\mu s$ [6.66]. The transverse relaxation time is transformed into pore size using Equation (6.29) after determining the surface relaxivity ρ, using Equation (6.28).

Experiments have been carried out by A. Bohris *et al.*, 1998, to follow the hydration process of cement pastes from 1 hour till 28 days [6.67]. Their results, presented in Figure 6.12, show three peaks at the age of 28 days, associated with chemically bound water (9 μs), gel water (80 μs) and the long component (350 μs) with capillary water. They calculate the pore sizes using Equation (6.29), by taking $T_{2,s} = 10\,\mu s$ (the relaxation time of the chemically bound water), and $T_2 = 8\,\mu s$, which gives a pore size of $d = 2.4\,nm$. A component with a very short spin–spin relaxation time T_2 represents the protons of the solid OH^- groups and the water of crystallization. A relatively long spin–spin relaxation time T_2 represents water in pores smaller than 2.5 nm and between the layers forming the C–S–H gel. The bulk water protons relax quickly due to the fast exchange between them and the bound water fraction, which defines the pore surface phase (see Figure 6.10). In the proton FID of Portland cement paste, this is shown by a component with a relatively long

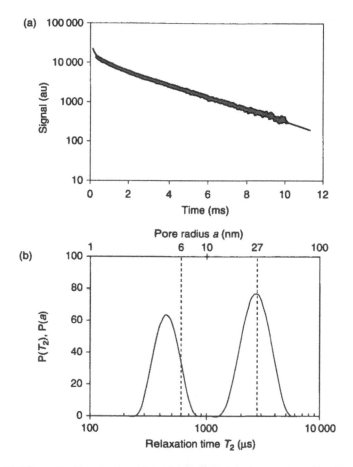

Figure 6.11 Magnetic resonance relaxation analysis of measurements in a cement mortar. (a) The NMR signal as a function of spin-echo time. The solid line represents a fit to the data with a double exponential. (b) Pore size distribution from the magnetization measurement in mortar assuming a normal (Gaussian) distribution.

Source: Reprinted from *Materials and Structures*, Vol. 34, R.M.E. Valckenborg, L. Pel, K. Hazrati, K. Kopinga, J. Marchand, "Pore water distribution in mortar during drying as determined by NMR," pp. 599–604, copyright 2001, with permission from RILEM.

transverse relaxation time T_2. The bound water is very effectively relaxed by the solid substrate through the strong coupling that exists between them. The proton relaxation times have also been studied for different water contents in cement pastes [6.69].

Other researchers have used the stretched exponential function to describe relaxation in cement pastes [6.68,6.70,6.30] and have found that during the hydration period the stretch parameter *a* changes, being close to 1 during the dormant period (corresponding to a single-exponential) and being reduced to 0.5–0.8 after 300 hours (=12.5 days) of hydration, depending on the type

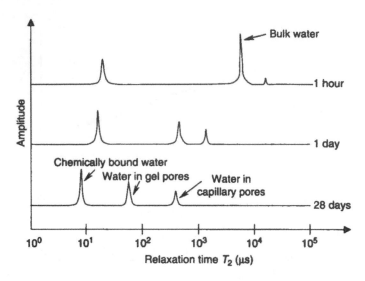

Figure 6.12 Proton T_2 relaxation times of cement paste with w/c = 0.50 at different hydration times, 1 hour, 1 day, and 28 days after mixing. The amplitudes of the peaks represent the amplitude of the different T_2 components. At 28 days, the peaks correspond to chemically bound water (9 μs), nano-pore water (80 μs, equivalent to a pore size of 24 Å) and capillary water (350 μs equivalent to a pore size of 150 Å). Short relaxation times correspond to confined water and long relaxation times to bulk water.

Source: Modified from *Magnetic Resonance Imaging*, Vol. 16, A.J. Bohris, U. Goerke, P.J. McDonald, M. Mulheron, B. Newling, B. Le Page, "A broad line NMR and MRI study of water and water transport in Portland cement pastes," pp. 455–461, copyright 1998, with permission from Elsevier Science, Amsterdam, The Netherlands.

of cement. Researchers believe that the stretch parameter *a* is related to the fractal geometry of the pore structure and is changing as the fractal dimension changes (decreases) during hydration, but have not correlated the exponent to the fractal dimension yet. The stretched exponential curve has been used to fit experimental data for both the longitudinal and transverse relaxation times. The value of the stretch parameter greater than 1 in the transverse relaxation has been explained by a superposition of relaxation and diffusion effects in internal magnetic field gradients [6.70].

Surface magnetic relaxation in porous media is usually attributed to hindered rotation and paramagnetic impurities. In cement pastes, however, it seems reasonable to attribute surface relaxation to dipole interactions in water of hydration on grain surfaces. The hydrate relaxation time is 17 μs, while the measured surface relaxation time is 30–40 μs [6.23].

6.4.2   NMR cryoporometry

It is well known that the physical properties of a liquid confined within small pores can be radically different from the properties of the bulk liquid; in particular, the freezing point is depressed in small pores. J. Strange *et al.*, 1993, employed the phenomenon of freezing point depression together with magnetic resonance to introduce NMR cryoporometry, a new method for pore size distribution measurements [6.71]. The technique was initially applied using porous silicas saturated with cyclohexane and was further developed by Strange's group to become a useful technique for pore size analysis on a variety of materials [6.72,6.59,6.34]. Several researchers have used cryoporometry since then to characterize many different porous materials, including silicas, porous carbons, rocks, cement pastes, and polymers.

NMR cryoporometry is very similar to Differential Scanning Calorimetry (DSC) Thermoporometry that was described in Chapter 5, and they both rely on melting point suppression to produce pore size distributions. NMR cryoporometry is based on the technique of freezing a liquid in the pores and measuring the melting temperature by NMR. It is necessary to vary the temperature of the saturated sample from about 50°C below to just above the bulk-liquid melting temperature. Since the melting point is depressed for crystals of small size, the melting point depression gives a measurement of pore size, and the change in liquid fraction as the temperature is increased reveals the distribution of pore sizes.

The combination of the two techniques, cryoporometry and relaxation analysis, can provide more information on the pore structure of a material than either of the two methods separately. The technique makes use of the fact that when lowering the temperature, the fluid in the large pores freezes first, which enables one to selectively "switch off" the contribution of these pores and study the relaxation of the hydrogen nuclei in the small pores separately. NMR can be used to measure the quantity of liquid within the sample, using the large discontinuity in spin–spin relaxation rime, T_2, at the freezing point to distinguish solid ice from liquid [6.43]. The sample is first cooled to ensure that all the pore water has frozen, and then as temperature is slowly increased, the signal intensity of a NMR spin-echo is recorded. Due to the very short spin–spin relaxation time of ice, any signal obtained can be assumed to originate from the liquid water inside the pores. The first signals detected on warming are therefore due to water in the smallest pores. Experiments have demonstrated that cement pastes have two principal pore structural components and the water identified is the capillary and gel pore water. The freezing-thaw process does not appear to have any irreversible effects on the pore structure [6.73,6.74,6.75].

As the frozen sample is warmed, the crystals formed within smaller pores will melt first. A minute crystal of linear dimension x will melt at a temperature T lower than the melting point T_m of the bulk-liquid by an amount $\Delta T(x)$, which is inversely proportional to the dimension of the crystal. The volume of liquid produced indicates the volume of pores with dimension less than or equal to x. The melting point distribution is determined by analyzing the NMR signal as a function of temperature. The lowering melting temperature of a liquid in a pore is generally attributed to the reduced crystal size in the pore and to the large surface-to-volume ratio, S/V. For a liquid confined within a pore in which a crystal is forming, assuming the contact angle between liquid, solid, and pore wall is 180°, the temperature reduction of the melting point is given by Equation (6.36) known as the Gibbs–Thompson equation [6.71,6.76]. Equation (6.36) can be simplified to Equation (6.37) for a particular liquid, by using a constant k_f that depends on the liquid:

$$\Delta T_m = T_m - T_m(x) = \frac{4\sigma T_m}{x\Delta H_f\rho} \tag{6.36}$$

$$T_m - T_m(x) = \frac{k_f}{x} \quad \text{with} \quad k_f = \frac{4\sigma T_m}{\Delta H_f\rho} \tag{6.37}$$

where ΔT_m is the temperature reduction of the melting point (K), T_m is the normal melting point of the bulk material (K), $T_m(x)$ is the melting point of a crystal of linear dimension x (K), σ is the surface energy at the liquid–solid interface (N/m), ΔH_f is the bulk enthalpy of fusion (J), ρ is the density of the solid (g/m^3), k_f is the melting point depression constant (K Å).

The melting point depression constant k_f is characteristic of the liquid and depends on the properties of the liquid and liquid/solid interface. The constant k_f must first be calibrated for the probe material using samples with known pore size distributions. It should be emphasized that the use of liquids other than water will change the parameter k_f and thus also the relative uncertainty in the derived pore radii. A large value for k_f implies a better sensitivity and a smaller uncertainty in the pore radius [6.77]. Two liquids that have been tested extensively in cryoporometry studies have been water and cyclohexane [C_6H_{12}]. The melting point depression constant k_f is 580 K Å for water and 1790 K Å for cyclohexane; these values do not change with pore surface structure or type of porous sample [6.72]. Cyclohexane has proved to be a very suitable liquid because its melting point depression constant has a large value and freezes to a plastic crystal that is soft minimizing any damage to the pore structure on freezing from 279 K down to 186 K [6.43]. In other studies it has been reported that the constant k_f is 20 K Å for bulk water and 6800 K Å for water contained in pores in the cement paste.

It should be noted here that Equation (6.37) is the same as Equation (5.20) that was mentioned in Chapter 5, with slightly different nomenclature used: the constant k_f is the equivalent of the constant A, and t_f is the equivalent of constant B. In addition, the surface energy, σ, even though not identical, is used as the equivalent to the surface tension, γ, and they both have the same units. The original symbols used in the technical literature referring to NMR cryoporometry are used in this section. However, the correspondence of parameters to the parameters used in thermoporometry is rather easy for the reader to identify.

R. Schmidt et al., 1995, modified Equation (6.37) to Equation (6.38) to be used for mesoporous materials (with pore radius greater than 10 Å) [6.77]:

$$\Delta T = \frac{k_f}{R_p - t_f} \tag{6.38}$$

where ΔT is the temperature change (K), k_f is the melting point depression constant (K Å), R_p is the radius of the pore (Å), t_f is the thickness of the non-freezable pore surface water (Å).

The parameter t_f is proposed to be identical to the thickness of a nonfreezing pore surface water and is taken to be $t_f = 3.49 \pm 0.36$ Å [6.77].

A porous material saturated with liquid, which is then cooled, will produce a distribution of melting temperatures that depends on the pore size distribution. Since supercooling is a common phenomenon, the true melting temperature characteristics are more reliably determined by raising the temperature after freezing all the liquid. Measurement of the fraction of liquid as a function of temperature yields the pore size distribution. The pore volume $V(x)$ is a function of the pore diameter x. The volume of pores with diameter between x and $x + \Delta x$ is $(dV/dx)\Delta x$. If the pores are filled with liquid, the melting temperature of the liquid $T_m(x)$ is related to the pore size distribution by Equation (6.39) [6.71]. By differentiating Equation (6.37) we obtain Equation (6.40), which introduced into Equation (6.39) gives Equation (6.41):

$$\frac{dV}{dx} = \frac{dV}{dT_m(x)}\frac{dT_m(x)}{dx} \tag{6.39}$$

$$\frac{dT_m(x)}{dx} = \frac{k_f}{x^2} \tag{6.40}$$

$$\frac{dV}{dx} = \frac{dVk_f}{dT_m(x)x^2} \tag{6.41}$$

The measurement of $dV/dT_m(x)$ can give dV/dx from Equation (6.41), provided the constant k_f is known for the liquid used. Then, a plot of dV/dx vs. x will give the pore size distribution.

Two ways of capturing the NMR cryoporometry data have been explored. One way is to ramp the sample temperature and capture multiple NMR spin-echoes, each at a slightly different temperature. This procedure produces a large number of data points, each with a significant associated noise level and provides a highly detailed, pore size distribution dependent on signal filtering, e.g. by the application of a smoothing function; therefore, care must be taken when interpreting the results. The second way is to step the temperature by a small amount and capture an entire CPMG decay curve (see Section 6.2.3.3). This procedure provides fewer number of points than the first procedure, but each point has a lower associated noise when repeated CPMG scans are averaged [6.72]. However, the CPMG method cannot be applied below 190 K because the signal amplitude is too small to form a sufficient number of echoes [6.73].

NMR cryoporometry has been applied to cement pastes by several researchers [6.78,6.73,6.58,6.33,6.34]. Pore sizes larger than about 100 nm cannot be identified by cryoporometry because the depression of the melting point is too small to be measured with sufficient accuracy. It has been found experimentally that from the decrease in the absolute size of the NMR signal upon freezing in cement samples, the volume of water in large capillary pores is less than 10% of the total evaporable water [6.34]. A sharp peak is observed due to water in nanopores and a broader peak due to water in the capillary pores. The sizes for the gel pores are in good agreement with the sizes obtained from other measurement techniques, but the capillary pore size is

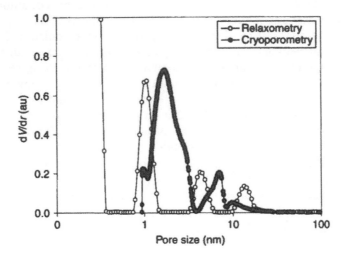

Figure 6.13 Comparison of pore size distributions of a cement mortar determined by cryoporometry and relaxometry at 0°C. The mortar specimens were made from white cement, and were 1-year-old at testing.

Source: Reprinted from *Journal of Physics D: Applied Physics*, Vol. 35, No. 3, R.M.E. Valckenborg, L. Pel, K. Kopinga, "Combined NMR cryoporometry and relaxometry," pp. 249–256, copyright 2002, with permission from the Institute of Physics.

somewhat smaller than expected. Since the transverse relaxation time T_2 of water inside gel pores is very short, its contribution to an NMR spin-echo at 200 µs is considerably attenuated [6.33]. It has also been reported in the technical literature that contamination of water within the cement pores from soluble ions do not change the melting temperature compared to distilled water [6.33]. An example of pore size distributions obtained in cement mortar using cryoporometry is shown in Figure 6.13, along with the pore size distribution obtained using relaxation analysis [6.75].

Researchers have also used a technique that is a combination of cryoporometry and imaging to study the pore structure of materials [6.43,6.79,6.78]. Unfrozen water content in concrete can be monitored as a function of position and temperature, and therefore, imaging in combination with NMR cryoporometry makes it possible to map pore size distributions within the porous material.

6.4.3 NMR imaging

Magnetic resonance imaging (MRI) started out as a tomographic imaging technique for routine use in medicine by producing an image of the NMR signal obtained from a thin slice through the human body. The technique was called MRI rather than NMRI because of the negative connotations associated with the word "nuclear" in the late 1970s. MRI was first used in porous materials by R. Gummerson *et al.*, 1979, to monitor water distribution during capillary flow [6.80]. Mapping of NMR data into spatial images (MRI and magnetic resonance microscopy) provides a nondestructive way of characterizing the internal structure of materials. Imaging techniques are now extended to applications in materials science beyond a 2-dimensional image to a 3-dimensional image from selected regions of a specimen [6.15]. As in medical imaging methods, MRI can probe structural features in porous solids containing water by providing images of the spatial distributions of the mobile fluid. A detailed description of the theory of MRI can be found in technical publications [e.g. 6.81,6.82].

In NMR spectroscopy, the applied magnetic field B_0 is required to be highly homogeneous, i.e. constant over the sample. On the contrary, in MRI a linear magnetic *field gradient* generated by a separate set of current-bearing coils is superimposed on the static magnetic field to relate the NMR resonance frequency to the spatial position within a sample [6.45]. The value of the magnetic field gradient will determine the resonant frequency of the nuclei; therefore, if a magnetic field is created with a different value at each point, excited nuclei at different positions in the sample will resonate at different radio frequencies. Resonant frequencies will then encode spatial information, and the frequency spectrum will describe a spatial distribution. By creating a magnetic field with three orthogonal field gradients, i.e. a field that varies in strength across all three dimensions of space, the resonant frequencies of a given atomic nucleus scanned reflect the nucleus' X, Y, and Z position. By sweeping across a range of RFs, and recording the strength of the signal emitted from the sample, the scanner builds a picture of the number of nuclei at each location [6.34,6.82].

MRI techniques can give a variety of information regarding pore structure in saturated materials. The simplest information obtained is the spatial variation of porosity, which can be obtained by careful selection of the applied pulse sequence and delays during the imaging experiment. Even though MRI is used extensively for biological applications, its use in porous materials, including cement pastes is not simple, because the structure and mineralogical composition of cement pastes perturb the proton resonance of water much more than do the constituents of a biological material. In a porous material, water molecules inside the pores have greatly reduced relaxation times, because of dipolar interactions with surface sites, and especially with any paramagnetic species that may be present. In addition, the strong contrast of magnetic susceptibility between the solid and liquid phases creates large local magnetic field variations within the pore structure, which affects the NMR signal causing broad resonance lines and provides poor resolution. As a result, standard MRI equipment is poorly suited to imaging cement-based materials.

Two of the techniques used to overcome the problem of broad resonance lines are the use of strong field gradients and the use of fast imaging methods. The magnetic field gradients can be made large enough to make the differences between neighboring nuclei much larger than the internal field variations. The electronics of the NMR instrument need to be able to deal with the rapidly varying signals that arise from the short relaxation times. Fast imaging methods narrow the resonance line artificially, either by the so-called magic-angle spinning, which requires the sample to be rotated rapidly, or by repeatedly applying RF pulses. Special pulses and gradients have been proposed and used by several researchers and some of these are briefly discussed in the following sections.

6.4.3.1 *Strong field gradients*

Stray-field imaging (STRAFI), exploits the extremely large and stable field gradient (as high as 20 T/m) that surrounds a strong superconducting magnet [6.83]. A RF pulse containing a narrow range of frequencies around the Larmor frequency is used to excite nuclei in a thin slice of the material at right angles to the gradient direction. The amplitude of the resulting signal is then proportional to the nuclear density in the slice. Since the gradient is static, the measurements have to be repeated for different positions of the sample to build up an image. The method has the advantage that the measurement is direct and no Fourier transformation is needed to produce the image [6.34]. It is well-suited to obtain 1-dimensional moisture distribution profiles and has been applied to several cementitious materials [6.67,6.68].

In *gradient-echo imaging*, inductive coils are used to apply strong field gradients that vary sinusoidally in time, and the coils can be tuned to impose large and rapid changes to the magnetic field [6.84]. Nuclear magnetic resonances are at a maximum when the magnetic field passes through zero, which ensures that the excitation pulses are least perturbed by the gradient and that the range of frequencies needed in the pulse is minimized.

In the *single point imaging* (SPI) sequence, a short, intense RF pulse is used to generate transverse magnetization [6.85,6.86,6.67,6.61,6.87,6.88,6.89]. Since the RF pulse is applied while the magnetic field gradient is on, duration of the RF pulse must be short enough that the bandwidth of the pulse covers the range of frequencies introduced by the gradient. Since no frequency encoding is employed, the acquired image is free of distortions resulting from magnetic field inhomogeneity, susceptibility variations, and chemical shifts. The resolution depends only on the strength of the gradient used. SPI is an appropriate technique to study systems with short relaxation time and has been used in cement-based materials, which have both short transverse and longitudinal relaxation times [6.89,6.85].

Despite its utility, SPI is time inefficient since one gradient switch and RF pulse is associated with each experimental point. To overcome these limitations, B. Balcom *et al.*, 1996, developed a version of the SPI sequence, which permits shorter data acquisition time, minimizes gradient vibration, and enables the introduction of the longitudinal relaxation T_1 constant, into a variety of images [6.85]. The pulse sequence is called SPRITE *(single point ramped imaging with T_1 enhancement)*, in which the data is stored in a multiple point manner, with small variations in the encoding time, and then an average of the Fourier transformed profile is taken. An important advantage of this technique is that the profiles are not distorted by chemical shifts or magnetic susceptibility variations, because the time evolution of the magnetization is not measured. SPRITE has been used as an imaging method in a number of studies on cement and concrete materials, including the measurement of water distributions during drying and in freezing [6.86,6.85,6.78,6.30,6.79].

The spatial resolution obtained in MRI depends on the material and on the magnetic field gradient strengths employed, but is rarely better than about 0.2 mm. MRI at its current state cannot image the pore structure in cement-based materials, but can image the aggregate particles. It can image liquids within the pore structure and therefore, is well-suited to mapping patterns or distributions that are imposed on a specimen by external processes, such as capillary water uptake and drying. Even though MRI on cementitious materials is predominantly water proton resonance imaging, it has also been used to observe the migration of sodium ions as an aqueous solution of sodium chloride is absorbed into bricks [6.74] as well as imaging chloride ion distributions.

6.4.3.2 Fast imaging methods

Fast imaging methods have been developed and can be applied to samples with good signal-to-noise ratios and long relaxation times. All imaging methods can also be performed with magnetic field gradients not in the B_0 field, but in the B_1 field; methods of this kind are referred to as *rotating-frame imaging* [6.82].

Magic angle spinning (MAS) is used to rotate a sample about an axis inclined at 54.7 (the magic angle) with respect to the magnetic field and in this way it averages away the anisotropic effects artificially [6.82,6.15]. The combined rotation and multipulse spectroscopy (CRAMPS) technique is an NMR technique which combines the MAS rotation of the sample with a multipulse sequence. It is therefore possible by proton NMR to distinguish in the spectra the ^1H lines corresponding to Ca–OH, Si–OH, and H–OH groups in the studied samples and to estimate their population.

A fundamental extension of MRI is *combined relaxographic imaging* (CRI), which was described by C. Labadie *et al.*, 1994 [6.90]. A relaxogram describes the probability distribution of spins having different values of relaxation times. Combined relaxography and imaging data can be displayed in two fundamental ways: one can display localized relaxograms from portions of an image, or make images from discrete portions of a relaxogram.

Magnetic resonance images depend on many variables and image quality is generally described in terms of the signal-to-noise ratio (SNR), spatial resolution, and contrast. The SNR determines how grainy the images appear: the less the SNR, the more grainy the image. Spatial resolution determines how sharp the image looks, and in MRI, resolution is bad. Structures of less than $50\,\mu m$ in solids and less than $5\,\mu m$ in liquids can hardly be resolved. The method has been successfully applied for pore sizes ranging from few nanometers up to about $10\,\mu m$. The image contrast is determined by the density of the observed nucleus, the relaxation times T_1 and T_2, the self-diffusion constant, the chemical shifts and indirect couplings of nuclear spins. Magnetization filters are given by pulse sequences and are used to generate contrast in imaging. Detailed information of filters applied can be found elsewhere.

6.4.3.3 Application to porous materials

MRI has found applications for materials characterization, to observe the spatial distribution of a liquid phase inside a porous material, such as building materials, soils, and oil reservoir rocks. The nucleus that is easier to observe is the proton ^1H, but images of other nuclei (^{23}Na, ^{19}F, ^{31}P) can also be made, but with a lower sensitivity. As compared to other tomographic techniques, MRI gives access not only to the 3-dimensional distribution of a liquid phase, but also to physico-chemical information through the NMR spectra and the NMR relaxation times.

Several researchers have used MRI for applications to cement-based materials. Considering the cement paste inhomogeneity and spin relaxation characteristics of water in porous media, the primary challenge of imaging concrete samples by MRI is overcoming the short FID time limitations. Some of the applications of MRI on cement-based materials have been the invasion of polymer coating material into a deuterated concrete using STRAFI, imaging water in concrete samples using SPI [6.68], and monitor chloride ion transportation in cement mortars [6.91].

Relaxation tomography has been proposed recently as a technique for obtaining basic structural information of porous materials [6.92,6.93]. The technique is a combination of relaxation analysis and imaging and is able to quantify the parameters connected to the local structure in the internal regions of a porous material saturated with water. Relaxation tomography can describe how porosity changes for different values of surface area–pore volume ratios of different materials, providing in this way information on the material's pore structure.

6.5   Advantages and limitations

Nuclear magnetic resonance is a fast, nondestructive technique, with no limitation on sample size. Prior to testing no drying is necessary which might alter the microstructure of cement pastes due to structure collapse. Actually, the sample must be water-saturated, if all the pores are to be tallied. In addition, the technique can be used to monitor evolution of the pore structure during hydration.

Measuring the spin–lattice relaxation time T_1 of a liquid contained in the pore space of a solid allows the determination of pore structure without making assumptions about the pore shape.

NMR has proved a very valuable tool for the investigation of both the pore size distribution within the cement paste using cryoporometry and also for monitoring the ingress of water and ions. However, the very short spin–spin relaxation times, and the very wide range of pore sizes encountered in cement pastes, introduce difficulties in analyzing the results.

The fact that the existing pore solution is not water, but concentrated alkali hydroxide, may influence the results. In addition, the internal porosity of the cement paste, with its complex and highly tortuous geometry may make the measurements difficult to interpret. In building materials, the presence of paramagnetic impurities due to high amounts of iron oxides result in very large internal field gradients, which may lead to a rapid dephasing of the nuclear spins and hence a rapid decay of the transverse nuclear magnetization during the diffusion in the pores. The effects of impurities complicate the analysis of the experimental data, since many of the theoretically developed models do not explicitly take these effects into account. Proton relaxation enhancement of pore water, due to the reduced mobility of water molecules near the solid surface, and the consequent linear relationship between the relaxation time and the pore dimension can be utilized to estimate the pore size.

An experimental constraint specific to cement-based materials, which present much stronger susceptibility inhomogeneity effects than most biological samples, is the short lifetime T_2^* of the transverse magnetization over excitation. The consequent magnetic field inhomogeneities cause a rapid signal decay, which can be partly compensated using Hahn spin-echoes.

The primary disadvantage of MRI is the complicated equipment, high associated costs and significant expertise in order to obtain reliable data. As a result, the pore structure determined using NMR has primary applications for studies of particular scientific interest rather than for regular studies on large numbers of samples. The use of MRI for detection of structural defects in materials is not likely to be very useful with the present state-of-the-art and limited resolution.

Cryoporometry is a nondestructive technique, and is suitable for spatial imaging of pore diameters in the range of a few nanometers to around 100 nm. Cryoporometry is a relatively rapid technique, it takes about 2 hours for one experiment, and gives reproducible and accurate results. Analysis of results can lead to determination of total pore volume and pore size distribution. Results are consistent with those of the conventional gas-desorption method. The method offers interesting possibilities, if combined with the spatially resolved methods of MRI to study the pore size variation in macroscopically inhomogeneous materials such as rocks and minerals. Cryoporometry appears to be independent of surface morphology for saturated samples. However, the technique is relatively new and experimental results are rather limited for meaningful comparisons.

The NMR technique and its associated variations is under continuous development and new pulses are used to compensate limitations of the technique during applications to porous materials, including cement pastes. In addition, combinations of techniques have proved useful in exploring the potential of the method.

References

6.1   Bloch F., Hansen W.W., Packard M. "Nuclear induction," *Physical Review*, Vol. 69, No. 3–4, 1946, pp. 127.

6.2   Bloch F. "Nuclear induction," *Physical Review*, Vol. 70, No. 7–8, 1946, pp. 460–474.

6.3   Purcell E.M., Torrey H.C., Pound R.V. "Resonance absorption by nuclear magnetic moments in a solid," *Physical Review*, Vol. 69, No. 1–2, 1946, pp. 37–38.

6.4   Ernst R.R., Anderson W.A. "Application of Fourier transform spectroscopy to magnetic resonance," *The Review of Scientific Instruments*, Vol. 37, No. 1, 1966, pp. 93–102.

6.5   Ernst R.R., Bodenhausen G., Wokaun A., *Principles of Nuclear Magnetic Resonance in One and Two Dimensions*, Clarendon Press, Oxford, 1987.

6.6   Lauterbur P.C. "Image formation by induced local interactions: examples employing nuclear magnetic resonance," *Nature*, Vol. 242, 1973, pp. 190–191.

6.7   Mansfield P., Grannell P.K. "NMR 'diffraction' in solids?," *Journal of Physics C: Solid State Physics*, Vol. 6, 1973, pp. L422–L426.

6.8   Atkins P.W. "Chapter 18: Magnetic resonance," *Physical Chemistry*, 4th Edition, W.H. Freeman and Company, New York, 1990, pp. 533–565.

6.9   Seligmann P. "Nuclear magnetic resonance studies of the water in hardened cement paste," *Journal of the PCA Research and Development Laboratories*, Vol. 10, No. 1, pp. 52–65 (January 1968), Reprinted as Bulletin 222, Portland Cement Association, 1968.

6.10 Atta-ur-Rahman. *Nuclear Magnetic Resonance, Basic Principles*, Springer-Verlag, New York, 1986.

6.11 Farrar T.C. *Pulse Nuclear Magnetic Resonance Spectroscopy; An Introduction to the Theory and Applications*, Farragut Press, Madison, WI, 1989.

6.12 Hore P.J. *Nuclear Magnetic Resonance*, Oxford University Press, Oxford, 1995.

6.13 Hore P.J., Jones J.A., Wimperis S. *NMR: The Toolkit*, Oxford University Press, Oxford, 2000.

6.14 Hornak J.P. *The Basics of NMR*, 1999, Rochester Institute of Technology, Rochester, NY (available at http://www.cis.rit.edu/htbooks/nmr/bnmr.htm).

6.15 Chmelka B.F., Pines A. "Some developments in nuclear magnetic resonance in solids," *Science*, Vol. 246, No. 4926, 1989, pp. 71–77.

6.16 Granger P. "Nuclear magnetic resonance basic principles," in *Application of NMR Spectroscopy to Cement Science*, P. Colombet, A.-R. Grimmer (eds), Gordon and Breach Science Publishers, New York, 1994, pp. 5–28.

6.17 Freeman R. *Magnetic Resonance in Chemistry and Medicine*, Oxford University Press, Oxford, 2003.

6.18 Leyden D.E., Cox R.H. "Chapter 3: Experimental and instrumental aspects of NMR," in *Analytical Applications of NMR*, John Wiley & Sons, New York, 1977, pp. 49–94.

6.19 Farrar T., Becker E.D. *Pulse and Fourier Tranform NMR-Introduction to Theory and Methods*, Academic Press, New York 1971.

6.20 Carr H.Y., Purcell E.M. "Effects of diffusion on free precession in nuclear magnetic resonance experiments," *Physical Review*, Vol. 94, No. 3, 1954, pp. 630–638.

6.21 Freeman R. *A Handbook of Nuclear Magnetic Resonance*, Longman Scientific & Technical, Essex, England, 1988.

6.22 Abragam A. "The Principles of Nuclear Magnetism," *The International Series of Monographs on Physics 32*, Oxford Science Publications, Clarendon, Oxford, 1989.

6.23 Watson A.T., Chang C.T.P. "Characterizing porous media with NMR methods," *Progress in Nuclear Magnetic Resonance Spectroscopy*, Vol. 31, 1997, pp. 343–386.

6.24 Pel L., Kopinga K., Brocken H. "Determination of moisture profiles in porous building materials by NMR," in *Proceedings of the International Symposium on Non-Destructive Testing in Civil Engineering*, September 26–28, 1995, Berlin, Germany, pp. 189–196.

6.25 Lahajnar G., Blinc R., Rutar V., Smolej V., Zupancic I., Kocuvan I., Ursic J. "On the use of pulse NMR techniques for the study of cement hydration," *Cement and Concrete Research*, Vol. 7, No. 4, 1977, pp. 385–394.

6.26 Gallegos D.P., Smith D.M. "A NMR technique for the analysis of pore structure: determination of continuous pore size distributions," *Journal of Colloid and Interface Science*, Vol. 122, No. 1, 1988, pp. 143–153.

6.27 Ackerman J.L., Garrido L., Ellingson W.A., Weyand J.D. "The use of NMR imaging to measure porosity and binder distribution in green state and partially sintered ceramics," in *Proceedings of the Conference on Nondestructive Testing of High Performance Ceramics*, Boston, MA, 1987, A. Vary, J. Snyder (eds), *The American Ceramic Society*, Westerville, OH, pp. 88–113.

6.28 Bhattacharja S., Moukwa M., D'Orazio F., Jehng J.Y., Halperin W.P. "Microstructure determination of cement pastes by NMR and conventional techniques," *Advanced Cement Based Materials*, Vol. 1, 1993, pp. 67–76.

6.29 Schreiner L.J., Mactavish J.C., Miljković L., Reeves L.W. "NMR line shape-spin-lattice relaxation correlation study of Portland cement hydration," *Journal of The American Ceramic Society*, Vol. 68, No. 1, 1985, pp. 10–16.

6.30 Prado P.J., Balcom B.J., Beyea S.D., Bremner T.W., Armstrong R.L., Pishe R., Gratten-Bellew P.E. "Spatially resolved relaxometry and pore size distribution by single-point MRI methods: porous media calorimetry," *Journal of Physics D: Applied Physics*, Vol. 31, No. 16, 1998, pp. 2040–2050.

6.31 Korringa J., Seevers D.O., Torrey H.C. "Theory of spin pumping and relaxation in systems with a low concentration of electron spin resonance centers," *Physical Review*, Vol. 127, No. 4, 1962, pp. 1143–1150.

6.32 Kleinberg R.L., Kenyon W.E., Mitra P.P. "Mechanism of NMR relaxation of fluids in rock," *Journal of Magnetic Resonance Series A*, Vol. 108, No. 2, 1994, pp. 206–214.

6.33 Leventis A., Verganelakis D.A., Halse M.R., Webber J.B., Strange J.H. "Capillary imbibition and pore characterisation in cement pastes," *Transport in Porous Media*, Vol. 39, No. 2, 2000, pp. 143–157.

6.34 McDonald P., Strange J. "Magnetic resonance and porous materials," *Physics World*, July 1998, pp. 29–34.

6.35 Hürlimann M.D., Helmer K.G., Latour L.L., Sotak C.H., "Restricted diffusion in sedimentary rocks. Determination of surface-area-to-volume ratio and surface relaxivity," *Journal of Magnetic Resonance Series A*, Vol. 111, No. 2, 1994, pp. 169–178.

6.36 Chapman R., Bloom M. "Nuclear spin-lattice relaxation of ^{3}He on neon surfaces," *Canadian Journal of Physics*, Vol. 54, No. 1, 1976, pp. 861–869.

6.37 Banavar J.R., Schwartz L.M. "Chapter 10: Probing porous media with nuclear magnetic resonance," in *Molecular Dynamics in Restricted Geometries*, J. Klafter, J.M. Drake (eds), John Wiley & Sons, New York, 1989, pp. 273–309.

6.38 Palmer R.G., Stein D.L., Abrahams E., Anderson P.W. "Models of hierarchically constrained dynamics for glassy relaxation," *Physical Review Letters*, Vol. 53, No. 10, 1984, pp. 958–961.

6.39 Peyron M., Pierens G.K., Lucas A.J., Hall L.D., Stewart R.C. "The modified stretched-exponential model for characterization of NMR relaxation in porous media," *Journal of Magnetic Resonance Series A*, Vol. 118, No. 2, 1996, pp. 214–220.

6.40 Kenyon W.E., Day P.I., Straley C., Willemsen J.F. "A three-part study of NMR longitudinal relaxation properties of water-saturated sandstones," *SPE Formation Evaluation*, Vol. 3, No. 3, 1988, pp. 622–636 and Erratum, Vol. 4, No. 1, 1989, p. 8.

6.41 Halperin W.P., D'Orazio F., Bhattacharja S., Tarczon J.C. "Chapter 11: Magnetic resonance relaxation analysis of porous media," in *Molecular Dynamics in Restricted Geometries*, J. Klafter, J.M. Drake (eds), John Wiley & Sons, New York, 1989, pp. 311–350.

6.42 Davies S., Packer K.J. "Pore-size distributions from nuclear magnetic resonance spin-lattice relaxation measurements of fluid-saturated porous solids. I. Theory and Simulation," *Journal of Applied Physics*, Vol. 67, No. 6, 1990, pp. 3163–3170.

6.43 Strange J.H., Webber J.B.W., Schmidt S.D. "Pore size distribution mapping," *Magnetic Resonance Imaging*, Vol. 14, Nos 7–8, 1996, pp. 803–805.

6.44 Gallegos D.P., Munn K., Smith D.M., Stermer D.L. "A NMR technique for the analysis of pore structure: application to materials with well-defined pore structure," *Journal of Colloid and Interface Science*, Vol. 119, No. 1, 1987, pp. 127–140.

6.45 Ewing B., Davis P.J., Majors P.D., Drobny G.P., Smith D.M., Earl W.L. "Determination of spatially resolved pore size information," *Studies in Surface Science and Catalysis, Vol. 62: Characterization of Porous Solids II, Proceedings of the IUPAC Symposium*, Alicante, Spain, May 6–9, 1990, F. Rodriguez-Reinoso, J. Rouquerol, K.S. Sing, K.K. Unger (eds), Elsevier Science Publishers, 1991, pp. 709–716.

6.46   Senturia S.D., Robinson J.D. "Nuclear spin-lattice relaxation of liquids confined in porous solids," *Society of Petroleum Engineers Journal*, Vol. 10, No. 3, 1970, pp. 237–244.

6.47   Loren J.D., Robinson J.D. "Relations between pore size fluid and matrix properties, and NMR measurements," *Society of Petroleum Engineers Journal*, Vol. 10, No. 3, 1970, pp. 268–278.

6.48   Brownstein K.R., Tarr C.E. "Spin-lattice relaxation in a system governed by diffusion," *Journal of Magnetic Resonance*, Vol. 26, No. 1, 1977, pp. 17–24.

6.49   Cohen M.H., Mendelson K.S. "Nuclear magnetic relaxation and the internal geometry of sedimentary rocks," *Journal of Applied Physics*, Vol. 53, No. 2, 1982, pp. 1127–1135.

6.50   Mendelson K.S. "Nuclear magnetic resonance in sedimentary rocks: effect of proton desorption rate," *Journal of Applied Physics*, Vol. 53, No. 9, 1982, pp. 6465–6466.

6.51   Mendelson K.S. "Nuclear magnetic relaxation in fractal pores," *Physical Review B*, Vol. 34, No. 9, 1986, pp. 6503–6505.

6.52   Brownstein K.R., Tarr C.E. "Importance of classical diffusion in NMR studies of water in biological cells," *Physical Review A*, Vol. 19, No. 6, 1979, pp. 2446–2453.

6.53   Davies S., Kalam M.Z., Packer K.J., Zelaya F.O. "Pore-size distributions from nuclear magnetic resonance spin-lattice relaxation measurements of fluid-saturated porous solids. II. Applications to reservoir core samples," *Journal of Applied Physics*, Vol. 67, No. 6, 1990, pp. 3171–3176.

6.54   Gallegos D.P., Smith D.M., Brinker C.J. "An NMR technique for the analysis of pore structure: application to mesopores and micropores," *Journal of Colloid and Interface Science*, Vol. 124, No. 1, 1988, pp. 186–198.

6.55   D'Orazio F., Bhattacharja S., Halperin W.P., Eguchi K., Mizusaki T. "Molecular diffusion and nuclear magnetic resonance relaxation of water in unsaturated porous silica glass," *Physical Review B*, Vol. 42, No. 16, 1990, pp. 9810–9818.

6.56   Halperin W.P., Bhattacharja S., D'Orazio F. "Relaxation and dynamical properties of water in partially filled porous materials using NMR techniques," *Magnetic Resonance Imaging*, Vol. 9, 1991, pp. 733–737.

6.57   Bryar T.R., Daughney C.J., Knight R.J. "Paramagnetic effects of iron(III) species on nuclear magnetic relaxation of fluid protons in porous media," *Journal of Magnetic Resonance*, Vol. 142, 2000, pp. 74–85.

6.58   Valckenborg R.M.E., Pel L., Hazrati K., Kopinga K., Marchand J. "Pore water distribution in mortar during drying as determined by NMR," *Materials and Structures*, Vol. 34, No. 244, 2001, pp. 599–604.

6.59   Allen S.G., Stephenson P.C.L., Strange J.H. "Morphology of porous media studied by nuclear magnetic resonance," *Journal of Chemical Physics*, Vol. 106, No. 18, 1997, pp. 7802–7809.

6.60   D'Orazio F., Tarczon J.C., Halperin W. P., Eguchi K., Mizusaki T. "Application of nuclear magnetic resonance pore structure analysis to porous silica glass," *Journal of Applied Physics*, Vol. 65, No. 2, 1989, pp. 742–751.

6.61   Beyea S.D., Balcom B.J., Bremner T.W., Prado P.J., Green D.P., Armstrong R.L., Grattan-Bellew P.E. "Magnetic resonance imaging and moisture content profiles of drying concrete," *Cement and Concrete Research*, Vol. 28, No. 3, 1998, pp. 453–463.

6.62   Halperin W.P., Jehng J.-Y., Song Y.-Q. "Application of spin-spin relaxation to measurement of surface area and pore size distributions in a hydrating cement paste," *Magnetic Resonance Imaging*, Vol. 11, No. 2, 1994, pp. 169–173.

6.63   Porteneuve C., Korb J.-P., Petit D., Zanni H. "Structure-texture correlation in ultra-high-performance concrete: a nuclear magnetic resonance study," *Cement and Concrete Research*, Vol. 32, No. 1, 2002, pp. 97–101.

6.64   Pel L., Kopinga K., Hazrati K. "Water distribution and pore structure in concrete as determined by NMR," 9. Feuchtetag, September 17–18, 1997, Weimar, Germany, pp. 294–300.

6.65 Pel L., Hazrati K., Kopinga K., Marchand J. "Water absorption in mortar determined by NMR," *Magnetic Resonance Imaging*, Vol. 16, Nos 5–6, 1998, pp. 525–528.

6.66 Hansen T.C. "Physical structure of hardened cement paste. A classical approach," *Materials and Structures*, Vol. 19, No. 114, 1986, pp. 423–436.

6.67 Bohris A.J., Goerke U., McDonald P.J., Mulheron M., Newling B., Le Page B. "A broad line NMR and MRI study of water and water transport in Portland cement pastes," *Magnetic Resonance Imaging*, Vol. 16, Nos 5–6, 1998, pp. 455–461.

6.68 Tritt-Goc J., Piślewski N., Kościelski S., Milia F. "The influence of the superplasticizer on the hydration and freezing processes in white cement studied by ^1H spin-lattice relaxation time and single point imaging," *Cement and Concrete Research*, Vol. 30, No. 6, 2000, pp. 931–936.

6.69 Englert G., Wittmann F. "Zum Studium der Eigenschaften des adsorbierten Wassers in hydratisiertem Tricalciumsilikat," *Zement-Kalks-Gips*, Vol. 24, No. 7, 1971, pp. 312–316.

6.70 Nestle N. "A simple semiempiric model for NMR relaxometry data of hydrating cement pastes," *Cement and Concrete Research*, Vol. 34, No. 3, 2004, pp. 447–454.

6.71 Strange J.H., Rahman M., Smith E.G. "Characterization of porous solids by NMR," *Physical Review Letters*, Vol. 71, No. 21, 1993, pp. 3589–3591.

6.72 Strange J.H., Mitchell J., Webber J.B.W. "Pore surface exploration by NMR," *Magnetic Resonance Imaging*, Vol. 21, Nos 3–4, 2003, pp. 221–226.

6.73 Jehng J.-Y., Sprague D.T., Halperin W.P. "Pore structure of hydrating cement paste by magnetic resonance relaxation analysis and freezing," *Magnetic Resonance Imaging*, Vol. 14, Nos 7–8, 1996, pp. 785–791.

6.74 Valckenborg R., Pel L., Kopinga K. "Cryoporometry and relaxation of water in porous materials," in *Proceedings of the Fifteenth European Experimental NMR Conference (EENC 2000)*, 12–17 June 2000, University of Leipzig, Germany.

6.75 Valckenborg R.M.E., Pel L., Kopinga K. "Combined NMR cryoporometry and relaxometry," *Journal of Physics D: Applied Physics*, Vol. 35, 2002, pp. 249–256.

6.76 Jackson C.L., McKenna G.B. "The melting behavior of organic materials confined in porous solids," *Journal of Chemical Physics*, Vol. 93, No. 12, 1990, pp. 9002–9011.

6.77 Schmidt R., Hansen E.W., Stöcker M., Akporiaye D., Ellestad O.H. "Pore size determination of MCM-51 mesoporous materials by means of ^1H NMR spectroscopy, N2 adsorption, and HREM. A preliminary study," *Journal of the American Chemical Society*, Vol. 117, No. 14, 1995, pp. 4049–4056.

6.78 Prado P.J., Balcom B.J., Beyea S.D., Bremner T.W., Armstrong R.L., Grattan-Bellew P.E. "Concrete freeze/thaw as studied by magnetic resonance imaging," *Cement and Concrete Research*, Vol. 28, No. 2, 1998, pp. 261–270.

6.79 Choi C., Balcom B.J., Beyea S.D., Bremner T.W., Grattan-Bellew P.E., Armstrong R.L. "Spacially resolved pore size distribution of drying concrete with magnetic resonance imaging," *Journal of Applied Physics*, Vol. 88, No. 6, 2000, pp. 3578–3581.

6.80 Gummerson R.J., Hall C., Hoff W.D., Hawkes R., Holland G.N., Moore W.S., "Unsaturated water flow within porous materials observed by NMR imaging," *Nature*, Vol. 281, 1979, pp. 56–57.

6.81 Callaghan P.T., *Principles of Nuclear Magnetic Resonance Microscopy*, Oxford University Press, Oxford, 1991.

6.82 Blümich B. *NMR Imaging of Materials, Monographs on the Physics and Chemistry of Materials: 57*, Oxford University Press, Oxford, 2000.

6.83 Samoilenko A.A., Artemov D. Yu., Sibel'dina L.A. "Formation of sensitive layer in experiments on NMR subsurface imaging of solids, *Journal of Experimental and Theoretical Physics Letters*, Vol. 47, 1988, pp. 417–419.

6.84 Cottrell S.P., Halse M.R., Strange J.H. "NMR imaging of solids using large oscillating field gradients," *Measurement Science and Technology*, Vol. 1, No. 7, 1990, pp. 624–629.

6.85   Balcom B.J., Macgregor R.P., Beyea S.D., Green D.P., Armstrong R.L., Bremner T.W. "Single-point ramped imaging with T_1 enhancement (SPRITE)," *Journal of Magnetic Resonance Series A*, Vol. 123, No. 1, 1996, pp. 131–134.

6.86   Beyea S.D., Balcom B.J., Prado P.J., Cross A.R., Kennedy C.B., Armstrong R.L., Bremner T.W. "Relaxation time mapping of short T_2^* nuclei with single-point imaging (SPI) methods," *Journal of Magnetic Resonance*, Vol. 135, No. 1, 1998, pp. 156–164.

6.87   Emid S., Creyghton J.H.N. "High resolution NMR imaging in solids," *Physica B*, Vol. 128, No. 1, 1985, pp. 81–83.

6.88   Bogdan M., Balcom B.J., Bremner T.W., Armstrong R.L. "Single-point imaging of partially dried, hydrated white Portland cement," *Journal of Magnetic Resonance Series A*, Vol. 116, No. 2, 1995, pp. 266–269.

6.89   Gravina S., Cory D.G. "Sensitivity and resolution of constant-time imaging," *Journal of Magnetic Resonance Series B*, Vol. 104, No. 1, 1994, pp. 53–61.

6.90   Labadie C., Lee J.-H., Vétek G., Springer C.S. "Relaxographic imaging," *Journal of Magnetic Resonance Series B*, Vol. 105, No. 2, 1994, pp. 99–112.

6.91   Cano F. de J., Bremner T.W., McGregor R.P., Balcom B.J. "Magnetic resonance imaging of ^1H, ^{23}Na, and ^{35}Cl penetration in Portland cement mortar," *Cement and Concrete Research*, Vol. 32, 2002, pp. 1067–1070.

6.92   Borgia G.C., Bortolotti V., Fantazzini P. "Changes of the local pore space structure quantified in heterogeneous porous media by H-1 magnetic resonance relaxation tomography," *Journal of Applied Physics*, Vol. 90, No. 3, 2001, pp. 1155–1163.

6.93   Borgia G.C., Bortolotti V., Fantazzini P., Gombia M., Zaniboni M. "Improved pore space structure characterization by fusion of relaxation tomography maps," *Magnetic Resonance Imaging*, Vol. 21, Nos 3–4, 2003, pp. 393–394.

7   Small-angle scattering

Small-angle scattering (SAS) techniques are powerful tools for characterizing complex microstructures and studying porous materials. A. Guinier first published his work on SAS studies in 1937 [7.1]. The fundamentals of SAS for both X-rays and neutrons have been described in detail by A. Guinier and G. Fournet, 1955 [7.2]. Several researchers have developed since then theoretically and experimentally the SAS technique [7.3,7.4,7.5,7.6] and good general introductions on SAS are given in the technical literature [7.7,7.8]. H. Ritter and L. Erich carried out the first small-angle X-ray scattering (SAXS) experiments on porous materials in 1948 [7.9]. The first small-angle neutron scattering (SANS) spectrometers were built in the 1970s in Germany and France. They proved very popular and they were soon built at most neutron scattering centers, including those in the United States. Due to their high cost of operation, the peculiarities of the experimental facilities, and the particular cycle of operation of a neutron source, a SANS experiment needs to be planned carefully and time at the facility be reserved well in advance. In scheduling time for a neutron scattering experiment, a scientist must arrange the most efficient use of the instrument for the benefit of all users.

Since 1980, SANS has developed into an important characterization method for the investigation of microstructure in solid state physics, materials science, metallurgy, chemistry, and biology. SANS has been used very successfully for the study of voids and radiation damage effects in various materials. The technique can probe inhomogeneities from the near atomic scale (1 nm) to the near micron scale (600 nm) in a variety of materials, such as polymers, ceramics, metals, and composites. The scattering through a small angle from the incident beam provides information about the state of the specimen over distances larger than few atomic diameters. SANS and SAXS give information on the size, shape, concentration, and surface area of inhomogeneities in materials including pores. SANS is particularly valuable for use with a material when X-rays cannot provide the desired information either because of lack of scattering contrast or because of too severe absorption from the material. Neutron absorption is low in most materials and samples few mm thick can be investigated [7.10]. Neutrons can penetrate thicker samples than X-rays or electrons, since they carry no electrical charge or field; however, neutrons are difficult to produce and the flux is low. Neutrons of 1 Å wavelength have kinetic energy

82 meV, which is much lower than the energy of X-rays and electrons of the same wavelength (12 keV and 150 eV respectively).

SAXS was systematically applied to cement paste for the first time by D. Winslow and S. Diamond in 1974 [7.11]. SANS was introduced to cement pastes by A. Allen *et al.*, 1982 [7.12] and has been used since then by several researchers. With SANS, information can be obtained at the fine end of the pore size distribution (less than approximately 30 nm diameter) on cement pastes without any specimen drying or pretreatment prior to the test.

In cement pastes, the scattering of both X-rays and neutrons is dominated by the interface between C–S–H and the pore medium (water or air) [7.13]. SANS has been used extensively in recent years to determine the surface and volume fractal dimensions of the C–S–H gel as a function of time and as a function of the content of mineral admixtures, such as fly ash or silica fume. In addition, SANS has also been applied for durability applications, e.g. to characterize silica aggregates for the potential of deleterious reactions in highly alkaline concrete [7.14].

7.1 Theoretical aspects

There are many similarities between the principles of small-angle scattering of neutrons, X-rays, and electrons (see Table 7.1) that would allow an experimentalist familiar with SAXS to contemplate a SANS experiment without too much hardship. X-rays have wavelengths 0.1–5 Å, neutrons 1–15 Å, and electron beams 0.05–1 Å. Interatomic distances are of the same order of magnitude as the wavelength of the incident beams. Neutrons can penetrate several centimeters deep into most materials, whereas electrons or X-rays of comparable wavelength are confined to the surface. Scattering arises from induced dipoles of atomic electrons. The scatters are approximately the same size as the incident wavelength, so a decay in the scattered signal occurs as the angle varies. Scattering from dislocations is weak and difficult to analyze in detail, but empirical relationships between SAS and deformation parameters may be useful in applied research.

The neutron was discovered by J. Chadwick in 1932. Neutrons and protons are the building units of the atomic nucleus. Unlike the proton, the neutron is a neutral particle, which possesses no electrical charge, has a spin of $\frac{1}{2}$, a substantial magnetic moment of -1.9132 nuclear magnetons, and weighs slightly more than the proton, i.e. 1.00867 atomic mass units [7.15]. Neutrons are

Table 7.1 Some characteristics of the different radiations employed in scattering methods [7.15]

Type of radiation	X-rays	Neutrons	Electrons
Wavelength range	0.1–5 Å	1–15 Å	0.1 Å
Source of scattering	Electron density	Density of nuclei	Electron cloud
Sample thickness	<1 mm	1–2 mm	100 μm
Problems	Absorption	Low fluxes	Low penetration

unstable particles, having a half life of approximately 12 minutes and decay into a proton, electron, and a neutrino. Neutrons can be produced either as a product of nuclear fission in a reactor, or by a spallation process, and have energies that depend on the temperature of the moderator used to reduce their energies (see also Section 7.2.2). The energies of neutrons are comparable to vibrational and diffusional energies of molecular systems and can be used to detect molecular motion occurring in the frequency range of 10^7–10^{14} Hz. Different types of neutrons can be produced depending on their energies, such as ultra-cold neutrons, slow neutrons, fast neutrons, epithermal neutrons etc. The types of neutrons used in SANS experiments are cold neutrons and thermal neutrons. Cold neutrons have energies that range between 5×10^{-5} eV and 0.025 eV; thermal neutrons have energies approximately equal to 0.025 eV. Neutrons also possess wavelengths that are ideally suited to allow measurement of structural information over a wide range, covering the wave function of hydrogen to structural morphology of macromolecules.

When an incident wave of X-rays or neutrons with a wavelength λ passes through a material, the intensity of the beam decreases as a result of its inter-action with the material. The decrease in intensity is due to two effects: (a) a change in the direction of the scattered particle, i.e. a scattering phenomenon, and (b) absorption, which is associated with the disappearance of the photons or neutrons [7.8]. Depending on the interaction of the beam with the speci-men, the resulting scattering can be elastic or inelastic. *Elastic* scattering means that there is no loss of energy in the process of scattering, and, therefore, the scattered beam of neutrons or X-rays will have a wavelength that is the same as the wavelength of the incident beam. When *inelastic* scattering takes place, the incident neutrons or X-rays lose or indeed gain energy by exchange with a quantum of vibrational energy. The inelastically scattered neutrons or X-rays will be of lower or higher energy, i.e. of longer or shorter wavelength respec-tively. A neutron of wavelength 1 Å has an energy of 0.08 eV, while an X-ray quantum of wavelength 1 Å has energy about 12 keV; this means that the energy of X-rays is 150 000 times greater than the energy of a neutron of the same wavelength [7.15]. When an X-ray is inelastically scattered, the change of energy or wavelength is too small to be detected experimentally. On the contrary, inelastic scattering of a neutron produces changes of energy and wavelength, which can be easily and accurately measured [7.16]. Therefore, neutrons, but not X-rays, are extremely powerful for investigating energy changes due to scattering.

Typically, the wavelength of the X-rays used in SAS is up to one order of magnitude smaller than neutron wavelengths. SAXS experiments generally use either CuK_α or MoK_α radiation (with wavelengths of 0.154 nm and 0.071 nm, respectively), or radiation of a similar wavelength at a synchrotron facility. SANS experiments on cement pastes typically use cold neutrons with wave-lengths of 0.5–0.8 nm. The scattering features at small angles correspond to structures ranging from tens to thousands of angstroms. The shorter the wave-length, the smaller the microstructural features are that can cause scattering at a given scattering angle. In SANS and SAXS studies, the effective maximum scattering angles and associated minimum microstructural dimensions studied are determined by the signal-to-noise ratio as the SAS signal decays away with increasing scattering angle [7.13]. The scatter is particularly noticeable at small angles, i.e. within a few degrees to the direction of the incident beam, and this is how the method received its name [7.8].

X-rays interact with orbital electrons of the atoms comprising the material through which they pass. The passage of X-rays through regions of differing electron density causes their scatter. Scatter at small angles occurs in addition to the Bragg diffraction at somewhat larger angles. Each electron in the scattering sample scatters like an induced dipole scatterer and emits X-rays. The X-rays scattered by different electrons travel through the sample and arrive at the detector with different phases. These phase differences must be taken into account when the amplitudes of the scattered waves are added together to calculate the resultant scattered amplitude [7.17]. Neutrons on the other hand, penetrate the electron cloud and interact directly with the atomic nucleus. Light scattering and SAXS studies rely on the differences in refractive indices or electron densities of the compound tested. Neutron scattering arises from differences in neutron scattering lengths. In both the SANS and SAXS cases, the scattering occurs at interfaces between two phases, and the data analysis for the two techniques is similar. The theory of X-ray SAS can be easily adapted to neutrons.

Neutrons are scattered either through collisions with atomic nuclei or by a magnetic field, if the atoms have unpaired electron spins. Since the neutron's interaction probability is small, the neutron usually penetrates well through matter making it a unique probe for investigating bulk condensed matter. Neutron scattering is therefore strongly dependent upon the phases present in a particular sample, upon the homogeneity of the various phases, and upon the phase contrast between the solid and pores. The force exerted by a nucleus on a neutron during the scattering process is described by a quantity which is known as the *scattering length*, l, and which depends on the particular isotope that causes the scattering. Different isotopes of the same element can have quite different neutron scattering lengths and there is no simple rule that expresses the dependence of the neutron scattering length on the atomic number. Scattering in cementitious materials is generally assumed to be dominated by the presence of small particles or fractal structures as opposed to pores or voids.

In the following sections, the SANS and the SAXS are going to be described together since the theory for the two methods is the same, and reference to neutron scattering is going to be made predominantly to facilitate the presentation of the topic. Some problems that may arise from different nomenclature that has been developed for the two radiations are pointed out. For example, for the scattering vector, k and Q have been used frequently in neutron scattering instead of h used in X-ray scattering. Another quantity treated differently in SANS and SAXS is the intensity of the diffracted beam: for X-rays, the intensity I is given in terms of electron units per atom, where the intensity per atom is compared to a number of diffracting electrons giving the same intensity. The corresponding quantity in neutron diffraction is the coherent differential cross section per atom, $d\sigma/d\Omega$, expressed in units of cm^2 per sterad per atom.

7.2 Experimental procedure

The equipment setup is different for the two techniques and are described in the following sections separately. The specimen used is rather small, usually a disc with 1 cm diameter and can be few mm thick for SANS experiments. However, in order to avoid the possibility of multiple scattering, very thin specimens are prepared (<1 mm). SAXS from cement paste is 10 times stronger than SANS and so is the absorption of X-rays from the material tested; therefore very thin samples are required to obtain X-ray data. Carbonation will occur very rapidly in the thin specimens involved, which alters the microstructure; for this reason, the specimen must be kept in a CO_2-free environment. It should also be noted that it is not required that the specimen be dried before the experiment.

7.2.1 Small-angle X-ray scattering equipment

In the SAXS experiment, an intense X-ray beam is focused on the sample. The beam then passes through the sample that is positioned at right angles to the radiation [7.18]. The sample's electrons scatter a portion of the incident flux and the scattered X-ray intensity is measured as a function of the scattering angle. An ionization counter, which is located about 0.5 m down-stream from the sample, detects the scattered X-rays. The detector is sequentially stepped along a path at right angles to the X-ray beam, and the intensities at various angles are recorded. A general experimental setup for X-ray scattering is shown in Figure 7.1 and includes the following components: conventional or synchrotron sources, monochromators, slits, focusing devices (mirrors and crystals), and detectors with position sensitivity [7.19].

X-ray sources A conventional X-ray source consists of the characteristic lines of the anode material and a broad spectrum, the so-called "brehmsstrahlung." The characteristic lines have intensities that are a few orders of magnitude more intense than the brehmsstrahlung and the SAXS instruments use the radiation from the characteristic lines. The most commonly used anode materials are Cu, Mo, and Cr, giving X-rays with wavelengths 1.54, 0.71, and 2.3 Å,

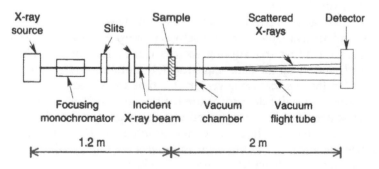

Figure 7.1 Schematic out-of-scale setup of SAXS equipment.

respectively. Long wavelengths make small values of the wave vector accessible for a certain collimation. The absorption of the X-rays in the sample increases as the wavelength increases, and this property often excludes the use of Cr radiation in SAXS experiments. The simplest source of X-rays is a sealed X-ray tube, which is cooled by circulating water.

Monochromators   A monochromator is used for energy selection of the synchrotron radiation. Depending upon the type of monochromator used, X-ray fluxes range from 7×10^{11} to 1×10^{12} photons per second. For conventional sources, the monochromatization procedure should pick out the characteristic line of the anode. Perfect crystals, like quartz, silicon, and germanium, are ideally suited for making focusing monochromators. One of the major problems when using synchrotron radiation is that the monochromator crystals and the mirrors can absorb a significant amount of heat, which can cause stability problems. Often the monochromator is water-cooled or cryogenically cooled.

Slits and collimation   The monochromatic radiation passes through a series of slits that form it into a beam with a rectangular cross section about $0.25 \, \text{mm} \times 20 \, \text{mm}$. Slits are employed to maintain reasonable intensities at high resolutions, however, slits cause a distortion or smearing of the observed pattern. The material for the slits should be strongly absorbing and therefore lead, tantalum, or tungsten are often used. The narrowly defined beam of X-rays i.e. produced impinges then on the sample. In theoretical considerations in the past, two shapes of the cross section of the incident beam of X-rays have been considered, with slight modification of the equations: (a) a long extremely narrow rectangular cross section, resulting from slit collimation, and (b) an extremely small circular cross section, resulting from pinhole collimation.

Focusing devices   In many pinhole cameras, the photon flux at the same position is increased by incorporating optical devices in the setup. The most obvious focusing device is an X-ray mirror usually made of fused quartz, which might be coated by a heavy metal to increase the critical angle.

The *sample* is contained in a special isolated cell with mica windows allowing transmission of the X-ray beam. The specimen is mounted transversely to the incident X-ray beam. A small fraction of the beam transmitted through the sample is scattered at angle 2θ with respect to the incident beam and is then registered by the detector.

Detectors   A great variety of detectors and detector systems have been developed and are available. For most setups that use Kratky and pinhole cameras the data for different scattering vectors can be recorded simultaneously, which greatly speeds up the data acquisition. For other cameras, the data have to be recorded for one scattering vector at a time in the sequential mode, and for these cameras a point detector is used. Photographic and electronic detection of the scattered radiation are used in SAS measurements. The scattering curve is obtained by recording the scattering intensity $I(Q)$ for several values of the scattering angle 2θ or of the scattering vector Q. Intensities are recorded over an angular range from $0°$ to about $2°$, which converts to Q values over a useful range from about 0.003 to $0.3 \, \text{Å}^{-1}$, i.e. about two orders of magnitude.

7.2.2 Small-angle neutron scattering equipment

One of the most striking observations about neutron scattering is the size of the apparatus: the distance from the moderator to the sample is typically between 10 and 120 m, while the distance between the sample and the detector could be between 0.5 and 40 m, depending on the type of the instrument [7.20]. SANS instruments, commonly include the following components: neutron sources, monochromators, slits and guides, and detectors (see Figure 7.2).

Neutron sources These are based on various processes that liberate excess electrons in neutron-rich nuclei, such as beryllium, tungsten, uranium, tantalum, and lead. Neutron sources can be of two types: a steady state fission reactor, or a spallation pulsed source. The techniques used for neutron scattering experiments at these two sources differ because the neutron spectra differ radically. At a steady state fission reactor, neutrons are produced continuously with a Maxwellian wavelength spectrum. At a spallation source, on the other hand, by bombarding a heavy metal target with energetic protons, neutrons are produced in short bursts ($<100\,\mu s$), 20–50 times per second, with a spectrum that extends to shorter wavelengths than the reactor Maxwellian. Neutrons are used more efficiently at spallation sources than at reactors. In order to have a low background environment, the SANS instrument is often placed in a guide hall, after a long neutron guide which can be curved or straight. The guide is often curved in order to prevent fast neutrons and gamma rays from the reactor core from reaching the detector [7.19]. Since the early days of neutron scattering, there has been an increasing demand for high neutron fluxes. Presently, the highest fluxes available are around a few $\times 10^{15}$ n/cm²/s [7.15].

Monochromators For instruments at steady state sources, a monochromator is a mechanical velocity selector, consisting of a rotating drum with helical

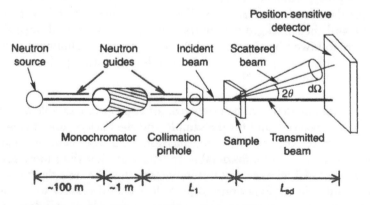

Figure 7.2 Schematic out-of-scale layout of the SANS experimental equipment. The distance between the monochromator and the sample, L_1, and the distance between the sample and the detector, L_{sd}, can be between 2 and 40 m, each. The angle between the direction of the incident beam and the scattered beam is 2θ and the solid angle is marked with $d\Omega$. The scattered beam and the detector are enclosed in a vacuum tube, not shown in the figure.

slots allowing only neutrons that have a wavelength and velocity that match the rotation speed of the monochromator to pass through it.

Slits and collimation   In SANS it is normal, but not necessary, to use an incident beam of neutrons with a circular cross section, with diameter usually between about 5 and 25 mm. Neutron beams cannot be focused, but small beam sizes can be achieved through the use of restrictive collimation. The neutrons are collimated by pinholes in a strongly neutron-absorbing material like boronated plastic or cadmium. The apertures can be rectangular or circular. The typical size for the first aperture is 2–5 cm and about 1 cm for the aperture in front of the sample [7.19]. The drawback with collimation is that the neutron flux after the collimator is reduced by a factor proportional to the cross-sectional area of the beam. Therefore, collimating the beam size from 12 mm down to 8 mm in diameter, reduces the flux at the sample by a factor of $36/16 = 2.25$. Increasing the beam size, however, degrades the scattering vector resolution, and so the choice of the beam size is always a compromise [7.20].

Sample container   Any kind of sample container can be used, either transparent or metallic. For typical experiments only about 100 mm^3 of sample is needed. Measurements of strong scatterers take about an hour to obtain at most neutron facilities. During the experimental procedure, light water inside the specimen can be replaced with heavy water, which is a much stronger neutron scatterer, thus giving higher contrast levels (see also Section 7.5) [7.21].

The *detectors* are typically multiwire two-dimensional and position-sensitive devices containing ^{3}He or BF$_3$ as the active gas. The incident neutrons are converted by nuclear absorption to charged particles, which can then be detected (e.g. ^{3}He is converted into triton ^{3}H$^-$, a proton H$^+$, and a gamma ray) [7.19]. The position-sensitive neutron detector covers an area of about 1 m^2 and is placed directly behind the sample. The spatial resolution of the detector is usually between 3 and 10 mm.

The *computers* control the acquisition and reduction of SANS data and record the data for further analysis. The data is collected and averaged by a computer and then background subtraction, desmearing, and normalization takes place, followed by Guinier and Porod plottings. Preliminary data treatment is a multi-step procedure and includes smoothing, desmearing and Fourier Transformation [7.7].

For every sample that is to be measured, an appropriate sample background must also be measured, and for the sample and the background it is necessary to measure both the neutron transmission and the neutron scattering. Transmission runs are of considerably shorter duration than most scattering runs, typically 5–10 minutes, but it is important that the time the transmission runs be included in the experimental plan. It is also necessary to measure the scattering and the transmission of a calibration sample or standard, and the transmission of a completely empty sample position.

During the experiment, a well-collimated beam of neutrons of known wavelength λ is allowed to impinge on the specimen. In practice, the wavelength λ may vary between a fraction of an Å and about 20 Å. Once a value of the wavelength λ has been selected, the scattering vector range of the instrument is fixed by the sample–detector distance, L_{sd}, and by the radial distance on the detector, r_{det}, at which neutrons are recorded. Neutrons arriving close to the center of the detector have been scattered to small scattering vector values. However, because the minimum and maximum values of the radial distance on the detector r_{det} are fixed, the corresponding range of the scattering vector is small. The only ways of expanding the range of the scattering vector are either by changing the wavelength λ of the neutron, or, more usually, by changing the distance between the sample and the detector, L_{sd}. Changing the wavelength λ will result in a change of neutron count rate, and this effect may be undesirable. By increasing the distance between the sample and the detector, neutrons scattered at a given angle appear at larger values of the radial distance on the detector, r_{det}; by moving the detector closer to the sample, the converse effect takes place. Therefore, the smallest values for the scattering vector are reached with the longest distance between the sample and the detector and the wavelength is the longest possible. Similarly, the largest values for the scattering vector are reached with the detector being as close to the sample as possible and with the shortest possible wavelength. In some instances it is also possible to move the detector off-axis by a few degrees, which has the same effect as changing the maximum and minimum limits of r_{det} [7.20].

7.2.3 Data collection

A beam of collimated neutrons of intensity I_0 and wave vector k_0, is incident at a sample having a small volume, V. Most neutrons are transmitted through the sample without any interactions, and some neutrons are absorbed. Some neutrons will be scattered and can be measured with a neutron detector placed in the direction of the scattered beam $\bar{k}_1$ [7.22]. During the scattering experiment, the neutron wave vector $\bar{k}_0$ suffers a change in direction (see Figure 7.3), which implies a change in the neutron momentum that must be exchanged with the sample. The momentum transferred to the sample is conventionally

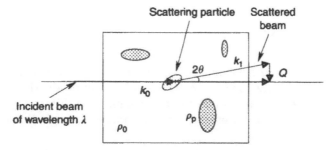

Figure 7.3 The geometry of the SANS experiment, showing the incident beam of neutrons with wavelength λ and the wave vector k_0. A fraction of the neutrons are scattered by a scattering particle through an angle 2θ. The wave vector of the scattered beam of neutrons is k_1, and the difference of the wave vectors of the incident and scattered beams is Q. The sample contains scattering particles of scattering length density ρ_p, which are dispersed in a matrix that has a scattering length density ρ_0.

described in terms of the corresponding wave vector Q. The neutrons scattered by the sample are measured as a function of the scattering angle 2θ. The scattering angle 2θ can be converted to the scattering vector Q by using Equation (7.1) [7.18]:

$$|\overline{Q}| = Q = \frac{4\pi}{\lambda} \sin\theta = [\overline{k}_0 - \overline{k}_1] \tag{7.1}$$

where $\overline{Q}$ is the scattering vector, λ is the wavelength of neutrons (or X-rays), 2θ is the scattering angle (degrees), $\overline{k}_0$ is the wave vector of the incident beam of neutrons, $\overline{k}_1$ is the wave vector of the scattered beam of neutrons.

The detector, or detector element, of dimensions $dx \times dy$ positioned at a distance L_{sd} from the sample and at scattering angle 2θ, records the intensity of neutrons, $I(\lambda,\theta)$, that are scattered into a solid angle element, $d\Omega$ (see also Figure 7.2). The scattered intensity $I(\lambda,\theta)$, can be expressed in general terms by Equation (7.2) [7.20]. The first three terms in Equation (7.3) are instrument specific, and the last three terms are sample-dependent:

$$I(\lambda,\theta) = I_0(\lambda)\Delta\Omega\eta(\lambda)TV\frac{d\sigma}{d\Omega}(Q) \tag{7.2}$$

where $I(\lambda,\theta)$ is the scattered neutron flux, $I_0(\lambda)$ is the incident neutron flux, λ is the wavelength of the neutron beam, $\Delta\Omega$ is the solid angle element defined by the size of a detector pixel, $\eta(\lambda)$ is the detector efficiency (sometimes also called the response), T is the neutron transmission of the sample, V is the sample volume illuminated by the neutron beam, $(d\sigma/d\Omega)(Q)$ is the microscopic differential scattering cross section, Q is the modulus of the scattering vector.

Sometimes, instead of the microscopic differential scattering cross section, $d\sigma/d\Omega$, the term $d\Sigma/d\Omega$ is used, which is known as the macroscopic differential cross section, and is expressed by Equation (7.3). Also, the product $N_P V_P$ is the same as the volume fraction of scattering bodies, ϕ_P [7.20]:

$$\frac{d\Sigma}{d\Omega}(Q) = N_P \frac{d\sigma}{d\Omega}(Q) \tag{7.3}$$

The objective of the SANS experiment is to determine the differential scattering cross section $(d\Sigma/d\Omega)(Q)$, so as to extract information on the shape, size, and interactions of the scattering bodies in the sample. The differential scattering cross section relates to parameters of the sample through Equation (7.4) [7.20,7.7]:

$$\frac{d\Sigma}{d\Omega}(Q) = N_P V_P^2(\Delta\rho)^2 P(Q)S(Q) + B_{inc} \tag{7.4}$$

where $(d\Sigma/d\Omega)(Q)$ is the differential scattering cross section, N_P is the number concentration of scattering particles, V_P is the volume of the scattering particle, $\Delta\rho$ is the difference in neutron scattering length density, $P(Q)$ is a function called the form factor or shape factor for each particle, $S(Q)$ is the inter-particle structure factor, B_{inc} is the incoherent scattering contribution to the total scattering (background signal).

A mixture of nomenclature is used in the technical literature treating SANS theory and experimental measurements. For example, the term $(d\sigma/d\Omega)(Q)$ is written as $(d\Sigma/d\Omega)(Q)$ or $I(Q)$. The term $P(Q)$ is expressed as $F(Q)$ or even $S(Q)$. And the scattering vector Q is represented also by the symbols q, h, k, or s [7.20].

A brief explanation of the parameters included in Equation (7.4) is given in the following paragraphs.

Differential scattering cross section The differential scattering cross section $[(d\Sigma/d\Omega)(Q)]$ is the dependent variable measured during a SANS experiment, and contains all the information on the size, shape, and interactions between the scattering particles in the sample. The variation of $d\Sigma/d\Omega$ with Q has dimensions of $(length)^{-1}$ is normally expressed in units of cm^{-1}, and is a quantity that can be routinely calibrated to yield absolute data*. In the case of isotropic SAS, $d\Sigma/d\Omega$ is a function only of the magnitude of the scattering vector Q [7.23]. The differential scattering cross section, $d\Sigma/d\Omega$, can be defined as the probability per unit time, sample volume, and neutron flux, that a neutron is scattered into a unit solid angle about a direction Ω with respect to the incident direction, assuming no absorption or other parasitic scattering process is taking place. It should be noted that the term $(d\sigma/d\Omega)(Q)$ is frequently and inaccurately referred to as the *intensity of scattering*, which is often represented by the symbol $I(Q)$.

Scattering length density A model commonly assumed for SAS studies involves two separate phases with very distinct properties. One phase is the matrix with a scattering length density ρ_0 and the other phase is a distribution of particles or inhomogeneities with scattering length density ρ_p, these particles are responsible for the scattering (see also Figure 7.3). The SANS technique uses the neutron contrast due to differences in the scattering power between small particulate regions or pores in a sample and the general background medium. When the neutron beam passes through a medium, the contrast between these particulate regions and the surrounding material causes a small component of the incident beam to be scattered through a small-angle with respect to the straight on direction. The width, intensity, and profile of the scattered component yields information on the mean size, the volume fraction, size distribution, and shape of the scattering particles or pores. The SANS of interest with the instrumental configuration used, occurs at less than 5° [7.24].

Scattering studies are most useful when the waves scattered by different scatterers in a system are coherent, i.e. when these waves have a definite phase relationship with each other. In this case, the scattered amplitudes (not intensities) are added to give the resultant amplitude from the system.

---

* Older texts may express the differential scattering cross section in units of barns/sterad/atom. To convert to cm^{-1}, one needs to divide by the average volume of an atom in cubic Å.

Since the wavelength λ of the neutron or X-ray beam is usually equal to a few Ångstrom, in the small-angle region Qd is almost always small compared to 1 for dimensions d of the scattering particles of the order of diameters of atoms or small molecules. Consequently, the rays scattered from different parts of an atom or small molecule, arrive at the detector almost completely in phase. The scattered intensity from atoms or small molecules is thus essentially independent of the scattering vector Q and so scattering measurements provide no information about structures on these length scales. A general property of scattering thus is that the scattering process cannot resolve (i.e. provide information about) the structure characterized by a length d when the following condition is fulfilled:

$$Qd = \frac{4\pi \sin(\theta)\, d}{\lambda} \ll 1 \tag{7.5}$$

where Q is the modulus of the scattering vector, d is the scattering length, 2θ is the scattering angle, λ is the wavelength of the neutron or X-ray beam.

One of the factors that determine the feasibility of a light-scattering experiment is the refractive index difference between the scattering particles and the matrix. If the refractive indices are the same, then there can be no light scattering from the scattering particles and it effectively becomes invisible. The same principle applies in SAXS and in SANS. In SAXS the factor is the difference in the electron density and in SANS the equivalent factor is the scattering length density, ρ [7.20]. The scattering length density ρ has dimensions of $(\text{length})^{-2}$ and is normally expressed in units of $10^{10}\,\text{cm}^{-2}$ or $10^{-16}\,\text{Å}^{-2}$. Selected values of ρ for cement pastes are given in Table 7.4 in Section 7.5.3 later [7.20]. The scattering length density varies irregularly between nuclei, and tables are available that give the scattering length for different elements [7.25,7.20]. Of particular significance is the difference in sign and magnitude between the scattering lengths of hydrogen and deuterium, and this has been used in studies involving cement pastes to determine the structure of C–S–H gel (see also Section 7.5.5 later) [7.20].

The scattering length density of a molecule is remarkably sensitive to the value of the bulk density used in its calculation, and careful consideration should always be given to this aspect [7.20]. Coherent neutron-scattering length density of a molecule ρ_p is defined as the product of the mean-scattering length per atom and the atomic number density, and is significantly different from the scattering length density of the surrounding medium ρ_0 [7.12]. Only coherently scattered neutrons carry structural information about the sample. All nuclei with non-zero spin also scatter neutrons incoherently.

Scattering contrast The parameter $(\Delta\rho)^2$ is more commonly called the scattering contrast and is simply the square of the difference in neutron-scattering length density between the scattering particles ρ_p and the surrounding medium or matrix, ρ_0. For the contrast, it does not matter if ρ_p is greater than or less than ρ_0 [7.12]. Although the scattering length density ρ can be negative, the scattering contrast $(\Delta\rho)^2$, and thus the differential scattering cross-section, $d\Sigma/d\Omega$, must either be positive or zero. When $(\Delta\rho)^2$ is zero, the scattering centers are said to be at contrast match [7.20].

Form factor   The parameter $P(Q)$ in Equation (7.4) is called the form factor or shape factor, contains intra-particle information and is dependent on and sensitive to both the size and the shape of the scattering particle. The form factor is a dimensionless function, and describes how the macroscopic differential cross section $d\Sigma/d\Omega(Q)$ is modulated by interference effects between neutrons scattered by different parts of the same scattering center [7.20].

Structure factor   The structure factor $S(Q)$ is related to the interparticle properties and tends to unity at high Q, as the concentration of scattering centers is reduced. Provided that the form factor $P(Q)$ is invariant with changes in concentration of the particles, the structure factor $S(Q)$ can be obtained from Equation (7.4) by measuring $d\Sigma/d\Omega(Q)$ at two different particle concentrations [7.20]. The structure factor $S(Q)$ takes into account the interference of waves associated with correlations in the positions of the component particles [7.23].

7.3   Plots obtained

SAXS and SANS have been employed in the investigation of many porous materials. From the scattering data, information can be obtained about the size of the pores and also about the properties of the pore-matrix boundary surfaces. The intensity of the scattered beam and the way in which the intensity varies with the scattering angle can be analyzed to obtain information about the structure of the sample. SAS data can be interpreted by using a set of approximations and models to derive microstructural parameters of interest. Most data analysis from multiphase systems is based on identifying specific trends. The main tools of analysis are the use of standard plots, and fitting of the scattering data to models [7.15]. Standard plots consist in assessing linear behaviors, when plotting functions of the intensity as functions of Q, in order to extract characteristic slopes and intercepts.

From its earliest development, SAS has been interpreted using a number of standard approximations and numerical methods to derive microstructural parameters of interest. The most commonly used approximations are the Guinier approximation for well-defined discrete inhomogeneities [7.2], and Porod's approximation for determining the surface area between two phases [7.3,7.26]. In addition, the Debye–Anderson–Blumberger model is used in research for randomly disordered microstructures on a single length scale [7.5]. Experimental and theoretical developments during recent years have made it possible to interpret the strong scattering observed by many materials whose microstructural disorder extends over many length scales, such as sandstones, cement pastes, clays etc. [7.27].

In general materials research, the obtained scattering curve, which is a plot of intensity $I(Q)$ vs. Q, may consist of a series of maxima, a single maximum, or a curve showing a gradual decrease in intensity (see Figure 7.4). The occurrence of discrete maxima (Figure 7.4a) may be viewed as a manifestation of a periodic or quasiperiodic structure in crystalline materials, or can result from widely separated spheres of the same size in a medium [7.28]. The occurrence of a single maximum (Figure 7.4b) is observed in a blend of phases separated

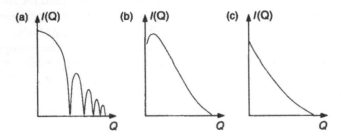

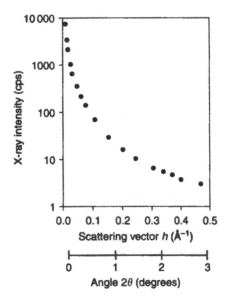

Figure 7.4 General types of SAS curves, plot of intensity $I(Q)$ vs. scattering vector Q, obtained from different types of materials. (a) Widely dispersed particles of the same size [7.28]; (b) phase-separated blend, like in metallic alloys or from porous glass [7.29]; (c) heterogeneous particles in a homogeneous medium, in porous and disordered systems, observed in natural stones [7.29].

Figure 7.5 Typical SAXS intensity curve for hardened Portland cement paste. The second abscissa scale represents the transformed scattering angle.

Source: Reprinted from *Journal of the American Ceramic Society*, Vol. 57, D.N. Winslow, S. Diamond, "Specific surface of hardened Portland cement paste as determined by small-angle X-ray scattering," pp. 193–197, copyright 1974, with permission from the American Ceramic Society, Westerville, OH.

from each other, such as in a metallic alloy, or in a glass [7.29]. The type of SAS that decreases in intensity with increasing angle in a continuous manner from a maximal value at $2\theta = 0$, shown in Figure 7.4c, is observed in porous and disordered systems, such as natural stones [7.29].

The intensity plot from cement paste shows a gradual decrease similar to case (c) in Figure 7.4, as it is shown in Figure 7.5, indicating a random structure with inhomogeneities [7.11]. Similar plots with the same shape have been obtained by several researchers, using both SAXS [7.30] and SANS [7.31,7.32,7.12,7.33].

The absolute value of the intensity at large values of the scattering vector, that is in the higher angle tail of the diffraction curve, depends on two parameters of the system [7.34]: (a) the difference in the electronic density (for SAXS) or scattering length density (for SANS) between the two phases, and (b) the total area S of the interface separating the phases. If the system examined is a sample containing only solid–gas interfaces, e.g. a dry porous material, then one would expect strong SAS, because the difference in scattering length densities between the two phases, $\Delta\rho$, is essentially equal to the electron density of the solid.

There are two limiting cases in a plot of the scattering curve at which useful information can be derived from the general behavior of the scattering function:

- The *Guinier* approximation, 1939, applies at low scattering angles, and consequently at small values of the scattering vector Q, such that the product of scattering vector and scattering length, is very small, i.e. $Ql \ll 1$. The pore size can be estimated from the measurement of the radius of gyration obtained at very small angles by using Guinier's law [7.35].
- The *Porod* approximation, 1951, is valid at large scattering angles, and large values of the scattering vector Q, such that the product of scattering vector and scattering length is large, i.e. $Ql \gg 1$. At large-scattering angles, or at the so-called Porod region, the shape of the curve is useful in obtaining information on the surface-to-volume ratio of the scattering objects and determine the surface area [7.3].

For some materials, e.g. polymers, other plots are also used, such as the Zimm plot, which is a plot of $(1/I)$ vs. Q^2, and the Kratky plot, which is a plot of the parameter $(Q^2 I)$ vs. Q [7.36,7.6]. With the Zimm plot, extrapolation to scattering vector equal to zero and to zero concentration yields the molecular weight of the material. The Kratky plot emphasizes the Gaussian nature of polymer chains [7.6].

The Guinier and Porod plots have been used quite extensively in the past to extract information that relates to the pore structure of a material, including cement pastes, and are described in the following sections.

7.3.1 *Guinier plot*

At small angles, the scattering intensity follows the approximation proposed by A. Guinier in 1939, which is a traditional method of analyzing scattering data [7.35]. For the Guinier approximation, the scattering particles need to satisfy the following conditions: (a) the sample must consist of particles of the same shape and be isotropic, i.e. composed of randomly oriented scatterers; (b) the parameter QR_g must not be much greater than 0.5; and (c) the scatterers in the sample must scatter independently, i.e. the scattered intensity from any scatterer must be unaffected by the presence of other scatterers. The scattering cross section for SANS and the scattering intensity for SAXS at small angles ($QR_g < 2.5$) are given by Equations (7.6) and (7.7), respectively:

$$\text{SANS} \quad \frac{d\Sigma}{d\Omega}(Q) = N_{PT} V_{PT}^2 (\Delta\rho)^2 \exp\left(\frac{-Q^2 R_g^2}{3}\right) \tag{7.6}$$

$$\text{SAXS} \quad I(Q) = I(0)\exp\left(\frac{-Q^2 R_g^2}{3}\right) \quad \text{where } I(0) = N(\Delta\rho)^2 V^2 \tag{7.7}$$

where $I(Q)$ is the SANS intensity at small angles as a function of the modulus of the scattering vector Q, N is the number of scatterers, $\Delta\rho$ is the contrast electron density, V is the volume of the scattering particle with radius of gyration R_g, $I(0)$ is the forward scattering amplitude, is also the extrapolated scattering intensity at $Q = 0$, R_g is the particle's radius of gyration, Q is the scattering vector.

By taking the natural logarithm of the Guinier approximation of Equation (7.7), we obtain Equation (7.8), with the parameters defined as previously:

$$\ln(I(Q)) = \ln I(0) - Q^2 \frac{R_g^2}{3} \tag{7.8}$$

Guinier's law states that a graph of $\ln I(Q)$ and Q^2 in the case of SAXS, and $\log (d\Sigma/d\Omega)$ vs. Q^2 in the case of SANS, should be linear at small values of Q^2, and that the slope of the linear portion of the Guinier plot will be proportional to a parameter called the *radius of gyration* of the scattering particles. The radius of gyration can be extracted from the slope $R_g^2/3$ of the plot and is related to the size of the scattering particles [7.7]. Interparticle effects always contribute to the radius of gyration R_g except in extreme cases, e.g. infinite dilution limit, which represents the case of an isolated particle. The usefulness of the Guinier plot stems from the fact that the obtained particle size R_g is independent of the absolute intensity $I(0)$ [7.15]. Regardless of the structure of a particle, the beginning of the Guinier scattering can be described with the help of two parameters, namely the intercept $I(0)$ characterizing the total amount of scattering matter, and the slope-related parameter R_g that bears information on the particle size [7.8].

An example of a Guinier plot obtained from cement pastes using SAXS is shown in Figure 7.6 [7.37].

The value of intensity $I(0)$ at zero angle cannot be measured experimentally, because of the interference of the incident beam, therefore, extrapolation to the zero value of the scattering vector Q must be used. The intensity $I(0)$ is characterized by the volume fraction of a porous material, which can be deduced from the intercept at $Q = 0$ [7.38]. In a similar way, when the chemical composition of a particle is known, the evaluation of $I(0)$ allows the molecular mass of a particle to be determined in both X-ray and neutron experiments [7.8].

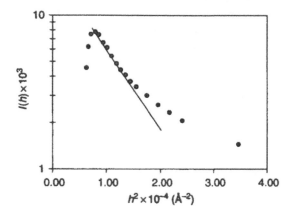

Figure 7.6 Guinier plot obtained on a cement paste using SAXS.

Source: Reprinted from D. Winslow, "The specific surface of hardened Portland cement paste as measured by low-angle X-ray scattering," PhD Thesis, Purdue University, copyright 1973, with permission from the author.

Guinier's approximation is not obeyed for a wide distribution of either pore sizes or pore shapes or both. Guinier's law is only valid when particle–particle correlations are absent, so that the measured intensity $I(Q)$ is purely due to an individual particle's size and shape. However, in a few cases where the particle density is not low, the Guinier approximation can still be valid as long as the particle–particle correlations are nearly a constant over the range of interest. When the parameter QR_g is small, the only particle–particle correlation is the nearest neighbor interaction for a connected network.

Radius of gyration   The radius of gyration is the root mean square of the distances of all the electrons from the electronic center of gravity of a particle. The radius of gyration is related to the size and shape of the particle. The process of obtaining R_g involves two steps: first, averaging all possible positions in the particle from which a vector r can start and be within the particle. Then, determining the probability that a randomly directed vector r from an arbitrary starting point in the particle will fall in the particle. The radius of gyration *for one particle* can be determined by Equation (7.9) [7.21,7.10,7.7,7.16]:

$$R_g^2 = \frac{1}{V_p} \int_{P_{V_p}} r^2 \, dr \tag{7.9}$$

where R_g is the radius of gyration, V_p is the volume of the particle, r is the distance from the electronic center of gravity.

The radius of gyration corresponds to the radius of inertia in mechanics, where the electron density, rather than the mass density, is used as the weighing factor. The concept of the radius of gyration is applicable to particles of any shape, but the range of the scattering vector where the radius of gyration can be identified may vary with different shapes. Examples of the radius of gyration expressed by the dimensions of simple uniform geometrical bodies are presented in Table 7.2. After having obtained an average radius of gyration from the Guinier plot, and by assuming a reasonable shape one may determine approximately the average size of the scattering particles [7.8,7.37].

Attention should be drawn that the measured average radius of gyration for particles with different sizes given by Guinier's law is not a simple arithmetic average of the radii of gyration of the individual pores. For a *distribution of particles*, the average radius of gyration is given by Equation (7.10) [7.2]:

$$\overline{R}^2 = \frac{\sum n_i R_i^8}{\sum n_i R_i^6} \qquad (7.10)$$

where $\overline{R}$ is the average radius of gyration of a collection of pores or particles, R_i is the radius of gyration of the ith class of pores or particles, n_i is the fraction of pores or particles in the ith class.

The accuracy of calculation of the radius of gyration, and consequently the particle (or pore) size, depends on a number of factors. Interparticle interference may somewhat influence the initial part of the scattering curve as it was mentioned earlier. Possible accumulation of particles also leads to distortions of scattering curves at very small angles, i.e. to a sharp increase in scattering intensity. Sometimes it is possible to detect visually the accumulation of particles from the Guinier plot (if several leading points deviate sharply from the straight line), and to determine the radius of gyration omitting these points.

Table 7.2 Radii of gyration of some homogeneous bodies [7.8,7.62]

Shape of geometrical bodies	Radius of gyration
Sphere of radius R	$R_g^2 = \frac{3}{5}R^2$
Spherical shell with radii $R_1 > R_2$	$R_g^2 = \frac{3}{5} \cdot \frac{R_1^5 - R_2^5}{R_1^3 - R_2^3}$
Ellipsoid with semiaxes a, b, and c	$R_g^2 = \frac{a^2 + b^2 + c^2}{5}$
Cylinder with radius r and height h	$R_g^2 = \frac{1}{4}\left(2r^2 + \frac{h^2}{3}\right)$
Elliptical cylinder with semiaxes a and b, and height h	$R_g^2 = \frac{a^2 + b^2}{4} + \frac{h^2}{12}$
Hollow circular cylinder with radii $R_1 > R_2$ and height h	$R_g^2 = \frac{R_1^2 + R_2^2}{2} + \frac{h^2}{12}$

7.3.2 Porod plot

At large values of the scattering vector Q, when the parameter $QR_g > 2.5$, the Porod approximation holds for SAS. G. Porod, 1951, demonstrated theoretically that a plot of the transformed function of the scattered intensity $h^3I(h)$ vs. h^3 for the case of SAXS, or a plot of the function $(Q^4 d\Sigma/d\Omega)$ vs. Q for the case of SANS, should achieve an asymptotic value, which should be proportional to the specific surface of the specimen producing the scatter [7.3]. A plot obtained from SAXS data on cement paste is shown in Figure 7.7 and from SANS data is shown in Figure 7.8 [7.11,7.23]. The plot obtained from SANS data shows a plateau reached that is used for Porod analysis. Similar plots on cement pastes have been obtained by other researchers [7.39,7.31].

In order to analyze the area of interest, the log-log plot of $(d\Sigma/d\Omega)$ vs. Q needs to be obtained, and gives a straight line with a gradient equal to -4 in the Porod-scattering regime [7.23]. A linear fit to the data in the Porod regime gives Porod's constant k_p as the slope of the line. The slope determined at high Q values in a sample depends on the shape of the particles (see also Figure 7.8) [7.3,7.20,7.40]. The Porod plot also yields information about the fractal dimension of the scattering objects.

- A slope of -4 represents smooth spherical particles in a multiphase system and polymers.
- A slope of -2 is a signature of Gaussian chains in a dilute environment or flat platelets.
- A slope of -1 points to rigid rods (cylindrical particles).

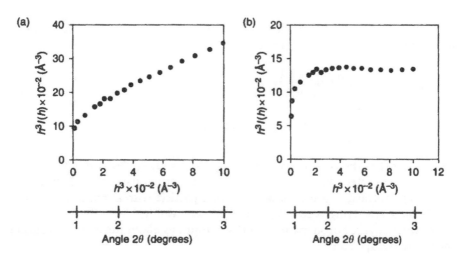

Figure 7.7 Initial (uncorrected) (a), and corrected (b) Porod plot for cement paste using SAXS.

Source: Reprinted from *Journal of the American Ceramic Society*, Vol. 57, D.N. Winslow, S. Diamond, "Specific surface of hardened Portland cement paste as determined by small-angle X-ray scattering," pp. 193–197, copyright 1974, with permission from the American Ceramic Society, Westerville, OH.

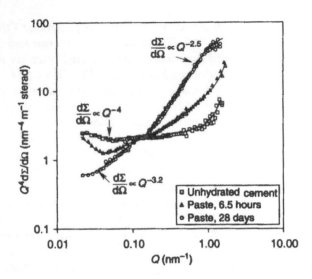

Figure 7.8 Plots of the parameter $Q^4 d\Sigma/d\Omega$ vs. Q for Portland cement pastes with w/c = 0.40 after 6.5 hours of hydration and at 28 days. The same plot for unhydrated cement is also presented for comparison. The lines are guides to the eye. The plateau region at approximately $Q = 1\,\text{nm}^{-1}$ is used for the Porod constant determination.

Source: Reprinted from Allen A.J., Oberthur R.C., Pearson D., Schofield P., Wilding C.R., 1987, "Development of the fine porosity and gel structure of hydrating cement systems," *Philosophical Magazine B*, Vol. 56, No. 3, pp. 263–288, copyright 1987, with permission from Taylor and Francis Ltd.

In the Porod region, the Equation for differential scattering cross section can be written as expressed by Equation (7.11) [7.23]:

$$\frac{d\Sigma}{d\Omega} = N_P V_P (\Delta\rho)^2 2\pi \left(\frac{S_P}{V_P}\right)\frac{1}{Q^4} \quad \text{or}$$

$$\frac{d\Sigma}{d\Omega} = k_p \left(\frac{S_P}{V_P}\right)\frac{1}{Q^4} \quad \text{with } k_p = 2\pi N_P V_P (\Delta\rho)^2 \tag{7.11}$$

where $d\Sigma/d\Omega$ is the differential scattering cross section, N_P is the number of scattering particles, V_P is the volume of scattering particles, $(\Delta\rho)^2$ is the scattering contrast, S_P is the surface area of a particle with a radius of gyration R_g, Q is the scattering vector, k_p is the Porod constant.

For the analysis, Porod made several assumptions about the material causing the scatter [7.3, 7.26]:

- The sample being investigated must be a two phase system. The electron densities for SAXS, or scattering length densities for SANS, of the two phases are different for each phase, but each phase has a uniform electron density. In the case that the Porod theory is used to determine the pore structure, the two phases are solid and air for a dried specimen, or solid and liquid in the pores for a saturated specimen. It is worth noting that Porod's development does not specify which of the two phases is the solid and which is the pore. The theory covers cases in which the pores may be filled with any liquid or gas, as long as the electron density in the pores is uniform.

- The boundaries between the two phases are assumed to be sharp, which means that the transition between one electron density and the other across the interface is abrupt.
- The particles are of the same shape but differ in size and are assumed to be oriented at random. The chance of finding a certain particle in the sample is independent of the position of the element relative to the position of the other particles.
- Multiple scattering can be neglected.

Porod's theory is exactly applicable only when the X-rays or neutrons are scattered purely by interfaces. Other sources of scatter in a specimen may be irregularities or vacancies in the lattice of the solid and scatter from liquid that may be in the pores. When non-interfacial scatter of this sort occurs, it causes the transformed plot to have a positive linear slope [7.11]. Characteristic deviations from the asymptotic Porod law may be caused by short range density fluctuations within the solid, or by roughness or a fractal character of the boundary separating pores and solid.

The asymptote in a plot of $h^3I(Q)$ vs. h^3 (or $Q^4d\Sigma/d\Omega$ vs. Q for SANS) is a function of the intensity of the incident X-ray beam, therefore, the intensity must be normalized before a surface area can be calculated. The normalization of the intensity is supposed to be done by measuring the intensity of the primary beam, which is experimentally very difficult, because of the extremely high intensity, millions of counts per second [7.37]. As an alternative normalization procedure, Porod developed a method by using the area under the entire transformed scattering curve as a measure of the intensity. Porod defined the area under the curve by a quantity Q_P that is called "Porod's invariant" and is given by Equation (7.13) (see also Figure 7.9). Porod showed that the quantity Q_P is an important integral characteristic of the scattering intensity, since its value does not vary with the state of subdivision, or with the specific surface of the specimen [7.11,7.7,7.8] The area under the curve of $hI(h)$ vs. h can be obtained by numerical integration [7.37].

According to Porod's law, 1952, for a system of two phases with volume fractions p and $(1-p)$, the total scattering surface area per unit volume, S/V, is determined directly from the Porod constant k_P, as expressed by Equation (7.12):

$$\frac{S}{V} = \frac{4p(1-p)k_p}{Q_P} \tag{7.12}$$

where $k_p = \lim_{h \to \infty} Q^3I(Q)$, and $Q_P = \int_0^\infty QI(Q)dQ \tag{7.13}$

where S/V is the area of the interface between the two phases per unit volume of irradiated sample, p is the volume fraction of one of the two phases, k_p is

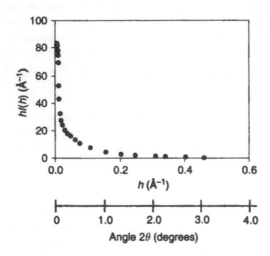

Figure 7.9 Plot used to obtain Porod's invariant based on SAXS measurements on cement pastes.

Source: Reprinted from *Journal of the American Ceramic Society*, Vol. 57, D.N. Winslow, S. Diamond, "Specific surface of hardened Portland cement paste as determined by small-angle X-ray scattering," pp. 193–197, copyright 1974, with permission from the American Ceramic Society, Westerville, OH.

the Porod constant, Q_P is the Porod's invariant, Q is the scattering vector, $I(Q)$ is the scattering intensity.

It is apparent from Equation (7.12) that in order to calculate the surface area, the porosity of the scattering specimen must also be known. Usually, the porosity is determined using other techniques, e.g. helium pycnometry [7.21].

Correct determination of the SANS surface area from the raw scattering data depends on several factors [7.37]. For example, choosing an appropriate range of Q values to calculate the Porod constant is very important. If the lower limit of Q is too low, the data does not fall into the Porod regime, and the resulting Porod constant value and surface area will be too low. While there is no theoretical upper limit on the Porod regime, the rapidly decreasing intensity as Q^{-4}, eventually makes it difficult to separate the signal from the background scattering. Differences in the data calibration of completely independent SANS experiments can result in estimated variations of up to 10% in the calculated Porod constant, while the various calibration procedures in use at different SANS facilities may lead to variations of up to 20% [7.40]. Porod's law is used to extrapolate the curve to very large angles without the need for making additional, time-consuming measurements.

7.4   Range of sizes

Small-angle scattering of X-rays or neutrons is a useful technique for learning about the structure of porous materials on a scale from about 1 to 100 nm [7.7]. This scale ranges from lengths slightly larger than the diameters of single atoms to distances almost large enough to be resolved by an ordinary optical microscope. Multiple neutron scattering in the small-angle regime enables the measurement of pores with size between 80 nm and 10 μm in size with 1–50% porosity. Materials characterized by high porosity (>50%) and pore sizes larger than 200 nm cannot be studied by traditional single particle techniques in SANS, since neutron scattering from these materials is dominated by multiple scattering. SAS techniques are particularly useful for powder characterization and for quantitative analyses of voids or particles that can be measured in a nondestructive manner. Both SAXS and SANS determine open and closed pores, as opposed to open pores only, that are accessible by other techniques, e.g. MIP, gas adsorption etc. SAS techniques have also been used to study the change in pore structure during hydration of cement, as it is discussed in Section 7.5.5.

There is no unique way to deduce the size distribution of particles of unknown shape from the measured scattering. For spherical particles, several attempts have been made to obtain experimentally the size distribution function or certain characteristics of it, but even under these simplified conditions, wide distributions are difficult to determine [7.10]. C. Vonk, 1976, developed theoretically two methods for analysis of SAXS to determine particle size distribution from intensity curves [7.41]. A similar analysis for obtaining particle size distributions from scattering intensity data has been provided by O. Glatter, 1977 [7.4]. Several researchers have obtained pore size distributions from cement pastes assuming spherical pores [7.12,7.24,7.31,7.42,7.43]. For example, D. Pearson and A. Allen, 1985, found a bimodal distribution of pores with a peak at around 5 nm in diameter, and another peak at around 10 nm for cement pastes with w/c = 0.6, stored in water or D_2O [7.24]. Other researchers have confirmed the observation of peaks at diameters about 5–10 nm [7.43,7.44].

7.5 Application to cement pastes

Significant amount of research has been published in the technical literature utilizing SANS to describe the internal surface and pore structure of Portland cement pastes [7.40,7.45,7.46]. Early applications of the SAS technique were dealing with the traditional methods of analysis, the Guinier and Porod plots, as it has already been mentioned. However, modern approaches for pore structure analysis include the use of fractals to characterize the cement paste microstructure. Therefore, current analysis of SAS data yields several parameters, which can provide a concise, quantitative description of the microstructure of cement pastes.

SAS from a heterogenous system such as cement paste is complicated by the fact that the scattering can originate at a number of different interfaces between the various phases present. However, it has been shown by research studies that SANS and SAXS are both dominated by the interface between the C–S–H gel and the pore fluid. The fact that the SANS is dominated by the gel/pore interface, allows the variation of the intensity over the scattering vector range to be related in a more general way to the C–S–H gel structure without having to consider contributions from other cement phases.

Considerable work by several researchers has demonstrated the usefulness of SANS for determining the total surface area of well-defined Portland cement paste systems, the water content of the C–S–H gel, and for interpreting the microstructure of C–S–H gel in terms of a scale-invariant microstructure. The open spaces in the C–S–H gel actually comprise the so-called "gel porosity" and perhaps part of the "capillary porosity" as well [7.13]. Even though the primary source of scattering is the C–S–H gel, the SANS from portlandite crystals [$Ca(OH)_2$] could also be significant. Therefore, it is necessary to consider all aspects of the scattering, i.e. Guinier and Porod regions, size distribution functions, intensity, when analyzing SAS data. It has also proven meaningful in cement paste studies to test identical samples stored under water (H_2O) and heavy water (D_2O), which give different contrasts between pores and the C–S–H gel, in order to confirm the source of SAS. The most important findings of analysis of SAS data in cement pastes are discussed in the following sections.

7.5.1 Guinier plot and radius of gyration

Cement paste is a poor system upon which to apply Guinier's law, because it is far from any of the ideal situations in which Guinier's law is closely obeyed. As it was already mentioned earlier, the radius of gyration represents the size of a scattering object. The average radius of gyration involves the ratio of the eighth power of the radius of gyration of the individual particles R_i to the sixth power of R_i; this means that with the broad pore size distribution of the cement paste, the average is dominated by the large pores and the small pores have virtually no impact upon the value of the radius of gyration. In this way, any interlayer spaces in the layered structure of the C–S–H gel do not contribute to the radius of gyration. Several shapes can be assumed for interpreting the scattering data that are particularly useful for minimizing the negative effects of multiple scattering. The sphere is an immediate choice due to its simplicity and the spherical pore model may be a reasonable assumption for calculating an approximate average pore size in cement pastes from the radius of gyration [7.37].

Guinier plots from cement pastes have prominent linear regions, which indicate a well-developed small particle size distribution. In SANS study of cement pastes a radius of gyration R_g equal to 2–2.5 nm has been determined and for a large particle size distribution $R_g = 5$ nm [7.21]. The "particles" have been identified as pores filled with water [7.21].

The effect of the degree of hydration, the water–cement ratio, and the pore solutions on the radius of gyration on cement pastes is shown in Figure 7.10. The radius of gyration changes significantly with time during hydration in the first 28 days for cement pastes with w/c = 0.27, and these structural changes are not completed at the end of the experimental investigations. In contrast, the sample with a water–cement ratio of 0.50 shows the end of structural changes already 7 days after the beginning of hydration. The researchers concluded that no significant microstructural changes take place in cement pastes in the size range of about 50–500 nm [7.44].

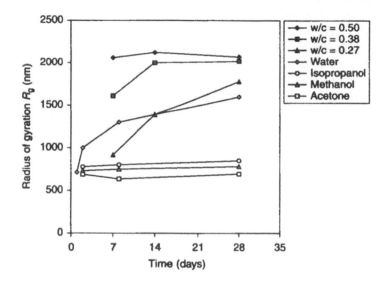

Figure 7.10 Change of the radius of gyration with time as a function of different water–cement ratios of cement pastes, and the liquid they are stored after mixing.

Source: Reprinted from *Cement and Concrete Research*, Vol. 20, F. Häußler, F. Eichhorn, S. Röhling and H. Baumbach "Monitoring of the hydration process of hardening cement pastes by small-angle neutron scattering," pp. 644–654, copyright 1995, with permission from Elsevier Science, Amsterdam, The Netherlands.

Using the equations in Table 7.2, a radius of gyration equal to 5 nm corresponds to a radius equal to 6.5 nm, assuming spherical pores, and to a radius of 20 nm assuming cylindrical pores with height equal to their radius.

7.5.2 Porod plot

The SANS due to large particles such as anhydrated cement particles should show perfect Porod scattering. When a continuous size distribution of scattering particles is present, the Guinier and Porod approximations are less apparent and may not even be applicable. From the scattering curves represented in the Porod plot [(dΣ/d$\Omega \times Q^4$) vs. Q], the Porod constant k_p can be obtained directly using Equation (7.14) [7.43]:

$$k_p = \frac{S}{m}2\pi(\Delta\rho)^2\rho = \frac{S}{V}2\pi(\Delta\rho)^2 \tag{7.14}$$

where k_p is the Porod constant, S is the surface area of the particles, V is the volume of the sample, m is the mass of the sample, ρ is the density of the sample, $(\Delta\rho)^2$ is the scattering contrast.

In the Porod plot, at small Q regions (about 8–1.1 nm^{-1}) smooth interfaces were observed.

Since the Porod constant is associated to the surface area of the scattering particles, or pores, heavy water is frequently used in studies with cement paste to determine the amount of water that is free and bound in the C–S–H gel. An example of the effect of heavy water is shown in Figure 7.11 [7.47]. In Figure 7.11, the values of the Porod constant were calculated as the intercept at $Q^4 = 0$, since in the Porod regime the intensity $I(Q)$ varies as k_p/Q^4.

SANS experiments have been carried out in the past by researchers on dry cement and on hydrated cement pastes and the Porod constants determined are reported in Table 7.3. A study on the effect of fineness of particles on the Porod constant has shown that as the particle size increased, the Porod

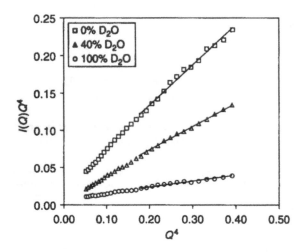

Figure 7.11 SANS data for cement paste specimens containing various levels of heavy water (D_2O). In the plot, $I(Q)$ is the normalized scattering intensity and Q is the scattering vector. The lines are the linear fits used to calculate the Porod constants.

Source: Reprinted from *Advanced Cement-Based Materials*, Vol. 7, J.J. Thomas, H.M. Jennings, A.J. Allen, "Determination of the neutron scattering contrast of hydrated Portland cement paste using H_2O/D_2O exchange," pp. 119–122, copyright 1998, with permission from Elsevier Science, Amsterdam, The Netherlands.

Table 7.3 Values for the Porod constant, k_p, of anhydrous and hydrated cementitious materials determined from SANS experiments

Sample	k_p ($10^{36}/m^5/sterad$)	Source
Anhydrous Portland cement	2.1	[7.23]
	3.23±0.14	[7.32]
	2.94	[7.48]
Anhydrous silica fume	3.10±0.11	[7.32]
Anhydrous blast-furnace slag	1.4	[7.23]
Portland cement paste	59.4	[7.23]
	122–197	[7.42]
	122–148	[7.31]

constant decreased [7.48]. The Porod's invariant Q_P has been found to range between $41 \times 10^{27}\,\text{m}^{-4}$ and $64 \times 10^{27}\,\text{m}^{-4}$ [7.42].

7.5.3 Scattering contrast

For cement-based materials, the surface area is almost entirely the area between the C–S–H gel and the pore water; therefore, for a saturated specimen the neutron scattering contrast can be expressed by Equation (7.15) [7.47].

$$(\Delta\rho)^2 = (\rho_{CSH} - \rho_{H_2O})^2 \tag{7.15}$$

where $(\Delta\rho)^2$ is the scattering contrast, ρ_{CSH} is the neutron scattering length density of C–S–H gel, ρ_{H_2O} is the neutron scattering length density of water.

The scattering length density of the C–S–H gel can be calculated from the average chemical composition and density values of each component using published results on the strength of the neutron interaction with each element [7.47,7.49]. However, this approach presents a problem specifically for the C–S–H gel, because it has a variable composition and a poorly defined structure, as it is not clear how much water should be included for the purpose of calculating the scattering length density, ρ_{CSH}. Some of the water in C–S–H gel is chemically bound into the structure, while there is additional water that is loosely bound in the interlayer spaces and within the gel pores. Since the calculated value of the scattering length density, ρ_{CSH}, is strongly sensitive to the amount of water assigned to C–S–H gel, it is important to establish this value accurately. The value of ρ_{CSH} used in many SANS results published in the technical literature has been calculated using a C–S–H formula of $C_{1.5}SH_{2.5}$ and a density of $2.18\,\text{g/cm}^3$, which are average values commonly quoted in the literature [7.23]. In recent years, an improved understanding of the nature of the C–S–H gel has led to the proposal of new formulae and densities, with variations predicted for different moisture contents within the sample [7.47,7.50]. From the coherent scattering lengths for neutrons by the various elements found in cement, the coherent scattering length density of the various phases with bound water (H_2O or D_2O) contents and the corresponding contrasts have been calculated in the past, and are reported in Table 7.4 [7.23]. It should be noted, however, that the neutron scattering values in international scattering databases have been updated and some of the currently reported values might be slightly different than the values reported in Table 7.4.

In principle, the correct neutron scattering contrast for hydrated cement paste can be determined experimentally by taking advantage of the large difference in the neutron scattering length densities between H_2O (water) and D_2O (heavy water). Experimental studies on the change in the scattering contrast with changing content in D_2O, which utilized the C–S–H gel composition, have established that when a hydrated cement paste is placed into heavy water D_2O, the D_2O exchanges fully with both the pore water and the water in the C–S–H gel, including structural water existing as OH^- groups, and this exchange greatly alters the scattering contrast [7.51,7.47,7.52]. Several researchers have studied the change in relative neutron scattering contrast of hydrated cement pastes as

Table 7.4 Values of scattering length densities, ρ, and scattering contrast $(\Delta\rho)^2$, for unhydrated cement phases, fully hydrated cement phases, and for deuterated cement phases determined by neutrons and X-rays [7.23]

Phase	Neutrons		X-rays	
	ρ $(10^{14}\,m^{-2})$	$(\Delta\rho)^2$ $(10^{28}\,m^{-4})$	ρ $(10^{14}\,m^{-2})$	$(\Delta\rho)^2$ $(10^{28}\,m^{-4})$
CaO	3.80	14.44	27.88	777
C_3S	3.94	15.52	26.94	726
H_2O	−0.60	—	9.58	—
$Ca(OH)_2$	1.61	4.88	19.20	92.5
C–S–H gel	2.29	8.35	19.04	89.4
D_2O	6.28	—	9.39	—
$Ca(OD)_2$	5.33	0.90	19.20	92.5
C–S–D gel	4.28	4.00	19.04	89.4

Notes
For the unhydrated phases CaO and C_3S, the contrast is with air, and for the hydrated phases in H_2O and D_2O, the contrast is with H_2O and D_2O, respectively. For the C–S–H gel, the stoichiometry $C_3S_2H_{2.5}$ is used.

a function of D_2O replacement in order to determine the stoichiometry of C–S–H [7.47,7.51]. The results indicate a C–S–H gel phase with composition $C_{1.7}SH_{2.1}$ and a density of $2.18\,g/cm^3$. These values are obtained when the interlayer water is included in the solid C–S–H phase but the gel pore water is not. The neutron scattering contrast between C–S–H gel and H_2O is $6.78 \times 10^{28}\,m^{-4}$.

It should be noted that the X-ray scattering contrast is not altered by exchange of water with heavy water D_2O, so the contrast variation technique cannot be used to obtain the SAXS contrast (see also Table 7.4) [7.47,7.40].

7.5.4   Surface area

Several researchers have used SANS to explain the source of the SANS/SAXS scattering considering a hydrating cement grain [7.48,7.43,7.40,7.53]. Portland cement paste consists of many crystalline and non-crystalline phases in various ranges of sizes. Since hardened cement paste has a large internal surface area many properties depend on surface interactions.

J. Thomas *et al.*, 1998, have used SANS to study the surface area development of hydrating cement pastes [7.40]. Their results supported the theory proposed that two different morphologies of the C–S–H gel form during cement hydration: a C–S–H morphology with high surface area, which rapidly fills the available pore space during the early hydration, and a morphology with a low surface area predominates at later ages of the cement paste (see also the Jennings–Tennis model in Section 1.2.1). The researchers reported that for Portland cement pastes at 28 days, the specific surface area ranges between $97\,m^2/g$ and $132\,m^2/g$ of dry paste for w/c = 0.35–0.60. The effect of relative humidity on the specific surface area has been determined for C–S–H gel using SANS [7.43]. The specific surface area is $36.6\,m^2/g$ for saturated samples, $17.2\,m^2/g$ for samples at 11% RH, and $8.3\,m^2/g$ for D-dried cement pastes, with cement particles of size less than 5 μm. The values were higher for cement pastes made of cement with larger cement particles.

D. Winslow and S. Diamond, 1974, have used SAXS to determine surface areas in cement pastes and found that the surface areas range from 427 m²/g for w/c = 0.30 to 618 m²/g for w/c = 0.60 [7.11]. Drying also affects the specific surface considerably: saturated cement pastes of w/c = 0.40 and hydrated by 86% had a surface area of 708 m²/g, which was reduced to 224 m²/g for D-dried samples. J. Völkl *et al.*, 1987, have investigated the specific surface area in cement pastes as a function of RH using SAXS [7.30]. The cement pastes had w/c = 0.50 and were stored in Ca(OH)₂ solution for 2 years. The specific surface area was about 150 m²/g at 0% RH, changed to increased 220 m²/g for 100% RH. During absorption of water, the specific surface area had its minimum value at 70% RH, and was 110 m²/g.

It is worth noting that the surface areas determined by SANS remain consistently much lower than the reported surface areas determined by SAXS. Values of the surface areas determined for cement pastes using SAXS and SANS are presented in Table 7.5 [7.13].

Table 7.5 Reported SAXS and SANS surface area values for saturated ordinary Portland cement (OPC) pastes, made from cement Type I or II

Technique	Specimen	Age	Surface area (m²/g of dry paste)
SAXS	Type I OPC, w/c = 0.30	1.5 years	427
	Type I OPC, w/c = 0.40	1.5 years	566
	Type I OPC, w/c = 0.60	1.5 years	618
	Type I OPC, w/c = 0.40	28 days	586
	Type I OPC, w/c = 0.40	1 year	605
	OPC, w/c = 0.50	2 years	215
	OPC, w/c = 0.40	6 months	210–235
	OPC, w/c = 0.60	6 months	280–320
SANS	Type I OPC, w/c = 0.40	28 days	83
	Type I OPC, w/c = 0.40	28 days	70
	Type I OPC, w/c = 0.40	28 days	131
	Type I OPC, w/c = 0.60	31 days	83
	Type I OPC, w/c = 0.40	11 days	63
	Type I OPC, w/c = 0.35	28 days	99
	Type I OPC, w/c = 0.40	28 days	97–102
	Type I OPC, w/c = 0.45	28 days	106
	Type I OPC, w/c = 0.50	28 days	123
	Type I OPC, w/c = 0.55	28 days	132
	Type I OPC, w/c = 0.60	28 days	147
	Type II OPC, w/c = 0.40	28 days	84
	Type I OPC, w/c = 0.40	5 years	93
	Type I OPC, w/c = 0.50	5 years	142

Source: Reprinted from *Concrete Science and Engineering*, Vol. 1, J.J. Thomas, H.M. Jennings, A.J. Allen, "The surface area of hardened cement paste as measured by various techniques," pp. 45–64, copyright 1999, with permission from Elsevier Science, Amsterdam, The Netherlands.

Note
The values have been compiled from data given in the technical literature with the references given in the original paper.

7.5.5 Structure of cement paste

Several researchers have used SANS to study the microstructural development of cement pastes [7.48,7.23,7.31,7.39,7.54]. A. Allen *et al.*, 1987, suggested the following three processes that influence SANS results during cement hydration: (a) the development of the C–S–H gel structure increases the SANS at high values of the scattering vector Q, while the shrinking unreacted cement clinker cores diminish the Q^{-4} Porod component at lower Q values; (b) the C–S–H gel grows and binds the cement together, and this may be responsible for the $Q^{-3.2}$ scattering in the intermediate Q range (see also Figure 7.8); (c) The growth of $Ca(OH)_2$ crystals and other phases may also be important and contribute to SANS scattering. The formation of the C–S–H gel is a nucleation- and diffusion-controlled process rather than a process limited by a gel/pore surface growth [7.23].

Allen *et al.*, proposed that SANS is from small globules of C–S–H about 2 nm radius that stick together to make up agglomerates of spheres with radii of about 36 nm. The formation of C–S–H gel from globules has been mentioned in Chapter 1 in the models describing the C–S–H gel structure (see also Figure 1.1). On similar analysis of SANS results, L. Aldridge *et al.*, 1995, concluded that scattering is a result of two scattering entities in the cement pastes, which they believed to be C–S–H globules resembling spheres with radius 40 nm and less than 5 nm [7.33].

A. Allen and R. Livingston, 1998, have suggested that the single most significant scattering feature in cement pastes is the fractal or scale-invariant microstructure of the C–S–H gel [7.39]. This property may be modeled based on a scattering power law which relates the SANS cross-sectional data to both the distribution of sizes of small spherical globules (between ~0.5 and 100 nm) of C–S–H gel, and the fractal agglomerates of the C–S–H globules. An alternative approach to interpreting SANS data has been proposed by T. Sabine *et al.*, 1995, who suggested that the scattering of hydrated cement pastes is due to the distribution of quasi-spherical objects as opposed to any fractal structure. The quasi-spherical objects are of a random size and can be related to the scattering vector by a simple empirical function [7.28]. J. Phair *et al.*, 2003, have used SANS for comparing and analyzing the microstructural characteristics of cement pastes made from a variety of cements and mineral admixtures [7.54]. Their study indicated that Portland cement and blast-furnace slag had the largest radius of gyration compared to other cementitious systems, like alkali-activated fly ash.

J. Tritthart and F. Häußler, 2003, used SANS experiments to study the state of water in cement pastes [7.31]. The researchers concluded from their study that the gel water is part of the structure, cannot be mobile, and does not fill pores. Their conclusions are in contrast with the proposed theory that part of the gel water can be mobile, based on ice formation.

7.5.6 Fractal dimensions

A general review of the fractal properties of porous and disordered systems, as determined by SANS and SAXS, has been given by P. Schmidt, 1991 [7.55].

A number of SAXS and SANS studies have shown that fractal gel microstructures form in cementitious systems [7.56,7.57,7.58,7.51,7.18,7.43,7.59]. Until about 20 years ago, it was not possible to interpret fully the scattering profiles obtained from complex materials with highly disordered microstructures such as cement paste. The relatively recent discovery that the microstructure of cement paste contains fractal, i.e. scale invariant, properties over large parts of the size range has increased significantly the amount of useful microstructural information that can be obtained from SAS measurements. For fractal analysis, it is required that the data extend over a range that is wide enough to encompass the SAS associated with all significant length-scales. Another experimental requirement is that the sample thickness must be sufficiently small to give negligible multiple scattering, e.g. cement paste samples should be no more than 1–2 mm thick for SANS experiments. It is also required that the data be absolutely calibrated so that the relative strengths of the SAS arising in different parts of the microstructure can be assessed properly [7.39].

Experimental studies have given evidence that the hydration products show fractal behavior, and are both surface fractals and mass fractals. The results have been confirmed using both SANS [7.43] and SAXS curves [7.59]. However, parts with a smooth surface could also be found, which might have been caused by nonhydrated clinker grains. SAXS curves from hydrating cement pastes showed two different decay exponents as demonstrated by the linear slopes in their log-log plots, related to the mass and surface fractal structure respectively of Portland cement. The mass fractal dimension, D_m, changed with hydration time from an initial value of 1.9 (at age of 1 day) to 2.8 (at 28 days). The surface fractal dimension, D_s, remained almost constant around 2.8. The high value of the surface fractal dimension, D_s, indicates a very rough and porous surface structure, which is largely conserved during hardening with only a slight trend to become smoother. A. Heinemann et al., 1999, 2000, found that the cement paste during hydration changes from a mass fractal with fractal dimension less than 3 to a surface fractal with a fractal dimension between 3 and 4 [7.28,7.58].

D. Winslow et al., 1995, used SAXS to determine the fractal types and dimensions of hydrated Portland cement pastes over a range of length scales [7.18]. They found that at the large scale, 200–1500 Å, at most degrees of water saturation the cement paste is a rough surface fractal, and at the small scale, 30–200 Å, the cement paste is a mass fractal. This geometry depends upon the level of saturation and was found to change gradually with drying, becoming a rough surface fractal as the saturation decreases to values less than about 50%. Completely oven-dried pastes were found to have a considerably altered geometry [7.18]. The change in slopes of intensities with water content and the change in fractal dimensions with Q (and size) are shown in Figure 7.12. A similar study has been reported elsewhere [7.60].

One especially important property of fractal systems is that many properties of these systems can often be described by quantities that are proportional to a power of another quantity. This relation is frequently called a power law. In particular, the intensity $I(Q)$ of the SAXS or SANS from many disordered systems has been found to be proportional to a negative power of the scattering vector Q, where 2θ is the scattering angle and λ is the X-ray or neutron

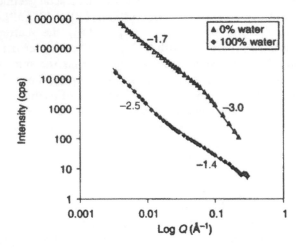

Figure 7.12 Typical results of a SAXS intensity curve plotted in a log-log scale, obtained from a cement paste with w/c = 0.40, at 28 days, and for different levels of water saturation. The values noted are the slopes of the lines.

Source: Reprinted from *Cement and Concrete Research*, Vol. 25, D. Winslow, J.M. Bukowski, J.F. Young, "The fractal arrangement of hydrated cement paste," pp. 147–156, copyright 1995, with permission from Elsevier Science, Amsterdam, The Netherlands.

wavelength. Usually the dependence of $I(Q)$ on a negative power of Q is observed only when Q is large enough to satisfy the condition $Ql \gg 1$, where l is the scattering length. This kind of scattering is often called power-law scattering and can be described by Equation (7.16) [7.55,7.61].

$$I(Q) = I_0 A_0 Q^{-a} \qquad (7.16)$$

where I_0, A_0 are constants, a is a constant determined by the structure of the sample, $I(Q)$ is the scattering intensity, Q is the scattering vector.

In research studies, effort is made to relate the exponent a to fractal dimension values. V. Castaño et al., 1990, determined the exponent a to be $a = 3.46 \pm 0.06$ for X-rays, and $a = 3.23 \pm 0.05$ for neutrons in cement pastes containing polymers [7.61]. They attributed the observed difference in the exponents from X-rays and neutrons to the fact that for neutron scattering there is a constant background scattering caused by incoherent scattering from the hydrogen in the water of hydration, or in the polymer added to the cements. There is no significant difference to the specimens between the specimens, which indicates no significant difference in the fractal dimension.

7.5.7 Factors affecting the results

Besides the experimental factors, e.g. the thickness of the specimen, several microstructural factors can affect the SAS results, such as the presence of mineral admixtures in cement pastes, and the drying techniques used. It should be noted here that specimen drying before the experiment is not necessary, however, SAS has been used to study the microstructural changes after drying the cement pastes by various techniques.

Intensity curves have been used to study the effect of the *sample thickness* on SANS data, using samples with thickness ranging from 0.5 to 12 mm [7.54]. For small thickness of the sample, the intensity was high and as the sample thickness increased, the intensity decreased.

Previous SANS studies on hydrating cement systems have shown that modifications in the microstructural evolution can be associated with the addition of *mineral admixtures*: silica fume, fly ash, or blast-furnace slag. When mineral admixtures are included in the mixture, then the interpretation of the SANS spectra must also consider scattering associated with the admixture/matrix interface [7.54]. Several researchers have used SANS to study microstructural development of cement paste having silica fume [7.48,7.39,7.23,7.32]. The addition of silica fume results in a parallel transition of the curves of $Q^4 d\Sigma/d\Omega$ vs. Q and the pastes with silica fume are mass fractals with fractal dimension between 2.5 and 2.7.

In the measured Q region the hardening cement paste does not show the Porod-like behavior of SANS curves. In contrast, the Porod's potential law holds for dry powder samples of clinker minerals and for silica fume. To calculate the surface area per unit volume, a SANS contrast between air and cement paste of $15 \times 10^{28}\,\text{m}^{-4}$, and air and silica fume of $17.5 \times 10^{28}\,\text{m}^{-4}$ were used [7.23,7.32].

The effect of *drying* has been shown by D. Winslow *et al.*, 1995, who determined that the fractal geometry of the completely oven-dried pastes is significantly different from that of pastes with evaporable water present [7.18]. Intensity plots from oven-dried cement pastes have slopes that are associated with mass fractals, i.e. slopes between 0 and -2 [7.18]. Cement pastes with high amount of water have slopes associated with surface fractals, i.e. slopes between -2 and -3.

Certain aspects must be taken into account when using SANS to investigate the properties of hardened cement pastes. A hydrated cement paste sample typically contains a variety of phases, such as unreacted cement grains, various hydration products, and electrolytic solutions. It is thought that even a single clinker grain consists of different phases and represents a dense packed system of heterogeneous elements. The neutrons traveling through the sample find various scattering objects with different scattering length densities. A selection of the most important scattering objects is necessary in order to effectively interpret the measured SANS data. In many samples, there are sources of electron density discontinuity other than interfaces. Two such sources are [7.37]:

- The presence of ions such as calcium and other ions found in pore water in cement paste
- The existence of atomic vacancies and other irregularities in the structure of the solid particles.

7.6   Advantages and limitations

SANS is a useful tool, because it is a nondestructive method and delivers results on a sample volume of about $0.5\,cm^3$, which can be considered as an average of all resolvable structural details of the whole cement paste. SANS may investigate samples made from cement paste with a thickness of several millimeters. SAXS demands disks with the thickness of only several micrometers.

Repeated measurements at the same sample are possible allowing, e.g. the observations of time-dependent changes. Using scattering techniques during the hydration process can be conducted in real time during cement hydration without interfering with the hydration process and this has allowed the development of the C–S–H gel structure and surface area to be monitored from the earliest hydration times of a few minutes. If the neutron flux is sufficiently high, experiments can be repeated continuously on the same specimen as it hydrates; such real-time investigations of early cement hydration have provided valuable information about the kinetics and the microstructure development during early hydration.

SAS is a nondestructive technique, and is amenable to a range of model assumptions regarding pore shapes and morphologies. Because SAS requires no destructive drying pretreatments and does not modify the microstructure itself, it can be used repeatedly in the real-time characterization of the microstructural evolution during cement hydration. SAS is an ideal technique for characterizing cement paste because specimens can be studied in their saturated state, thus avoiding possible problems associated with drying the C–S–H gel. SANS can be used to probe a sample volume several millimeters in diameter and $\approx 1\,mm$ thick, enough to probe the undisturbed, statistically representative microstructures in cement pastes.

The scattering techniques are nondestructive and non-invasive. SAS can be used to detect blocked or closed pores inaccessible to adsorptives. SAXS can study the meso- and micro-pore size distributions of many materials. X-ray measurement would be expected to measure all pores in the system, irrespective of any encapsulation. The pores are measured whether they are open or closed. Results such as these are entirely consistent with the idea of the pore size distribution that emerges from mercury porosimetry, but definitely in conflict with the trend of the capillary condensation results. As X-rays can readily penetrate the cement paste, they are capable of registering pores which might be inaccessible to particular molecules, such as mercury or nitrogen. So the measurement of SAXS should be more accurate at the fine end of the pore size distributions.

Neutrons are well suited for investigating polycrystalline materials as the wavelength can be chosen sufficiently long (for steady state instruments) to prevent double Bragg scattering. Furthermore, the absorption of neutrons is quite small in most materials and quite thick samples (1–5 mm) can often be used; this makes the sample preparation quite easy.

Neutrons are also nondestructive allowing interpretation of data without fear of radiation damage effects. This means that time-dependent behavior can be followed on the same sample along a reaction coordinate. Neutrons also possess a spin, and consequently a magnetic moment, which can interact with any unpaired electrons in the sample, although this is not widely exploited. Also, neutrons are highly penetrating, allowing not only bulk properties of materials to be probed under normal atmospheric conditions, but also when held within complex environments such as furnaces, cryostats, pressure, and shear cells. Neutrons are sensitive to magnetic spins, allowing to resolve magnetic structures and also to light elements, in particular hydrogen/deuterium, which remains invisible in TEM and SAXS.

X-ray scattering has the great advantage that conventional X-ray sources are relatively cheap. It is therefore possible to have an in-house X-ray instrument, which can be used 24 hours a day for collecting data. For pinhole cameras, there is practically no instrumental smearing of the scattering curve, which makes the data analysis easier than for long-slit cameras and neutron scattering experiments. However, attention should be drawn to the hazards of radiation and all necessary precautions should be taken when using the equipment.

SANS has an important limitation that it is expensive, non-portable and is signal-limited. Neutron sources are huge, relatively inaccessible, and the neutron facilities are expensive to build and run. The number of these facilities around the world is small, and it costs millions of dollars annually to operate a nuclear research reactor, and it costs that much in electric bills alone to run a pulsed neutron source.

Neutron sources are characterized by relatively low fluxes compared to X-ray sources (synchrotrons), and have limited use in investigations of rapid time-dependent processes.

A related sample preparation issue concerns the state of saturation. Conventional tests often specify oven-drying at 105°C to achieve a standard saturation in the pores. This may not be necessary for neutron scattering measurements, but omitting this step may make it impossible to compare results between conventional and neutron scattering measurements. SAXS and SANS measure the properties of non-dried paste, may be difficult to interpret, and only yield information on pores at the lower end of the size range. They are valuable non-contact methods, but they yield only general features of the network structure not including a pore size distribution.

Another issue is time resolution: under certain conditions, it typically takes 1 or 2 hours to acquire a data point by neutron scattering. This may be too long a time interval to study phenomena occurring in the very early hydration period. The counting time may be reduced in the future as high flux neutron sources come on line. In the meantime, it may be necessary to trade off precision vs. counting time.

References

7.1 Guinier A. "Rayons X. Dispositif permettant d' obtenir des diagrammes de diffraction de poudres cristallines très intenses avec un rayonnement monochromatique," *Comptes Rendus Hebdomadaires des Séances de l' Académie des Sciences, Paris,* Vol. 204, 1937, pp. 1115–1116.

7.2   Guinier A., Fournet G. *Small-Angle Scattering of X-rays*, Wiley, New York, 1955.

7.3   Porod G. "Die Röntgenkleinwinkelstreuung von dichtgepackten kolloiden Systemen. I. Teil," *Kolloid Zeitschrift*, Vol. 124, No. 2, 1951, pp. 83–114.

7.4   Glatter O. "A new method for the evaluation of small-angle scattering data," *Journal of Applied Crystallography*, Vol. 10, 1977, pp. 415–421.

7.5   Debye P., Anderson H.R. Jr, Brumberger H. "Scattering by an inhomogeneous solid. II. The correlation function and its application," *Journal of Applied Physics*, Vol. 28, No. 6, 1957, pp. 679–683.

7.6   Kratky O., Porod G. "Röntgenuntersuchung Gelöster Fadenmoleküle," *Recueil Des Travaux Chimiques Des Pays-Bas – Journal of The Royal Netherlands Chemical Society*, Vol. 68, No. 12, 1949, pp. 1106–1122.

7.7   Glatter O., Kratky O. *Small Angle X-ray Scattering*, Academic Press, New York, 1982.

7.8   Feigin L.A., Svergun D.I. *Structure Analysis by Small-Angle X-Ray and Neutron Scattering*, 1986, translated by G.W. Taylor, Plenum Press, New York, 1987.

7.9   Ritter H.L., Erich L.C. "Pore size distribution in porous materials. Interpretation of small-angle X-ray scattering patterns," *Analytical Chemistry*, Vol. 20, No. 7, 1948, pp. 665–670.

7.10   Kostorz G. "Small-angle scattering and its applications to materials science," in *Treatise on Materials Science and Technology, Vol. 15: Neutron Scattering*, Kostorz G. (ed.), Academic Press Inc., New York, 1979, pp. 227–289.

7.11   Winslow D.N., Diamond S. "Specific surface of hardened Portland cement paste as determined by small-angle X-ray scattering," *Journal of the American Ceramic Society*, Vol. 57, No. 5, 1974, pp. 193–197.

7.12   Allen A.J., Windsor C.G., Rainey V., Pearson D., Double D.D., Alford N.M. "A small-angle neutron-scattering study of cement porosities," *Journal of Physics D: Applied Physics*, Vol. 15, No. 9, 1982, pp. 1817–1833.

7.13   Thomas J.J., Jennings H.M., Allen A.J. "The surface area of hardened cement paste as measured by various techniques," *Concrete Science and Engineering*, Vol. 1, 1999, pp. 45–64.

7.14   Livingston R.A., Neumann D.A., Allen A., Rush J.J. "Application of neutron scattering methods to cementitious materials," in *Symposium Proceedings Vol. 376: Neutron Scattering in Materials Science II*, D.A. Neumann, T.P. Russell, B.J. Wuensch (eds), Materials Research Society, Warrendale, PA, 1995, pp. 459–469.

7.15   Hammouda B. *A Tutorial On Small-Angle Neutron Scattering From Polymers*, National Institute of Standards and Technology, Gaithersburg, MD, June 1995, p. 73.

7.16   Bacon G.E. *Neutron Scattering in Chemistry*, Butterworths, London, 1977.

7.17   Gerold V., Kostorz G. "Small-angle scattering applications to materials science," *Journal of Applied Crystallography*, Vol. 11, 1978, pp. 376–404.

7.18   Winslow D., Bukowski J.M., Young J.F. "The fractal arrangement of hydrated cement paste," *Cement and Concrete Research*, Vol. 25, No. 1, 1995, pp. 147–156.

7.19   Pedersen J.S. "Instrumentation for small-angle scattering," in *Modern Aspects of Small-Angle Scattering*, H. Brumberger (ed.), Kluwer Academic Publishers, Dordrecht, 1995, pp. 57–91.

7.20   King S.M. "Chapter 7: Small-angle neutron scattering," in *Modern Techniques for Polymer Characterization*, R.A. Pethrick, J.V. Dawkins (eds), John Wiley & Sons, New York, 1999, pp. 171–232.

7.21   Pearson D., Allen A., Windsor C.G., Alford N.McN., Double D.D. "An investigation on the nature of porosity in hardened cement pastes using small-angle neutron scattering," *Journal of Materials Science*, Vol. 18, No. 2, 1983, pp. 430–438.

7.22   Price D.L., Sköld K. "Chapter 1: Introduction to neutron scattering," in *Methods of Experimental Physics Vol. 23: Neutron Scattering Part A*, K. Sköld, D.L. Price (eds), Academic Press Inc., New York, 1986, pp. 1–95.

7.23 Allen A.J., Oberthur R.C., Pearson D., Schofield P., Wilding C.R. "Development of the fine porosity and gel structure of hydrating cement systems," *Philosophical Magazine B*, Vol. 56, No. 3, 1987, pp. 263–288.

7.24 Pearson D., Allen A.J. "A study of ultrafine porosity in hydrated cements using small-angle neutron scattering," *Journal of Materials Science*, Vol. 20, No. 1, 1985, pp. 303–315.

7.25 D.R. Lide, *CRC Handbook of Chemistry and Physics*, 84th Edition, CRC Press, Boca Raton, FL, 2003–2004.

7.26 Porod G. "Die Röntgenkleinwinkelstreuung von dichtgepackten kolloiden Systemen. II. Teil," *Kolloid Zeitschrift*, Vol. 125, No. 1, 1952, pp. 51–57 and 108–122.

7.27 Allen A.J. "Time-resolved phenomena in cements, clays and porous rocks," *Journal of Applied Crystallography*, Vol. 24, Part 5, 1991, pp. 624–634.

7.28 Sabine T.M., Bertram W.K., Aldridge L.P. "A method for interpreting small angle neutron scattering data from quasi-spherical objects," in *Symposium Proceedings Vol. 376: Neutron Scattering in Materials Science II*, D.A. Neumann, T.P. Russell, B.J. Wuensch (eds), Materials Research Society, Warrendale, PA, 1995, pp. 499–504.

7.29 Kostorz G. "Small-angle scattering studies of phase separation and defects in inorganic materials," *Journal of Applied Crystallography*, Vol. 24, 1991, pp. 444–456.

7.30 Völkl J.J., Beddoe R.E., Setzer M.J. "The specific surface of hardened cement paste by small-angle X-ray scattering effect of moisture content and chlorides," *Cement and Concrete Research*, Vol. 17, No. 1, 1987, pp. 81–88.

7.31 Tritthart J., Häußler F. "Pore solution analysis of cement pastes and nanostructural investigations of hydrated C_3S," *Cement and Concrete Research*, Vol. 33, No. 7, 2003, pp. 1063–1070.

7.32 Häussler F., Eichhorn F., Baumbach H. "Description of the structural evolution of a hydrating Portland cement paste by SANS," *Physica Scripta*, Vol. 50, No. 2, 1994, pp. 210–214.

7.33 Aldridge L.P., Bertram W.K., Sabine T.M., Bukowski J., Young J.F., Heenan R.K. "Small-angle neutron scattering from hydrated cement pastes," in *Symposium Proceedings Vol. 376: Neutron Scattering in Materials Science II*, D.A. Neumann, T.P. Russell, B.J. Wuensch (eds), Materials Research Society, Warrendale, PA, 1995, pp. 471–479.

7.34 Eanes E.D., Posner A.S. "Chapter 33: Small-angle X-ray scattering measurements of surface areas," in *The Solid-Gas Interface*, E. Alison Flood (ed.), Vol. 2, Marcel Dekker Inc., New York, 1967, pp. 975–994.

7.35 Guinier A. "La diffraction des rayons X aux très petits angles: Application a l' étude de phénomènes ultramicroscopiques," *Annales de Physique*, Vol. 12, 1939, pp. 161–237.

7.36 Zimm B. "The scattering of light and the radial distribution function of high polymer solutions," *The Journal of Chemical Physics*, Vol. 16, No. 12, 1948, pp. 1093–1099.

7.37 Winslow D.N. "The specific surface of hardened Portland cement paste as measured by low-angle X-ray scattering," PhD Thesis, Purdue University, August 1973.

7.38 Hutchings M.T., Windsor C.G. "Chapter 25: Industrial Applications," in *Methods of Experimental Physics Vol. 23: Neutron Scattering Part C*, K. Sköld, D.L. Price (eds), Academic Press, Inc., 1987, pp. 405–482.

7.39 Allen A.J., Livingston R.A. "Relationship between differences in silica fume additives and fine-scale microstructural evolution in cement based materials," *Advanced Cement Based Materials*, Vol. 8, No. 3–4, 1998, pp. 118–131.

7.40 Thomas J.J., Jennings H.M., Allen A.J. "The surface area of cement paste as measured by neutron scattering – evidence for two C–S–H morphologies," *Cement and Concrete Research*, Vol. 28, No. 6, 1998, pp. 897–905.

284 *Small-angle scattering*

7.41 Vonk C.G. "On two methods for determination of particle size distribution functions by means of small-angle X-ray scattering," *Journal of Applied Crystallography*, Vol. 9, 1976, pp. 433–440.

7.42 Häußler F., Palzer S., Eckart A., Hoell A. "Microstructural SANS-studies of hydrating tricalcium silicate (C_3S)," *Applied Physics A: Materials Science and Processing*, Vol. 74 (Suppl.), 2002, pp. S1124–S1127.

7.43 Häußler F., Hempel M., Baumbach H., Tritthart J. "Nanostructural investigations of hydrating cement pastes produced from cement with different fineness levels," *Advances in Cement Research*, Vol. 13, No. 2, 2001, pp. 65–73.

7.44 Häußler F., Eichhorn F., Röhling S., Baumbach H. "Monitoring of the hydration process of hardening cement pastes by small-angle neutron scattering," *Cement and Concrete Research*, Vol. 20, No. 4, 1990, pp. 644–654.

7.45 Berliner R., Popovici M., Herwig K., Jennings H.M., Thomas J. "Neutron scattering studies of hydrating cement pastes," *Physica B*, Vol. 241, 1997, pp. 1237–1239.

7.46 Berliner R., Popovici M., Herwig K.W., Berliner M., Jennings H.M., Thomas J.J. "Quasielastic neutron scattering study of the effect of water-to-cement ratio on the hydration kinetics of tricalcium silicate," *Cement and Concrete Research*, Vol. 28, No. 2, 1998, pp. 231–243.

7.47 Thomas J.J., Jennings H.M., Allen A.J. "Determination of the neutron scattering contrast of hydrated Portland cement paste using H_2O/D_2O Exchange," *Advanced Cement Based Materials*, Vol. 7, No. 3–4, 1998, pp. 119–122.

7.48 Häußler F., Eichhorn F., Baumbach H. "Small-angle neutron scattering on hardened cement paste and various substances for hydration," *Cement and Concrete Research*, Vol. 24, No. 3, 1994, pp. 514–526.

7.49 Sears V.F. "Neutron scattering lengths and cross sections," *Neutron News*, Vol. 3, No. 3, 1992, pp. 26–37.

7.50 Taylor H.F.W. *Cement Chemistry*, Thomas Telford Services Ltd, London, 1997.

7.51 Häussler F., Hempel M., Baumbach H. "Long-time monitoring of the microstructural change in hardening cement paste by SANS," *Advances in Cement Research*, Vol. 9, No. 36, 1997, pp. 139–147.

7.52 Allen A.J., Baston A.H., Wilding C.R. "Small-angle neutron scattering studies of pore and gel structures, diffusivity, permeability and damage effects," in *Symposium Proceedings Vol. 137: Pore Structure and Permeability of Cementitious Materials*, L.R. Roberts, J.P. Skalny (eds), Materials Research Society, Warrendale, PA, 1989, pp. 119–125.

7.53 Winslow D.N., Bukowski J.M., Young J.F. "The early evolution of the surface of hydrating cement," *Cement and Concrete Research*, Vol. 24, No. 6, 1994, pp. 1025–1032.

7.54 Phair J.W., Schulz J.C., Bertram W.K., Aldridge L.P. "Investigation of the microstructure of alkali-activated cements by neutron scattering," *Cement and Concrete Research*, Vol. 33, No. 11, 2003, pp. 1811–1824.

7.55 Schmidt P.W. "Small-angle scattering studies of disordered, porous and fractal systems," *Journal of Applied Crystallography*, Vol. 24, Part 5, 1991, pp. 414–435.

7.56 Kriechbaum M., Degovics G., Laggner P., Tritthart J. "Investigation on cement pastes by small-angle X-ray scattering and BET: the relevance of fractal geometry," *Advances in Cement Research*, Vol. 6, No. 23, 1994, pp. 93–100.

7.57 Heinemann A., Hermann H., Häussler F. "SANS analysis of fractal microstructures in hydrating cement paste," *Physica B*, Vol. 276–278, 2000, pp. 892–893.

7.58 Heinemann A., Hermann H., Wetzig K., Häussler F., Baumbach H., Kröning M. "Fractal Microstructures in Hydrating Cement Paste," *Journal of Materials Science Letters*, Vol. 18, No. 17, 1999, pp. 1413–1416.

7.59 Kriechbaum M., Degovics G., Tritthart J., Laggner P. "Fractal structure of Portland cement paste during age hardening analyzed by small-angle X-ray scattering," in

Progress in Colloid and Polymer Science, Vol. 79: Trends in Colloid and Interface Science III, 1989, pp. 101–105.

7.60 Beddoe R.E., Lang K. "Effect of moisture on fractal dimension and specific surface of hardened cement paste by small-angle X-ray scattering," *Cement and Concrete Research*, Vol. 24, No. 4, 1994, pp. 605–612.

7.61 Castaño V.M., Schmidt P.W., Hörnis H.G. "Small-angle scattering studies of the pore structure of polymer-modified Portland cement pastes," *Journal of Materials Research*, Vol. 5, No. 6, 1990, pp. 1281–1284.

7.62 Mikhail R.S., Turk D.H., Brunauer S. "Dimensions of the average pore, the number of pores, and the surface area of hardened Portland cement paste," *Cement and Concrete Research*, Vol. 5, No. 5, 1975, pp. 433–442.

8 Microscopic techniques and stereology

The optical microscope (OM) is much older than other characterization instruments and uses the visible or near visible portion of the electromagnetic spectrum to observe objects of interest. OMs are the cheapest "modern" instrument and take up little physical space. They are capable of handling almost every type of sample and can easily provide high magnifications up to 1400×. Even though it is not quite certain who invented the microscope, its origin can be traced in the Netherlands between 1590 and 1610 and the first well-known microscopists were A. van Leeuwenhoek and R. Hooke. Since the mid-1800s, the OM has been used to view virtually all materials, regardless of their nature or origin. In 1882, H. Le Chatelier published the results of his work on cement chemistry, in which he identified various constituents of Portland cement clinker using microscopic examination in polarized light [8.1].

The development of a microscope utilizing electrons as its source of radiation was conceived after L. de Broglie's suggestion in 1924, that electrons have a wavelength that is significantly shorter than the wavelength of light, a theory that was confirmed experimentally by C. Davisson and L. Germer in 1927 [8.2,8.3]. The possibility of constructing an electron microscope offered the prospect of forming images with considerably better resolution than images from a light microscope. By 1931, M. Knoll and E. Ruska showed experimentally that an electron beam could be focused using magnetic fields and two years later, the first electron microscope was constructed, capable of a magnification of 12000×. A brief review of the development of the electron microscope can be found in the technical literature [8.4]. In 1939, O. Radczewski et al., published the first electron microscopy studies on cement, including his work on hydration of tricalcium silicate, and tricalcium aluminate [8.5,8.6]. Scanning electron microscope (SEM) imaging using secondary electrons (SEs) was first applied on cement pastes by S. Chatterji and J. Jeffrey in 1966 to study fracture surfaces [8.7].

Around 1930, air-entrained concrete was introduced to improve durability of concrete to freeze–thaw and various parameters were proposed by T. Powers, 1954, to characterize the air void system in concrete [8.8]. Since 1950, many papers have appeared in the literature describing methods using the OM for calculating the air void parameters in concrete in order to determine concrete durability to freeze–thaw. These methods had been developed mathematically and statistically since the early 1900s and constitute the science of stereology. For practical applications, analysis methods have been used for a long time by petrographers for examination of rocks, and by metallurgists for identification of different phases in an alloy. In the microscopic technique for air void analysis, the ground surface of a concrete specimen is obliquely illuminated and viewed through a microscope using reflected light. In early applications for air void analysis using the optical microscope, the air voids were revealed by the shadows of their leading edges on their hollows, i.e. by the optical contrast difference they create upon illumination.

Examination of thin sections in the OM has become a well-proven technique for examining defects and components in concrete, limited only by the wavelength of visible light. Grinding and polishing techniques for the production of thin sections were first used in 1858 by H. Sorby for the microscopical study of rock minerals [8.9]. The fluorescent thin sections were introduced around 1980 in concrete science and are particularly useful for the study of porosities and microcracking in concrete specimens.

The microstructural characteristics of a cement-based material can be investigated by direct observation using an optical or an electron microscope. Qualitative analysis is the object of cement chemistry, hydration processes etc. In pore structure determination, it is the quantitative analysis of the features observed that are of interest. The determination of spatial size distributions from data that can be obtained from sections or projections is possible by using mathematical and statistical analysis techniques that constitute the science of stereology. Direct observations of pore casts, polished sections, and thin sections in concrete with image processing have developed into a powerful technique for pore structure analysis. Image analyzers were introduced around 1970, and image analysis has been used to provide a quantitative evaluation of the number, size, and shape of pores. With the development of modern image analysis techniques, stereology is playing an increasingly significant role.

In this chapter, the principles of optical and scanning electron microscopy are briefly discussed and the use of image analysis to quantify the features observed on a plane section using microscopy are presented. The principles of stereology that help extract volume information from information obtained on a section are also briefly reviewed, with emphasis given on the use of the techniques to study the microstructure and the air void structure in cement-based materials.

8.1 Optical microscopy

The OM magnifies small objects, enabling the observer to directly view structures that are below the resolving power of the human eye, which is approximately 0.1 mm. Microscopy is the most definitive method to observe and measure particles or features. Even though it becomes highly subjective, if the basic principles of sampling, preparation, and counting are followed, a precise count can be made with a thorough understanding of the nature of the particles being studied. The basic principles of optical microscopy are briefly reviewed in the following sections. The reader can find detailed information in technical publications treating the subject (e.g. [8.10,8.11]).

8.1.1 Types of microscopes

Different types of microscopes are available based on the direction of light during operation and the type of image the microscopes provide. The direction of light during operation may be reflected or transmitted (see Figure 8.1).

Reflected light microscopy can be performed on highly polished surfaces, on unpolished (e.g. fracture) surfaces, and on polished thin sections.

Transmitted light microscopy is performed on thin sections only, which have a thickness of about 25–30 µm (see also Chapter 2). The light source is located below the specimen; the light emitted from the source is collected and is directed towards the specimen after passing through a number of filters, diaphragms, and lenses. Light is shone through the specimen, collected by an objective lens, and directed towards the eye piece (ocular). The light is controlled in several ways, both below and above the specimen, so that light interacts

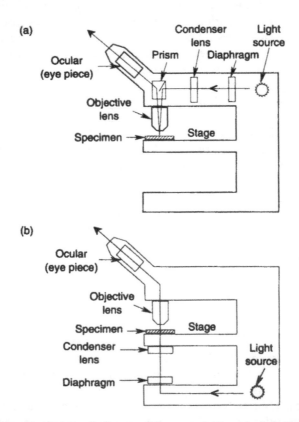

Figure 8.1 Schematic out-of-scale layout of the two microscopes most commonly used in concrete science. (a) Reflected light microscope, and (b) transmitted light microscope. The arrow indicates the direction of the light beam.

with the specimen in a certain way that is understood and can be used to help interpret the resulting images.

Microscopes can also be divided into compound microscopes and stereo-microscopes, based on the type of image they can provide.

Compound microscopes are the most common type of microscopes and provide a two-dimensional image to the observer. They have a number of objectives of varying magnifications mounted in a rotating nosepiece. Typically the range of magnification on a compound microscope is between 40× and 1000×, although some microscopes are capable of higher or lower magnifications. Because only one objective is used at a time, the viewer sees a two-dimensional image of the specimen, and the image is reversed and upside-down. The compound microscope consists of four components:

- The light source
- The condenser lens that directs light rays through the specimen
- The objective lens, that is nearest to the object
- The eyepiece (or ocular).

The condenser lens collects light rays from the light source, which are focused onto the specimen. The objective lens forms a real, magnified image of the object in the intermediate image plane.

Stereomicroscopes give the viewer an erect upright and unreversed three-dimensional (stereoscopic) image, which is particularly useful for viewing voids in concrete. The stereomicroscope is a specialized form of the compound microscope and gives the operator a real sense of depth when observing specimens. Most stereomicroscopes are used at magnifications from 5× to 50×, but with the proper accessories, magnifications up to 400× can be achieved. Mechanically, stereomicroscopes are arranged so that image focusing is adjusted by raising or lowering the entire microscope body to or away from the specimen, using a rack and pinion mechanism. On modern high power microscopes, however, focusing of the microscope is achieved most frequently by raising or lowering the specimen stage.

Fluorescent light microscopy Because of the porous nature of concrete, it is possible to fill the voids and pore spaces with resin that has been mixed with a fluorescent dye, which fluoresces under ultraviolet illumination (as it was described in Chapter 2). Fluorescent microscopy requires a strong light source, an exciting filter inserted before the condenser to ensure that only the shorter wavelengths reach the specimen, and a suppression filter placed between the specimen and the ocular to exclude the shorter wavelengths and isolate the fluorescence for viewing. A deep blue (UV) filter is inserted before the light beam enters the thin section, and the resultant ultraviolet light causes the fluorescent dye in the thin section to fluoresce yellow. By inserting a yellow filter after the light has passed through the specimen, all excessive blue light is filtered out and only the yellow light emitted from the fluorescent epoxy in the specimen is allowed to pass through the ocular lens and to the observer. As the epoxy is located in original voids and cracks in the concrete, including the capillary pores of the cement paste, this method highlights strongly the presence of pores in concrete [8.12]. For observation of porosity on thin sections, the concrete is required to have been impregnated with a thin epoxy containing 0.8–1.0% of fluorescent dye [8.13].

Materials with a crystalline structure cannot be studied using ordinary white light, principally because the light vibrates in all directions and consists of a range of wavelengths; this results in a composite of information, which is analytically useless. The polarized light microscope can overcome these problems. By placing a polarizer in the light's path before the sample, light is allowed to vibrate in one direction only, which enables the microscopist to isolate specific properties of a material in specific orientations. *Plane-polarized light* is polarized in an east–west direction, when it passes through the specimen and is not repolarized before reaching the eye. *Cross-polarized light* is polarized before passing through the specimen, and is re-polarized after leaving the specimen, and before entering the eyepiece [8.14].

8.1.2   Characteristics of the microscope

As the sophistication of investigations has increased over the years, modern OMs have evolved into instruments that have superior spatial resolution or depth of focus. An OM includes the following general characteristics: numerical aperture (NA), depth of field, maximum resolving power (resolution), and magnification [8.10].

The *numerical aperture* is a measure of the light-gathering capacity of the objective. The major function of an objective lens is to collect the maximum quantity of light coming from any point on the specimen, and combine it into an image. The NA of the objective is a measure of the light collection function: the higher the NA, other things being equal, the better the objective is able to separate the details of the specimen in forming an image, and the brighter the image. Objectives with higher NA are usually more expensive. The NA of a specific objective lens is a constant and its value is usually engraved on the lens mount along with its magnification or focal length.

Depth of field is the vertical distance in the specimen measured from above and below the exact plane of focus that yields an acceptable image. The depth of field depends on the NA: the higher the NA, the shallower the distance.

Maximum resolving power (usually mentioned as resolution) is the ability of the objective to yield an image that clearly distinguishes separate points lying close together in the specimen. The shorter the distance between the points that can be separated, the better the resolving power of the objective. The typical unaided human eye can resolve points that are about 120 μm apart; this means that two discrete object points with a separation of 120 μm or greater can be resolved as two separate and distinct entities. The maximum resolving power of a compound light microscope is approximately 0.2 μm and of a stereoscopic microscope about 8–10 μm. The resolving power is determined by the wavelength of illumination, and by the NA of the objective lens: the higher the NA, the better the resolving power.

Magnification is the ratio of a line obtained on a photo micrograph to a similar line located in reality on the specimen. From a statistical viewpoint, lower magnifications enable measurement of large areas, lessening the influence of sample heterogeneity. Magnification must be high enough so that the features to be characterized are easily resolved. On the other hand, if the magnification is too high, each field may show only a few features and more fields are required to be observed in order to obtain a certain degree of accuracy. Therefore, the magnification is selected as a compromise between the requirements for resolution and for statistical sampling.

The OM is used to identify and count particles with size ranging from 0.5 to 100 μm in diameter. The resolving power of an OM is optimally about 0.25 μm, and can be improved down to 0.1 μm by using ultraviolet light and quartz optics. Capillary pores can be studied at 350×, a magnification that is sufficient to resolve pores wider than 1 μm. The theoretical limit for resolution with an OM is of the order of 0.10 μm, but for practical purposes 10.0 μm is a more reasonable limit [8.13].

8.2 Scanning electron microscopy

The SEM is an instrument designed primarily for studying the surfaces of solids at high magnification. The images are qualitatively similar to the images obtained by an OM, but the SEM possesses much greater resolution and depth of field. The use of electron beams has certain benefits in image formation compared to light beams since the interaction of electrons with solids is more diverse than the interaction of photons with solids. In addition, SEM can provide information on crystal orientation, chemical composition, magnetic structure, or electric potential in the specimen [8.15].

The SEM has a relatively high instrument cost (US$300 000–500 000) but can provide a wide range of information in a timely manner; in addition, the testing cost is about US$100 per hour. For this reason, SEM often replaces the OM as the preferred starting tool for microstructural studies. The images obtained by SEM can be magnified up to 300 000×, because the wavelength of electrons is not as limiting as the wavelength of visible light, and therefore features with size of a few nm become resolvable. Sample requirements are more stringent for the SEM than for the OM: the specimens must be coated with a thin conducting layer and must be vacuum compatible, since for the conventional SEM a pressure less than 1.33×10^{-3} Pa is maintained. The depth examined inside the specimen varies from few nm to a few μm, depending upon the accelerating voltage and the type of electrons used for obtaining the image, i.e. backscattered or secondary electrons.

Specimens up to a few mm in each dimension can easily be accommodated inside the SEM and even larger specimens can be used with some restrictions to movement. Electrically insulating materials such as cement pastes accumulate electrical charge from the primary electron beam, which deflects both the beam and the trajectories of the collected electrons, and results in a grossly distorted image. To avoid this problem, the specimen is most often coated with a thin layer of conducting material, usually carbon. As an alternate solution, the specimen may be examined at a reduced accelerating voltage, when the electron emission from the surface can balance the rate of arrival of electrons in the beam.

The specimen sits in a vacuum chamber close to the probe-forming lens, with sufficient clearance that it can be moved and tilted freely about two axes. The specimen is surrounded mostly from above by a variety of detectors that are sensitive to a variety of signals. The specimen should always be placed in the SEM with the surfaces normal to the electron beam so that the magnification, which changes with working distance, will be the same on all areas of the viewing screen. The signal is captured and viewed on a screen. Particle counting can be done directly from the viewing screen, from photographs, or by using an image analyzer.

8.2.1   Design and physical basis of operation

The principles, electron optics, imaging modes, and other aspects of scanning electron microscopy are covered extensively in many texts (e.g. [8.16,8.17,8.18]). In this section, only a brief overview is presented.

A schematic diagram of a typical SEM is shown in Figure 8.2. Even though details of the arrangement vary between different manufacturers, the SEM has the following essential features [8.15]:

- An electron gun that produces a narrow beam of electrons accelerated through a potential difference of up to about 50 kV.
- Two or three lenses that focus the electron beam to a spot with size 0.5–10 nm on the specimen surface.
- A system to deflect the beam over the raster on the specimen.
- A specimen stage that holds the specimen and permits movement, tilt, and rotation of the specimen.
- A unit for collecting and amplifying the emitted electrons.
- Cathode-ray tubes (CRTs) to display the image.

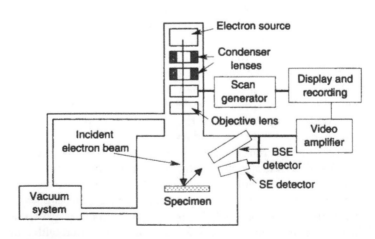

Figure 8.2 Schematic out-of-scale setup of a typical SEM.

- Electronic circuits to supply and control the electron gun, accelerating voltage, lens currents, scan generator, and signal amplification.
- A high-capacity vacuum system to maintain a pressure below 1.33×10^{-3} Pa and permit rapid evacuation after changing the specimen.

SEM images are formed by a quite different mechanism from the mechanism in an OM. In SEM, no objective lens is used, but instead, images are built up point by point, in a way similar to that used in a television display. Electrons are emitted from a cathode, and are accelerated by the application of a voltage difference between the cathode and anode. The electrons are focused in vacuum into a fine probe with size 5–10 nm, if a thermionic electron gun is used, and 0.5–2 nm, if a field emission gun is used. As the electrons penetrate the surface of the specimen, a number of interactions occur that can result in the emission of electrons or photons from the surface of the specimen or through the specimen. To build up the complete image, the electron beam in the microscope is scanned over an area on the specimen surface and the pattern of scan is called a raster. In an SEM, the raster is normally square and is covered in a line-by-line scan of up to 1000 lines (as compared with 625 lines in most television sets). Appropriate detectors can collect a reasonable fraction of the electrons emitted, and the output can be used to modulate the brightness of a CRT. The beam in the CRT display is scanned in synchronism over a geometrically similar raster: every point that the beam strikes on the sample is mapped directly onto the corresponding point on the screen. The image dimension is therefore measured in pixel dimensions along the x and y axes, typically being 512×512 pixels for lower resolution images and more for higher resolution images. A shade of gray is allocated to each pixel depending on the intensity of the signal received from the specimen surface. The intensity of the CRT is modulated by any of the signals resulting from the electron–specimen interactions. The magnification is increased by scanning a smaller area on the sample while keeping the image size on the CRT constant. The image on the CRT is a map of the intensities of the electron emission from the specimen surface, in a similar way as the image in an OM is a map of the light reflected from the surface.

8.2.2 The performance and characteristics of the SEM

The important characteristics of a SEM are the depth of field, which is a purely instrumental factor, the image noise, on which the specimen has some influence, the resolving power (resolution), on which the specimen has a very great influence, and the contrast, which depends on the specimen [8.15].

Depth of field is the distance along the microscope axis through which the specimen can be moved without blurring the image. The depth of field is sometimes wrongly called depth of focus. The SEM has about 100–300 times the depth of field of an OM.

The SEM *resolving power* depends on the operating conditions, the imaging mode, and the nature of the specimen. The image resolution from SEs is 10 nm or better, while the image resolution from backscattered electrons (BSEs) is about 0.4 μm, a size that corresponds to the size of coarse capillary pores in cement pastes. The resolution of a BSE image is comparable to the resolution of optical microscopy, however the better image contrast in BSE images allows better definition of the constituents than with optical microscopy imaging [8.19]. In some cases, the resolution attainable is limited by the electron penetration depth, and can be used to count particles ranging in size from 1 μm to 0.1 mm. Particles smaller than 0.1 μm usually have too low a contrast with the background to be counted efficiently.

The resolution depends on the probe size and on the information volume that contributes to the signal, and thus on the specimen and the mode of operation. The magnification is given simply by the ratio of the side-lengths of the display and specimen rasters, and is normally variable from about 20× to 100 000×. The resolution, however, is the most important quantity, and with an ideal specimen it is at best equal to the diameter of the electron beam where it strikes the specimen surface (the beam cannot in practice be focused to a perfect point). In current high performance instruments, this can be as small as 5 nm, compared with a resolution of about 300 nm for an OM. However, the resolution in the SEM depends critically upon the nature of the specimen and the mode of operation of the instrument, and 10–15 nm is a more typical figure [8.15]. For a resolution of about 10 nm and considering that the SEM is capable of very low magnification from about 10× up to about 50 000×, it can be used to identify particles ranging in size from 1 mm to 0.1 μm.

The *contrast* of the SEM images depends more on the topography of the sample than on differences in atomic number. Particle counting can be done directly from viewing the screen, from photographs, or by using an automatic image analyzer. Areas impregnated with epoxy resin appear almost black and provide an easy identification of cracks, air voids, and porosity content of the concrete. Particles smaller than 0.1 μm usually have too low contrast with the background to be identified efficiently.

8.2.3 *Electron–specimen interactions and principal images*

In the SEM, the interaction of the electron beam with the specimen is complex. The energy of the incoming electrons is dissipated by a series of scattering events over a fairly large volume of the specimen, which is known as the interaction volume (see Figure 8.3). The electron penetration depth is typically of the order of a micrometer. In some cases, the resolution attainable is limited directly by the electron penetration depth. In cement-based materials the interaction volume may be about 2–3 μm in depth and diameter [8.20].

The signals from the specimen include: SEs, BSEs, elastically scattered electrons, low-loss electrons, characteristic X-rays, and Auger electrons. The principal images produced in the SEM are of three types: SE images, BSE images, and elemental X-ray maps. SEs are detected from a near-surface region of the interaction volume, while BSEs are detected from both the near surfaces and from much deeper areas of the specimen. Primary electrons have high energy (up to 40 keV), SEs have low energy (less than 50 eV), and BSEs have energy ranging from 50 eV to that of the primary electrons. Therefore, SEs and

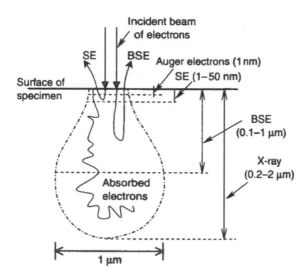

Figure 8.3 Volume of interaction of an electron beam with the specimen. The depths inside the specimen that generate the BSEs, SEs, X-rays, and Auger electrons are indicated in the figure (values in parentheses). The total depth of interaction in the specimen can be up to 5 μm, including X-ray emission and secondary fluorescent emission, which are not shown in the given figure.

BSEs are conventionally separated according to their energies because they are produced by different mechanisms. An additional electron interaction of major importance in the SEM occurs when the primary electron collides with and ejects a core electron from an atom in the solid. The excited atom will decay to its ground state by emitting either a characteristic X-ray photon or an Auger electron. The X-ray emission signal can be sorted by energy in an energy dispersive X-ray detector (EDX) or by wavelength with a wavelength spectrometer. These signal distributions are characteristic of the elements that produced them and SEM can use these signals to produce elemental images that show the spatial distribution of particular elements in the field of view. The ability of SEM to obtain elemental images is used for determining different compounds in the cement paste for structural characterization. Due to collision of electrons inside the specimen, the spatial resolution of an elemental image is rarely better than 0.5 μm.

Once generated, an electron must escape from the specimen in order to be detected, and the escape depth also increases with increasing energy. Reducing the accelerating voltage reduces the depth of the sample imaged by BSE, and so increases the potential spatial resolution; however the atomic number contrast is also reduced. The effect of accelerating voltages between 10 and 20 kV on the BSE image has been investigated by researchers, and has been found that 15 kV gives the best resolution with adequate atomic number contrast for cement pastes and are commonly used [8.20].

8.2.3.1   Secondary electrons

SEs result from the inelastic scattering of the incoming electrons. These have much lower energies than the incident electrons, so they escape from a thin layer at the surface of the specimen and are collected by a detector. The intensity of SEs depends on the local inclination of the specimen surface, and they are used for imaging of fracture surfaces and for porous materials and provides surface topographic data. Intensity variations on the screen image exhibit a three-dimensional character of the surface as a consequence of the high depth of field and current contrast.

Many hundreds of SEs can be excited within the specimen by a single primary electron, but only the SEs that are excited within approximately 10 nm of the surface can escape into the vacuum and be detected. Thus, SEs can be subdivided into electrons that escape from the specimen at the point where the primary electrons enter the specimen, and electrons that are excited at the point where the electrons leave. In most cases, the primary-excited SEs are capable of higher resolution than the SEs excited at other locations. SEs produced much deeper in the material suffer additional inelastic collisions, which lower their energy and trap them in the interior of the solid. The output of SEs varies according to the accelerating voltage of the incident electron beam, charge accumulation on the specimen's surface, the local density of the specimen, and, in particular, to the angle that the incident beam forms with respect to surface features.

8.2.3.2   Backscattered electrons

BSEs are incident electrons that have been elastically scattered through a wide angle so that they re-emerge from the specimen. BSEs have energies similar to the energy of incident electrons and they are emitted and detected from greater depths than the SEs (see also Figure 8.3). The terms re-diffused and reflected are also used to describe BSEs.

The intensity with which electrons are backscattered depends principally on the local atomic number of the specimen, but it is also affected by topography. Backscattering is more likely to occur for higher atomic number of a material. Thus, as a beam passes from a low atomic number to a high atomic number area, the signal due to backscattering, and consequently the image brightness, will increase. Other factors affecting the BSE intensity are the angle between detector and specimen, as well as on the angle between specimen and electron beam, and the atomic number of the irradiated specimen area. In contrast to the SEs, BSEs, with their higher kinetic energy, reach the detector system in straight lines with negligible electromagnetic deflection; therefore, the contrast due to shadowing in BSE images is more intense than in SE images.

SEM microstructural research in cement-based materials has concentrated mainly on BSE imaging of plane polished surfaces, usually of epoxy-impregnated specimens. BSE images of a fracture surface are also of high quality and suitable for analysis due to the high depth of field of the SEM (see Figure 8.4). SE images are often at high contrast, but are not as rich in detail as many BSE images.

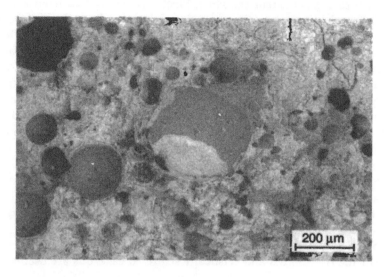

Figure 8.4 SEM image obtained using BSEs at low magnification (100×) on a fracture surface of air-entrained concrete. The air voids are clearly visible. A Philips XL30 ESEM was used, operating at an accelerating voltage of 20 keV.

8.2.4 Environmental scanning electron microscopy

SEMs require a high vacuum to operate and consequently specimens are observed in a dried state. Various attempts have been made to minimize the effects of drying on specimens that contain water, either by the use of very low temperatures or by the use of environmental microscopes. The environmental scanning electron microscope (ESEM) was developed by G. Danilatos, 1979, 1981, and uses a differential pumping system that enables the sample chamber to be maintained at relatively low vacuum (around 2.67×10^3 Pa) [8.21,8.22]. The gaseous layer around the specimen becomes ionized and suppresses the accumulation of charge, therefore no conductive coating is necessary to avoid problems with accumulation of charge on the specimen. The gas is used as the detection medium, so by using water vapor as the imaging gas, water-saturated conditions can be produced by controlling the temperature of the sample.

The ESEM has become popular with cement pastes because it allows observation of the microstructure without the associated alterations introduced by drying techniques. The technique has proved very useful in studying hydrating cement pastes and giving insight to the processes and mechanisms involved in the hydration of cement.

8.2.5 Transmission electron microscopy

The transmission electron microscope (TEM) compared to the SEM is analogous to transmitted light microscope compared to reflected OM. The principles of SEM are to a great extent the same. Electrons are emitted from a cathode and accelerated by a voltage between the cathode and anode. The beam of electrons is focused down to a fairly small spot by a first condenser lens. A second condenser lens projects the beam at the specimen and enables the convergence angle and area of illumination to be controlled. An aperture controls the number of electrons allowed into the beam and hence helps control the intensity of the illumination. The electron beam illuminates the specimen, which must be very thin, typically 5 nm–0.2 μm for 100 keV electrons, depending on specimen density and elemental composition. The electrons may be undeflected (transmitted with no interaction), deflected with no loss of energy (elastically scattered), or deflected with a significant loss of energy (inelastically scattered). Particular groups of electrons may be selected to contribute to the final image by the use of apertures. The electron-intensity distribution behind the specimen is magnified and imaged with a three- or four-stage lens system onto a fluorescent screen.

In a SEM, only BSEs and SEs are used for imaging, while in a conventional TEM only the transmitted and elastically scattered electrons are used to form images.

8.3   Scanning acoustic microscopy

Acoustic microscopes have gained much popularity in Materials Science due to the little effect they have on the material that is used during testing. One very popular acoustic microscope is the Scanning Acoustic Microscope (SAM), which has progressed rapidly during the 1980s to become a commercially available laboratory tool [8.23]. In 1936, S. Sokolov first described the possibility of utilizing acoustic waves for microscopy. The development of the pulsed acoustic systems, which enabled the reflections of the acoustic beam from the specimen to be separated from spurious reflections, and the creation of images by raster-scanning the specimen under a focused beam have led to the designs of SAMs available today.

Scanning acoustic microscopy is a high resolution, high frequency ultrasonic imaging technique, most commonly used at a high magnification mode, where it is employed as a counterpart to conventional optical and electron microscopy, to see fine detail at and below the surface of the sample. SAMs use focused ultrasonic waves and operate at frequencies around 1–2 GHz to produce images with a resolution comparable to that of the conventional OM. Low frequency ultrasound (in the range 2–10 MHz) gives spatial resolutions in the millimeter range and can obtain greater depths of penetration in the material. The method uses ultrasound waves to image the local impedance at each point of the material. In an acoustic microscope, two-dimensional images are built up by scanning a point transducer in the X and Y directions. The same aspects of the material structure that determine the mechanical properties of the specimen, influence the propagation of mechanical waves in the specimen. The principles of operation for SAM are described in detail by A. Briggs, 1992 [8.23,8.24].

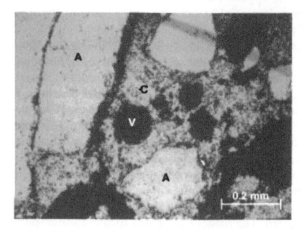

Figure 8.5 Image obtained on the polished surface of a concrete specimen using a Scanning Acoustic Microscope. The aggregate particles (A), cement paste matrix (C), and the air voids (V) can be identified easily due to differences of their density. It is worth noticing the interface between the aggregate particles and the cement paste.

The technique was introduced to concrete science in 1994 by K. Aligizaki *et al.*, for observations of cracks and pores on the polished surface of hardened concrete specimens [8.25]. It was later used by R. Livingston and colleagues for studying the microstructure of cement pastes [8.26]. Due to the difference in density between cement paste, aggregates, and voids, the constituents of a concrete specimen can be easily identified (see Figure 8.5). The technique can provide images very fast and does not require special preparation techniques. However, it involves the use of expensive equipment and trained personnel for its operation. For this reason, its use at the current state could only be limited for scientific studies in association with OMs or SEMs.

8.4   Image analysis

Image analyzers have become very popular in materials science since the 1970s. Image analysis has been used successfully for a long time in metallography, mineralogy, and biology, and in connection with stereology represents a method for describing the size distribution of pores in cement-based materials. Image analyzers can be interfaced with an optical, an electron, or an acoustic microscope to permit direct analysis of the image captured. The main objective of image analysis is to aid quantifying the parameters of interest using contrast differences in two-dimensional images [8.27].

The techniques of image analysis evolved very rapidly thanks to progresses in image acquisition and the development of algorithms and software, for general or specific applications. Various semiautomatic and fully automatic devices have been developed and are currently available that permit rapid data collection and analysis.

Semiautomatic instruments make use of a human operator to detect and distinguish the objects to be analyzed. The features of interest are input into a computer, which then provides the analysis and displays the numerical results.

Fully automatic instruments replace the human operator with automatic detection and identification of the features of interest. Fully automatic instruments view the image with a television scanner, digitize the image into a large number of picture elements (pixels), each with an associated gray shade, and transfer the digitized image into the computer. Then, the digitized image analyzer can be processed by appropriate software to extract the geometrical information and display the results.

In this section only a brief overview of the main characteristics of image analysis is presented. The reader can obtain further information from technical publications that treat the topic in great detail (e.g. [8.28,8.29]).

8.4.1   *Image analysis steps*

The most important component of an image analyzer is the scanner, because the resolution capability of the scanner is the limiting factor in the image analysis system. The primary component in image resolution is the number of resolvable picture elements (or pixels) in the image, which depends on the quality of the tube used and the magnification. The scanner must exhibit high differential sensitivity to permit distinction and identification of constituents having small gray-level differences.

Image analysis consists of several steps including area selection, image acquisition and digitization, image processing, feature recognition, and data analysis and output [8.30]. Each step must be controlled properly to obtain accurate, reproducible results. Sample selection must be systematic and well-planned to ensure that the samples analyzed are representative of the whole section. Pixel densities of 512×512 and 1024×1024 are the most popular densities used.

8.4.1.1   *Preparation*

The area selected for analysis must exhibit adequate contrast to allow the researcher to distinguish the various components of interest on the surface examined. In order to identify and analyze the pore structure in cement-based materials, porous regions, and solid regions in the image must be differentiated by selecting the gray level that defines the boundary between pore and solid matrix. Image processing then creates a binary image in which the black areas are pores and the white areas are solid phases of the cement paste [8.31]. It is advised that the cavity space of the pores be filled with a synthetic resin, and that after grinding the surface be smoothened by polishing to provide better contrast and reduce noise due to surface anomalies [8.32].

8.4.1.2 Detection and measurement

The primary mode of feature detection is gray-level thresholding. Several procedures, ranging from automatic to manual thresholding, are available for feature detection. For optimum detection accuracy, the features of interest must be treated to have a contrast range that is as narrow as possible. Setting the threshold to detect only gray levels within specific ranges enables the selective detection of constituents.

8.4.1.3 Digitizing

The image captured from the microscope is digitized by the computer into an array of square pixels whose locations in the array are stored in the computer. Image processing refers to the manipulation of the detected image to improve the accuracy of measurements, e.g. separating adjoining particles before counting.

White has a digitized intensity value of 255 (2^8 bits $= 256$) and black the value 0; in this way, all gray levels in a digital image between black and white are represented by 256 values between 0 and 255. The human eye can only discern between 40 and 60 different values of gray levels. Any image can be described in terms of the histogram of its gray levels [8.30]. Each pixel is assigned a gray level corresponding to the average brightness of the image at its particular location. Typically, in modern commercial image analyzers there are about 10^5–10^6 pixels and 10^2–10^3 gray levels in an image array.

The first image processing routine is usually a smoothing filter that evens out minor undesirable variations in the image due to noise. However, any smoothing also reduces resolution of image detail, therefore, little or no processing is sometimes preferred in order not to alter the image. Image enhancement procedures are becoming prevalent and produce a more attractive picture, especially when insufficient contrast, artifacts, and/or distortions prevent straightforward feature analysis. Image analysis involves the operations that usually quantify some aspects of the image, such as area fraction or diameter of a feature or particle.

8.4.1.4 Image processing and analysis

During the analysis of objects in images it is essential that we can distinguish between the objects of interest and the remaining of the image, which is referred to as the background. The techniques that are used to find the objects of interest are usually referred to as *segmentation* techniques because they segment the foreground from the background. Segmentation is the most critical image processing step. Two of the most common segmentation techniques are thresholding and edge finding. *Thresholding* is the process of differentiating phases on the basis of their gray level. The operator selects a range of gray levels that corresponds to the objects of interest. In the binary image, the pixels within the segmented gray-level range are on, while the rest are off. In most cases, the threshold is chosen from the brightness histogram of the region or image that we wish to segment. *Edge finding* is a procedure that finds the pixels that belong to the borders of the objects. The number of pixels in the image determines the resolution and an image with 512×512 pixels uses all the resolution that an OM can offer, which is about 5 µm. In this way, pores with diameters of about 5–200 µm are identified.

A scientist trying to quantify microstructural features using image analysis needs stereological relationships in order to relate the parameters observed on the image to the parameters in the three-dimensional material. In the following sections, a brief review is given on the mathematical relationships between two- and three-dimensional features in order to give the reader a better understanding of the methods used and reported by the different authors.

8.5 Stereology

Stereology, sometimes also mentioned mistakenly quantitative microscopy, is the science of reconstructing a three-dimensional picture of an object from measurements made on a two-dimensional plane of polish. The term stereology was proposed for the interdisciplinary branch of science dealing with the body of methods that facilitate the investigation of solid bodies when only sections and projections of these bodies are amenable to direct observations. Stereology deals with the geometry of microstructures and with quantitative characteristics of points, lines, and areas. For several decades quantitative microstructural analysis has been an important component in the characterization of materials. In the present context, the specific tasks that the concrete scientist faces are to evaluate, as far as may be possible, the porosity, specific surface, and pore size distribution of the porous cement-based material, given information of a two-, one-, or zero-dimensional character, i.e. of areas, lines, and points, respectively.

All the pioneering work in volume analysis application on materials, including the discovery of the principles and the development of equipment has been done by petrographers. Stereological methods have been theoretically developed for all kinds of particles. Different shapes that can be analyzed are spheres, convex particles, ellipsoids, lamellar spacings. The methodology is very well developed in the case of spherical particles because the sphere is the simplest shape to treat, since it possesses maximum symmetry. Entrained air voids in concrete are spherical and most stereological mentions in the technical literature that relate to concrete science have been for spherical pores. In recent years, however, with the developments of image analysis and observations on microstructure of cement pastes, reference is also made to other shapes.

A section through the solid exposes the internal objects in a very specific way: the section will cut across some of the objects, but will miss others. There are three methods of obtaining data about the features observed on a plane section: point counting, lineal analysis, and section analysis, which is also mentioned as areal analysis. Each method requires two types of information for data analysis:

- Basic stereological counting measurements that are obtained experimentally (e.g. from the plane section through the solid).

- Fundamental theoretical relations in stereology, which are used to calculate the geometric properties of the three-dimensional microstructure from the counting measurements. By application of the principles of mathematical probability, equations have been derived relating the surface parameters to the actual parameters in the solid and are mentioned in several technical publications [8.33,8.34,8.35].

In general, more measurements improve the accuracy of the results, however, in practice, the researcher must balance the accuracy desired with the amount of effort that may logically be expended in making the measurements.

Generally, the parameters needed to characterize a microstructure are the mean diameter, D, the standard deviation, $\sigma(D)$, and the total number of particles per unit volume, N_V. However, in the case of air voids in concrete a size distribution is needed to characterize the distribution of air voids in concrete. In Table 8.1 a comparison is made of some commonly mentioned methods for obtaining volume fraction and size distribution of systems of particles mainly with circular shape (spheres). Detailed information about

Table 8.1 Some of the most commonly mentioned stereology analysis techniques developed by researchers for determining pore volume and pore size distributions in a three-dimensional solid from measurement done on a plane (two-dimensional)

Method	Parameters determined			
	Volume	*Size distribution*		*Application*
Point count	Thomson, 1930	—		—
	Glagolev, 1933	—		—
Lineal analysis	Rosiwal, 1898	Spektor, 1950	Analytical	—
		Lord, Willis, 1931	Graphical	Sphere
		Cahn-Fullman, 1956	Analytical	Lamellar
		Bockstiegel, 1966	Analytical	Sphere
Section analysis	Diameters —	Wicksell, 1925	Coefficients	—
		Scheil, 1931	Coefficients	—
		Schwarz, 1934	Coefficients	—
		Saltykov, 1958	Coefficients	—
	Areas Delesse, 1848	Johnson, 1946	Coefficients	—
		Saltykov, 1958	Analytical	Spheres, convex particles

Note
An analytical method provides equations for calculations, and a method with coefficients uses values given in tables.

these methods of analysis or others applicable to various shapes is given elsewhere (e.g. [8.36]).

Stereological measurements are fundamentally statistical in nature. A measurement (or series of measurements) is made on a field from the sectioned area and recorded. Then a new field of view is revealed, the same measurements are made on that field, and the procedure is repeated. The number of fields required depends on the magnification used and the features examined. The mean value of each set of measurements and the standard deviation is computed to characterize the precision of the measurements [8.34].

Certain assumptions are made for systematic identification and counting of particles, voids, or minerals in a finely ground surface or thin section in order to estimate volume proportions. These assumptions are the following [8.13,8.37]:

- The particles of various sizes and shapes are distributed randomly throughout the specimen without any regular packing.
- The surface or section is cut through randomly orientated particles.
- The sectioned area is large enough to be representative of the material.

In the following sections, the stereological techniques used are briefly described, and the relationships that relate the features obtained on a plane section to the pore parameters in the solid are presented. Detailed information about the principles and derivations can be found in the technical literature [8.35,8.36,8.37,8.38,8.39,8.40].

8.5.1   Point counting

Even though point counting is the simplest way of measurement, somewhat surprisingly, it did not come into use until many years after the lineal and areal methods, which are described in Sections 8.4.2 and 8.4.3. The first application to structure analysis appears to have been made in 1930 by Thomson [8.41]. In 1934, Glagolev introduced a one-dimensional point count performed by advancing the specimen in discrete steps under the cross hairs of a microscope; this method was taken up by F. Chayes, in 1949, who devised a simple modification of a standard stage involving the addition of click stops on the traversing screws, which facilitated the advance of the specimen in constant increments [8.42].

If a point is placed at random within a volume, the probability that it will fall within a given constituent is equal to the volume fraction of that constituent. Consequently, the expected value P_P of the fraction of points within a constituent is given by Equation (8.1). Similarly, if points are placed at random on a plane section the expected value of the fraction of points are given by Equation (8.2), and for a random dispersion of points along a line are given by Equation (8.3) [8.36]:

$$P_P = V_V \tag{8.1}$$

$$P_P = A_A \tag{8.2}$$

$$P_P = L_L \tag{8.3}$$

where P_P is the fraction of points falling within the particles of interest, V_V is the volume fraction of the particles of interest, A_A is the areal fraction of the particles of interest, L_L is the line fraction of the particles of interest.

The areal fraction can be estimated by determining the fractional number of points in a random or nonrandom array that fall within the boundaries of the constituent on the plane of polish. The analysis using a nonrandom array of points is called systematic point count. In such a count, the points are distributed in a regular fashion, e.g. at the intersections of a square grid of lines [8.36]. In the case of point counting on the polished surface of cement pastes or concrete, one examines a sufficiently large number of points on a surface and notes the total number of points, and the points that fall in the pore space (either air voids or microstructural pores) (see Figure 8.6). The importance of determining the minimum number of points to produce a valid result is critical, because counting is time-consuming and sometimes visually difficult. The analysis time increases if many points are counted. Practically, a series of traverse lines are superimposed on the sample surface and at regularly spaced stops the constituent directly under the cross hairs is identified. The stops form a rectangular grid of points. The point count is a unitless fraction. Points falling on the edge between two features (e.g. air and paste), can complicate the procedure. It can take 2–6 hours of microscopic observation to complete a point counting analysis on concrete to determine the air void parameters using a stereoscopic microscope and following the standardized method [8.33, 8.34]. The method can only estimate total pore volume, or air volume for air void analysis, without giving information about the pore size distribution.

The precision of point counting methods depends on the number of points counted, the sizes of the component particles, and the size of the area traversed, which must be representative of the volume to be sampled.

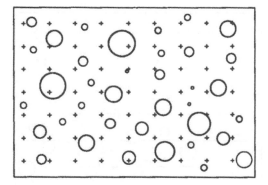

Figure 8.6 Schematic diagram of the modified point counting procedure on a plane surface received from a section through a solid containing spherical particles, such as air voids in concrete. The circles represent the cross sections from spheres, and the crosses represent the points used for the measurements. The points that fall inside the circles and the total number of points are counted.

8.5.2   Lineal analysis

Lineal analysis was introduced by the geologist A. Rosiwal in 1898 and is sometimes referred to by his name [8.43]. A. Delesse, 1848, had also stated that the volume fraction could be determined by a lineal analysis approach, but he did not develop such a method [8.44]. A lineal analysis involves traversing the surface along a series of straight lines, preferably, but not necessarily, parallel and uniformly spaced (see Figure 8.7) [8.45].

The traverse lines intercept a fraction of the depressions on the plane surface. The intercepts are random chord lengths or chord intercepts. The individual intercept lengths, i.e. the lengths of the portions of the lines passing through the pore space are measured and their frequency within a size group is obtained. Also, the total intercept length is obtained by summing the lengths of all intercept lengths as well as the total length of the traverse lines at the end of the analysis (see also Figure 8.7) [8.33,8.36]. This is a laborious process and as a result lineal analysis did not come into general use until the development of microscope stages that would automatically integrate the distances traveled in various constituents. The procedure is currently known in concrete analysis as the *linear traverse method.*

The method has been used on hardened concrete for measuring air content using thin sections, as well as with polished sections observed by reflected light. Experience has shown that magnifications of 30× to 40× are adequate for air void analysis. It has been found in the past that perception of the air voids is greatly facilitated by use of the binocular (stereoscopic) microscope [8.45]. A typical air void analysis test using a stereoscopic microscope can take 2–6 hours of microscopic observation to complete. Void size distributions are usually neglected in favor of determining the specific surface. The technique is used to measure both pore volume and pore size distribution as it is described in the following sections.

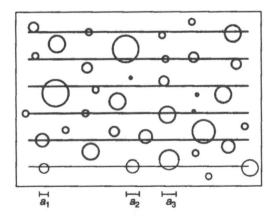

Figure 8.7 Schematic diagram of the lineal analysis (linear traverse procedure). The total intercept length is the summation of lengths $a_1 + a_2 + \cdots$. During the measurement, the total intercept length, and the total length are measured, as well as the distribution of sizes of the chords.

8.5.2.1 Pore volume

A. Rosiwal, 1898, showed that the volume fraction of the constituents in a material could be obtained from measurements along a random line through the volume of the material [8.43]. The lengths of segments or chords intercepted are summed separately for each constituent. Then, the proportion for any constituent in the volume is the summation of all chords for that constituent divided by the summation of the chords of all constituents [8.45]. If l is the total length of intercept of a test line with particles of one phase in the structure (e.g. the air voids), and L is the total length of a line within a solid volume, then the fraction of this test line occupied by the particles is l/L [8.36]. If the test lines are uniformly distributed over the area, then the probability of selecting a test line in any range of positions is the same for all positions. The total volume of a phase in the cubic sample is given by Equation (8.4), which simply states that the lineal fraction occupied by a phase in the volume of the sample is an unbiased estimator of the volume fraction of that phase:

$$V_V = \frac{V_p}{V} = L_L \quad \text{where} \quad L_L = \frac{l}{L} \tag{8.4}$$

where V_V is the volume fraction of a phase in the solid volume, V_p is the total volume of the particles in the solid, V is the volume of the solid, L_L is the length fraction of a line within the volume of the solid, l is the total length of the segments falling in a constituent, L is the total length of the line.

If the line is placed randomly on a plane containing two-dimensional features instead of placed randomly in a volume element, it can be proven that Equation (8.5) is valid.

$$L_L = A_A \tag{8.5}$$

where A_A is the areal fraction of the particular two-dimensional feature, L_L is the length fraction of a line within a feature.

Equations (8.4) and (8.5) indicate that the linear fraction of a constituent measured on a plane surface is equal to the volume fraction of the constituent in the volume.

8.5.2.2 Pore size distribution

The section chord methods are based on the determination of chord or intercept length distributions obtained with three-dimensional objects. Such a distribution would be obtained if a representative sample were intersected with uniformly distributed parallel straight lines [8.46]. The chord length distribution is obtained by intersecting the circles obtained in the sections with uniformly distributed parallel straight lines. It has been proven by F. Dullien *et al.*, 1969, that the chord distribution obtained in this manner is identical to the distribution obtained by directly intercepting the spheres in the solid with straight lines [8.47].

Analyses for determination of the pore size distribution from a distribution of chords, have been provided by several researchers using different approaches [8.37]. For example, G. Lord and T. Willis, 1951 used a mathematical–geometrical analysis, and J. Cahn and R. Fullman, 1956 provided a differential notation.

The method by G. Lord and T. Willis, 1951, was developed for air void analysis in concrete [8.48]. The researchers have given both a graphical and a formal mathematical solution. The graphical technique for determining air void size distribution presumes a distribution curve that has a decreasing slope from the origin to some maximum point. The inability of the method to predict the size distribution of the smaller void sizes is a serious drawback, because there is a significant amount of small air voids, which are also the most effective voids to freeze–thaw durability. A continuous function of the exponential type could be used to approximate the discrete frequency distribution function. To determine the air void size distribution, it is necessary to know two things: the relative frequency distribution of the spherical air voids, and the total number of the voids.

The approach by J. Cahn and R. Fullman, 1956, gives the total number of intersections per unit length of secant with chord lengths between l and $l + dl$ to be expected from a distribution $N_V(D)$ and is given by Equation (8.6) with the solution given by Equation (8.7) [8.49]:

$$N_L(l)\, dl = \frac{\pi}{2} l\, dl \int_l^\infty N_V(D)\, dD \tag{8.6}$$

$$N_V(l) = \frac{2N_L(l)}{\pi l^2} - \frac{2}{\pi l}\frac{dN_L(l)}{dl} \tag{8.7}$$

where $N_L(l)$ is the number of intersections per unit length of secant, l is the chord length, $N_V(D)$ is the distribution of spheres with diameter D, $N_V(l)$ is the number of spheres in the solid with chord length l, $dN_L(l)/dl$ is the slope of the distribution.

The frequency distribution curve of any variable is constructed by plotting the frequencies with which each magnitude of the variables occurs, against the magnitudes. It is an inherent characteristic of such a curve that the area underlying the curve is proportional to the total number of occurrences of all the magnitudes of the variable.

8.5.3   Section analysis

Section analysis is sometimes referred to as the Delesse analysis, because it was introduced by A. Delesse in 1848 and it entails measuring and then summing the areas of a constituent on a plane section [8.44]. Delesse measured the areas of a constituent by cutting up a tracing of the structure and weighing the pieces. The section analysis can be used to determine both total pore volume and pore size distribution.

8.5.3.1 Pore volume

If A is the area of intersection of a particular test plane with all particles of interest, in our case the pores or air voids, then the fraction of the area of plane surface occupied by the particles is given by Equation (8.8) (see Figure 8.8) [8.37].

$$A_A = \frac{A_p}{A} \tag{8.8}$$

where A_A is the fraction of the area of the plane surface occupied by the particles of interest, A_p is the total area of particles, A is the total area analyzed.

If the test planes are uniformly distributed in position, so that the probability of finding a test plane in any interval is the same as that for any other interval, then it can be shown that the total volume of the air voids in the volume V is given by Equation (8.9) [8.33,8.36].

$$V_V = \frac{V_p}{V} = A_A \tag{8.9}$$

where V_V is the volume fraction of a phase of interest, V is the volume of the solid, A_A is the area fraction of the plane surface occupied by the particles of interest.

Section methods are based on the fact that a spherical particle is cut by a random plane with equal probability at any distance from its center [8.37]. If the sample is large enough and the spheres are uniformly distributed in space, one intersecting plane will result in the same distribution of circles in the plane of sectioning like a set of uniformly distributed parallel planes [8.46].

Several methods are available for obtaining the spatial size distribution of spheres from the size distribution of their planar sections. The three main types of measurements that can be made on planar sections are: section diameters, section areas, and section chords. The analysis using section chords was already discussed in Section 8.5.2.2.

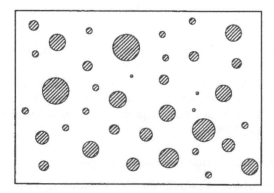

Figure 8.8 Schematic diagram of the areal (section) analysis measuring the areas of the circular cross sections. The total rectangular area, the total area from the summation of the circular cross sections, and the distributions of diameters of the circular cross sections are measured.

8.5.3.2   Pore size distribution from section diameters

Early analyses of pore size distributions from section diameters provide tables for determining the pore size distributions and were described by several researchers, E. Scheil, 1931, H. Schwartz, 1934, and S. Saltykov, 1958, as it was mentioned in Table 8.1.

Saltykov's method is based on improvements of the previous analysis methods and is considered superior to earlier approaches. All these analyses have in common that the experimentally determined distribution of section diameters is not considered as a continuous function, but is broken up into discrete size groups (class intervals). Then the particles within each interval are considered to have the same diameter. Scheil originally chose 12 subdivisions, Schwartz 10, and Saltykov recommended between 7 and 15 subdivisions [described in 8.37].

The number and sizes of the sections on the plane of polish are measured. To facilitate the calculations, the distribution of section diameters is broken down into discrete size groups. Then the probable number and size of the spheres that give rise to the observed section distribution can be deduced from this information using table coefficients [8.37]. In a polydispersed system of spheres intersected by a plane when the number of sections of the same diameter is counted, allowance is made for the fact that some of the sections derive from spheres of larger diameters. For this reason, a successive subtraction process is followed to account for the large spheres.

Instead of providing tables for obtaining pore size distributions, a probability density function is given by Equation (8.10) and was derived for the first time by S. Wicksell, 1925 [8.50]. This equation presents the relation between the size distribution of spheres and the size distribution of sections, and its solution is given by Equation (8.11):

$$f(r) = \frac{r}{R} \int_r^{R_{max}} F(R) \frac{dR}{\sqrt{R^2 - r^2}} \tag{8.10}$$

$$F(R) = \frac{2R\overline{R}}{\pi} \int_r^{\infty} \frac{1}{\sqrt{r^2 - R^2}} \cdot \frac{d}{dr}\left(\frac{f(r)}{r}\right) dr \tag{8.11}$$

where $f(r)$ is the size distribution of the sections, R_{max} is the maximum size of the spheres, $F(R)$ is the size distribution of spheres.

Equation (8.11) is a general solution of the integral equation but more practical solutions have also been presented, in order to derive a size frequency distribution of spheres from a size frequency distribution of profiles measured on random sections [8.39].

8.5.3.3   Pore size distribution from section areas

Analysis procedures have been published by W. Johnson, 1946, and S. Saltykov, 1958 to determine size distribution of particles in a solid from section areas [described in 8.37]. Johnson's derivation is applicable only to single phase structures. Saltykov's improvement of Johnson's method applies to distributions of particles as well as to grain sizes. Although Saltykov has improved

Johnson's method, the essential ideas introduced by Johnson are retained. The procedures follow a similar approach to the approach followed using section diameters, i.e. categorizing the distribution of section areas into discrete group sizes and use coefficients to calculate the size distribution in the solid.

8.5.4 Comparison of stereological methods

The three stereological techniques mentioned and used i.e. the point count, the lineal analysis and the section analysis, vary in terms of accuracy and ease of use. Studies have been carried out in the past by researchers comparing the different analysis techniques. A brief discussion on how these techniques compare is presented in the following sections.

8.5.4.1 Comparison for pore volume

In order to determine total pore volume and porosity, a researcher can use either point-count, lineal analysis, or section analysis of the structure. The choice depends solely on the preference of the experimenter. All the procedures require a sampling of the structure and will therefore be subject to a statistical error. The choice between the procedures available must be made on the basis of their efficiencies, the most efficient being the one requiring the least effort for a given accuracy. Besides the statistical errors, other sources of error in a volume fraction analysis are bias introduced during preparation of the specimen, and errors resulting from the limited resolution of the microscope and the experiments [8.36]. After selecting a method, the experimenter must decide which of several available analysis techniques is to be used for the analysis.

J. Hilliard and J. Cahn, 1961, have published a very thorough theoretical analysis of the statistics associated with the various methods for estimating volume fraction and have concluded that the systematic point count method is the most efficient of all methods proposed [8.51]. The lineal analysis and the areal analysis show approximately the same variance. For a given number of observations, the systematic point count has a lower variance than both the areal and lineal analyses. Point counting has the advantage of being free from instrumental errors. Equally spaced traverses give an appreciably lower variance than non-equally spaced traverses, as would be expected [8.36]. From the standpoint of ease of gathering the experimental data using a conventional OM, the chord methods are preferable to the areal or diametral methods. Moreover, the counting and sizing of chord lengths by automatic electronic scanning devices promise to be especially rapid [8.52]. However, by using an image analyzer with currently available software, any parameter can be acquired with the same ease.

8.5.4.2 Comparison for pore size distribution

Point count cannot give pore size distribution, therefore it is the linear traverse method vs. the areal analysis method or different calculation procedures that can be compared with each other. The analysis of the data to develop a three-dimensional particle size distribution is considerably simpler when the distribution of lineal intercepts (chords) is measured as opposed to the distribution of diameters, which is essentially an area measurement. The mathematical steps involved using the lineal analysis are fewer than the steps involved using the diameter analysis. As a result, determination of size distribution based upon the measurement of lineal intercepts is, in theory, less susceptible to error and to the accumulation of experimental differences than is the method based upon diameter measurements. Furthermore, it is easier to measure the distribution of lineal intercepts, which involves a scan along a test line than to measure diameters or principal axes, which involves a feature to feature scan in the area. It has been shown that the procedures of measurement of distribution of section diameters (Scheil, Schwartz, Saltikov), section areas (Johnson, Saltikov), and chord lengths (Spektor) are approximately equal in precision and in good agreement with one another. However, the accuracy of results depends also on the specific analysis procedure used.

F. Dullien *et al.*, 1969, studied and compared two statistical methods of particle size distribution measurement for spheres: Scheil's method, using the distribution of circle diameters and the Cahn–Fullman method based on the distribution of chord lengths [8.47]. With the help of a computer, groups of spheres were generated, which had known, pre-assigned size distributions. The spheres were sectioned to obtain distributions of circles and chords. Then the original and known sphere diameter distribution was back-calculated from the measured distribution of circle diameters and chords. A great deal of valuable information was obtained regarding the effect of sample size on the accuracy of these methods. All the distributions used were of the normal (Gaussian) type. By using planes or lines, diameter distributions and chord distributions were obtained. The researchers found that the section diameter method requires far fewer data for comparable accuracy than the section chord method. It appears that in their analysis the two consecutive sectioning operations performed in the section chord method magnify the error of the data, much as two consecutive differentiations of a function do. Therefore, it is apparent that the section chord method requires a great deal more data than the section diameter method for comparable accuracy.

8.6   Application of microscopic techniques to cement paste microstructure analysis

Pore structure in cement pastes is usually identified and quantified using SEM or OM at high resolutions. Image analysis techniques coupled with stereological methods have been used extensively in recent years in order to quantify the pore structure parameters and compare them to results obtained by indirect techniques.

Capillary pores in cement pastes made with w/c = 0.35 have been studied by researchers in the past using optical microscopy and image analysis at 100× and thin sections with UV illumination [8.53]. However, due to the limitations in resolution and magnification of the OM, the features of the cement paste microstructure are usually examined using SEM. Many modern SEMs can examine samples with dimensions as large as 150 mm × 150 mm × 30 mm [8.13]. Therefore, the dimensions of the specimens do not pose a limitation to the observations. Several studies have been reported in the technical literature on cement-based materials for analysis of clinker, cement, and concrete microstructures, and in the characterization of concretes damaged by alkali-silica reaction and by sulfate attack [8.19].

It has been found that BSE images are well-suited for microstructure imaging, because contrast in BSE images is produced by variations in atomic number within the specimen; this makes it possible to distinguish the different cement phases, anhydrous material, other hydration products (e.g. C–S–H gel, $Ca(OH)_2$), pores and aggregates [8.54]. The pores being an absence of matter do not scatter electrons and so appear uniformly black or as the darkest areas of a digital image of microstructure (see also Figure 8.4). If the incident electrons are not scattered, they will continue down into the specimen and will eventually be scattered by the material at the bottom of the pore. After thresholding, the image processing computer creates a binary image in which the black areas are pores and the white areas are solid phases of the cement paste. Some of the electrons backscattered from the bottom may re-emerge from the pore and be detected. If the pores are filled with resin, similar behavior will occur because resins usually contain organic compounds that absorb and scatter electrons weakly [8.20].

Analysis of BSE images has often been used to quantify the microstructure of cement pastes, including the determination of total porosity [8.55,8.56,8.57,8.58,8.59], and pore size distribution [8.58,8.60].

BSE images enable the study of the arrangement of pores with respect to other microstructural features and their analysis can be extremely valuable for studying the pore structure in cement pastes. However, there are limitations to the information that can be obtained in relation to the fine pores and to the three-dimensional structure of the pore network [8.20]. Other researchers have used BSE images of cement pastes obtained at a magnification of 1500× in order to resolve capillary pores at the micron scale [8.61]. High magnifications also generate problems in use, because it is difficult to maintain consistent image quality. Qualitative analysis includes identification of phases and choice of proper magnification to correctly identify and characterize the features of interest. Quantitative analysis is conducted easily at present times using image analysis. BSE images have been used by several researchers to study the pore structure of cement pastes. K. Scrivener and P. Pratt, 1987, first processed BSE images of polished sections of cement pastes to identify the various phases, such as anhydrous phases, calcium hydroxide, calcium silicate hydrate, and porosity on the basis of the gray level [8.54].

The selection of magnification during SEM analysis is an important factor that affects accuracy of results and time. The choice of magnification is a compromise between resolution of the phases and the need to obtain a representative sampling without undue effort [8.55]. The microstructure of cement paste is not homogeneous, therefore, the area fraction occupied by any component will vary from field to field. For this reason, a sufficient number of fields must be measured, so that the average area fraction of each component is representative of the bulk cement paste. The higher the magnification used, the greater will be the natural variation between fields, and therefore more fields need to be measured in order to obtain representative results. For example, one field at 400× covers the equivalent of 25 fields at 2000×, so it would be very time-consuming to conduct analysis of microfeatures at 2000× in a statistically satisfying manner. On the other hand, at low magnifications it is difficult to resolve the different components.

Several researchers have tested the effect of magnification on accuracy of results in order to identify the optimum value determined from the number of fields needed for accuracy, and therefore time, and resolution of the feature of interest [8.20,8.62,8.58,8.61,8.63,8.60]. K. Scrivener, 1986, used image analysis of BSE images obtained from polished sections of cement paste at 2000× and at 400×. She found that the difference between the porosity values determined at the two magnifications is small, with the porosity determined at 400× being generally slightly higher than the porosity determined at 2000×. Consequently, a magnification of 400× was regarded as satisfactory for the porosity measurements in her study [8.55]. In general, magnifications at 400× and 1200× are mentioned by researchers for pore structure analysis [8.60]. For evaluating the area fraction of anhydrous remnant cement grains, a magnification of 200× has been found sufficient in order to correctly quantify the area fraction of anhydrous cement grains [8.63].

Quantitative analysis of pore structure is facilitated with the use of image analysis. Due to the shape of capillary pores and their irregular arrangement in the pore network, a section through the specimen of study will reveal a distribution of features with irregular shapes and different sizes. Porosity is described by only one image-based parameter: the area fraction of pores (pore fraction). Pore fraction may be of primary importance, but alone it does not provide insight into shape and structure of porosity. Several image-based analyses have been used to extract meaningful quantities, which characterize pore structure. Sizing, two-point correlation, and fractal analysis techniques are some of the approaches used to describe several spatial characteristics of porosity [8.58].

S. Diamond and M. Leeman, 1994, used SEM image analysis to determine the size distribution of capillary pores in hardened cement pastes with water–cement ratio 0.25 and 0.40, and of ages ranging from 1 to 28 days [8.60]. They assumed that the area fraction of the whole field occupied by the pore feature is equal to the fraction of the volume of the whole specimen occupied by the pore. They compared the pore size distribution results from SEM images to the results obtained using mercury intrusion porosimetry (MIP). They found that estimates of pore diameters by MIP are two orders of magnitude smaller than the sizes revealed by the image analysis. Diameters of air voids are even more drastically underestimated by mercury intrusion. The authors attributed the differences to the limitations of MIP. In a similar study, D. Lange *et al.*, 1994, used image analysis techniques to test the nature of pore structure as observed in BSE images of polished sections of plain cement pastes, pastes with silica fume, and mortars [8.58]. They sorted the pore clusters by size and obtained plots of cumulative pore area vs. pore size. They compared pore size distributions they obtained using MIP to pore size distributions they derived from BSE images and concluded that images capture more information than MIP. Further information and discussion is given in Section 9.4.

F. Dullien and collaborators, 1974, 1981, used Wood's metal in sandstone samples and applied the principles of quantitative stereology using SEM images obtained from polished sections of the material that had been saturated by Wood's metal [8.46,8.64]. To avoid deterioration by oxidation of the polished Wood metal surfaces, micrographs were usually prepared of the polished sections and the analysis was carried out on the micrographs rather than be obtained directly from the surface of the specimen. The pores penetrated with Wood's metal appear as white areas, while the rest of the material appears as black, and in this way contrast is enhanced, and image analysis is extremely facilitated.

K. Willis *et al.*, 1998, used Wood's metal as the intruding liquid in cement pastes, and after hardening of Wood's metal inside the pores, they used BSE images obtained at 1500× for analysis [8.61]. Image analysis techniques were applied to the SEM images obtained from a polished surface of the sample after intrusion. The pore volume fraction from Wood's metal samples was calculated as the area fraction from the binary images and the results were compared to pore volume results obtained using MIP. The volume of intruded mercury reported by MIP was converted to volume fraction by gravimetric conversions using the sample mass and specific gravity. The sizes of the pores were determined using the minor and major axes of each pore section. A pore size distribution was determined from the areas of the sections and was compared to the pore size distributions obtained using MIP. The MIP curve shows much higher percentages of intrusion occurring at smaller pore sizes than the intrusion indicated by the Wood's metal intrusion porosimetry (WMIP) curve.

Attention is needed, however, when trying to calculate pore size distributions. Even though volume fraction is equal to the area fraction, as it was mentioned in Section 8.5.3.1, the pore size distribution and the size distribution of the areal fractions are different; the size distribution of the sections obtained on the plane section is related but is not identical to the pore size distribution of the pores in the volume of the material (see also Section 8.5.3.2).

In a similar study, K. Tanaka and K. Kurumisawa, 2002, used gallium to examine the pore structure in hardened cement pastes using image analysis [8.65]. The visual distinction of cement hydrates from gallium in the visual field is extremely difficult with the use of OM or SEM. For this reason, the researchers used elemental mapping by electron probe microanalyzer (EPMA*), and the locations and shapes of pores were discussed. They compared the porosity determined using area fraction measurements from image analysis results to the volume of gallium intruded in the specimen by calculating the mass difference of the specimen before and after intrusion. The porosity results in their study compared well, suggesting that the gallium image represents reasonably well the quantitative information. Similar studies have

---

* EPMA is an apparatus that irradiates electron beams onto a sample to observe the surface using the secondary and reflected electrons, while obtaining information on the elements in a microarea of the sample using simultaneously generated X-rays.

been carried out using Newton's metal as the intruding metal and analyzing images using TEM and SEM (see also Chapter 2) [8.66].

8.7   Air voids analysis using optical microscopy

The two most commonly used methods for analysis of air voids in concrete are the Chayes point count method and the Rosiwal linear traverse techniques that were described in Section 8.4. Until recently, the point count method and the linear traverse method were the only methods used systematically for air void analysis. These methods have been standardized by ASTM and are described in detail in ASTM C 457. In order to provide satisfactory results, the standard requires that for an aggregate of maximum nominal size of 19 mm the determination be carried out with a total number of points of at least 1350 and a minimum length of traverse of 2286 mm of a surface greater than 71 cm^2. These methods have been used extensively due to the simple equipment they require for analysis, the stereoscopic microscope. The parameters determined are the void content, spacing factor, and specific surface that were defined in Section 1.4.2.

Significant research has been carried out in order to standardize the two techniques, the point count method and the lineal traverse method, and to determine the total air content, spacing factor and specific surface of air voids, as they were defined by T. Powers [8.8]. In recent studies, research has focused mainly on the following areas:

- Identify the parameters that affect the measurements with traditional methods
- Use different mathematical parameters
- Use of image analysis
- Use of section analysis
- Use air void size distribution.

In the following sections, a review and discussion of developments in these areas are presented, intended to highlight some aspects of interest and some trends and developments that have become effective lately due to equipment advances.

8.7.1   Factors affecting results

Several factors affect the results of air void analysis and extensive studies have been carried out in the past to determine the effect of these parameters.

8.7.1.1   Specimen preparation

A good grinding technique is of paramount importance in order to obtain a good surface for observation and minimize artifacts due to scaring. A sample of hardened concrete is ground to provide an extremely smooth and plane section suitable for microscopic inspection. Unsatisfactory surfaces lead to higher spacing factors making the concrete appear to be in worse condition than it is.

The better the lapping, the better the measurement will be because the degree of subjective interpretation by the operator is proportionately reduced.

Faulty or incomplete specimen preparation has been cited as a source of error by many researchers. Specimen preparation can skew the air content results obtained using the ASTM C 457 test by as much as 3% [8.67]. Concrete with low air content requires 1–3 hours per specimen for the grinding procedure, while concrete with high air content needs more time for grinding, 1.5–4.5 hours [8.68]. Polished sections of concrete that has high air content, or is weak due to other reasons (e.g. decomposition due to chemical or physical attack), present problems during surface preparation. At very high air contents, the air voids tend to coalesce, and the thin septa and ridges between the voids can easily be damaged during grinding, therefore, these factors impair the accuracy of the measurements. The surfaces ground for a short time are of poor quality. It is essential that the specimens be thoroughly cleaned after each grinding operation and before proceeding to the next grinding step. It has been found experimentally that an unsatisfactory grinding technique always results in a high spacing factor.

8.7.1.2 *Magnification*

The magnification recommended by ASTM C 457 for air void analysis in concrete is 50× and should not be less than 50×. At high levels of magnification the operator can see and measure the small voids [8.68,8.69]. Therefore, at high magnifications, one can determine a higher air content and a smaller spacing factor than at low magnifications. Small voids have little influence on total air content, they have significant impact on specific surface, and therefore on spacing factor. The influence of high magnifications is more evident with a large spacing factor: a magnification of 100× instead of 50× can reduce a spacing factor of 0.17 mm by 10–13%, and a spacing factor of 0.25 mm by at least 10%. The spacing factor is used to determine the adequacy of air voids in concrete and thus the frost resistance of concrete. A low spacing factor (lower than 0.25 mm) indicates a frost resistant concrete and a high magnification might therefore lead to results indicating a concrete with good quality. For this reason, high magnifications should be linked with stricter requirements. R. Pleau *et al.*, 1996, have suggested a minimum magnification of 100× and a maximum of 125× [8.70].

8.7.1.3 *Operator objectivity*

During air void analysis, the human operator makes the decisions about the individual measurements of chord length, and whether the constituent under the crosshairs is paste, coarse aggregate, fine aggregate or an air void. Therefore, operators must be well-trained and well-experienced. The operator subjectivity as the cause of variability in the calculated spacing factor has been documented experimentally: as an example, an inexperienced operator can report a spacing factor of about 0.145 mm, while an experienced operator can report a spacing factor of about 0.210 mm on the same specimen [8.70].

8.7.1.4   Inclusion of entrapped voids

ASTM C 457 states that no provision be made for distinguishing among entrapped air voids, entrained air voids, and water voids. Any such distinction is arbitrary, because the various types of voids integrate in size, shape, and other characteristics. A comparative study between laboratories has shown that some testing agencies consider it useful to distinguish between the entrapped air voids and the entrained air voids [8.68]. By omitting the entrapped (irregular) air voids and measuring only the entrained (spherical) air voids, the standard deviation for the spacing factor can be reduced by 50% for concretes with low air content, and by 20% for concretes with high air content. It is recommended that the entrapped and the entrained air void be counted separately and that the spacing factor be calculated from the entrained air voids only. With a length of traverse of about 2.5 m, limits of confidence of ±0.01 mm can be expected. If the precision of the normal method is considered satisfactory, the length of traverse can be reduced accordingly by omitting the entrapped air voids. The omission of the entrapped air voids reduces the required spacing factor. A spacing factor of 0.25 mm calculated from all the air voids corresponds to a spacing factor calculated from the entrained air voids of only about 0.22 mm [8.68].

8.7.1.5   Methods of measurement

The modified point count method is easier and much more rapid than the linear traverse method. In an international comparative study between laboratories and techniques to determine the air void system of hardened concrete, it has been found that the method of measurement (Rosiwal, modified point count, and three methods which were modifications of the Rosiwal and modified point count) do not have a significant influence on the results [8.68].

8.7.1.6   Thin sections vs. polished sections

The area of a typical thin section does not exceed 25 mm × 25 mm and can be examined in both transmitted and reflected light. Sections as large as 150 mm × 100 mm may be prepared to represent the full cross-section of a core. Large area thin sections are more difficult to manipulate, therefore many concrete petrographers work with 75 mm × 75 mm sections and obtain acceptable results [8.13].

T. Aarre, 1995, compared the results of air void analysis using thin sections and polished sections [8.71]. The analysis was done using the linear traverse method according to ASTM C 457. The air content measured on thin sections was measured generally lower than the air content on plane sections using a probability theory approach by approximately 15%. The finite thickness of thin sections leads to a systematic error in the determination of the air content of concrete. The researcher explained that the air content measured on thin sections is smaller than the air content measured on plane sections, because it is the smallest diameter visible that is measured in a thin section. In her studies, she found no systematic deviation between the specific surface measured on polished plane sections and on thin sections.

8.7.2 Different mathematical parameters

Spacing equations have been proposed by R. Philleo, 1983 [8.72], E. Attiogbe, 1993 [8.73], R. Pleau and M. Pigeon, 1996 [8.70]. Each of these equations attempts to characterize the spacing among air voids, and estimate the distance that water must travel to reach the nearest air void. It is generally accepted that the assumptions made by T. Powers to determine the spacing factor, on which the standard measurement and analysis is currently based, do not represent realistic conditions. However, even though better approaches are proposed, it is going to be a significant time before any new parameters be extensively tested, accepted, and standardized for use.

8.7.3 Image analysis on air voids

The image analyzers started being in use for air void analysis in the late 1970s. Computer programs are becoming increasingly sophisticated in manipulating data derived from images and can now be used on a personal computer with sufficient capacity. Additional contrast enhancement is needed in the case of image analysis and several researchers have proposed methods, where the air voids on the ground surface of concrete are filled with contrast material [8.74]. As it was mentioned in Section 8.4 earlier, image analysis techniques can be semi-automatic or fully automatic. The semi-automatic method is preferred for its higher objectivity in identifying correctly the elements of interest. A comparison of an image analysis method vs. a traditional manual method has reported a reduction of measurement time by 83–88%, and a reduction in the coefficient of variation of the measurement by 75% when using an image analysis method [8.75].

The research carried out in the past using image analyzers had as its purpose to compare the results obtained by linear traverse and point counting, to the results obtained by the same methods when the OM is used, in order to identify limitations and problems associated with the use of image analysis. More recently, the use of section analysis, has been explored because of its ease to conduct with an image analyzer, and in that aspect image analysis is discussed in Sections 8.7.4 and 8.7.5 [8.76,8.77].

S. Chatterji and H. Gudmundsson, 1977, have used the image analyzing technique and the chord method to determine the air void parameters in concrete [8.78]. The results obtained by image analysis in their study are identical to those obtained by using the standard (Rosiwal) technique. Also, the air content measured by the automatic image analyzing technique in hardened concrete compares well to the air content measured in fresh concrete. The researchers reported a reduced testing time using image analysis, taking about 1.5 hours for sample preparation, and 0.5 hour for the measurements. In a similar study, J. Cahill *et al.*, 1994, explored the possibility of using image analysis in conjunction with the linear traverse method on polished samples of hardened concrete, which had previously been assessed by the standard linear traverse method [8.74]. From the distributions of chord lengths, the distributions of diameters of air voids were calculated for both methods and are presented in Figure 8.9. The data illustrate the similarities between the two methods at the upper end of the size range, while highlighting the significant differences below around 70 μm, with the image analysis recording many smaller voids.

A study presented summarized results from 13 laboratories in Europe, where a manual method and an image analysis system were used for air void analysis in concrete [8.79]. Image analysis methods included measuring chord lengths, diameters, section areas, and the perimeters of the air voids. The study showed that image analysis techniques statistically lie in the group of all measurement results. However, the image analysis methods can be problematic when a high amount of porous sand grains is present in the concrete, because they affect the image and therefore alter the results. Therefore, sample preparation is of crucial importance for obtaining reliable and comparable results using image analysis. In addition, automated methods show difficulties to measure the paste content of concrete specimens. In an effort to quantify the cement paste content, K. Peterson *et al.*, 2001, used a flatbed scanner for air void analysis of hardened concrete and scanned the surface three times in order to identify the different phases, i.e. cement paste, air voids, and aggregates [8.80]. The researchers used the modified point count technique to obtain results both manually and by using analysis of the digital output image, in order to determine the air content in the cement paste. In their approach, the contrast between the hardened cement paste and aggregates was enhanced by staining

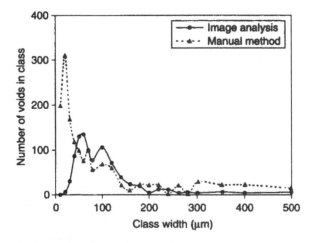

Figure 8.9 Size distributions of air voids obtained using the linear traverse method by ASTM C 457 (manual method) and by image analysis. Both analyses were done by measuring the length of chords.

Source: Reprinted from *Petrography of Cementitious Materials*, S.M. DeHayes, D. Stark (eds), ASTM STP 1215, J. Cahill, J.C. Dolan, P.W. Inward, "The identification and measurement of entrained air in concrete using image analysis," pp. 111–124, copyright 1994, with permission from the American Society for Testing and Materials, West Conshohocken, PA.

the cement paste pink using a solution of phenolphthalein in alcohol. Phenolphthalein turns the cement paste pink, and it is important that staining is done shortly after cutting the specimen so that no carbonation takes place that will hinder the staining of the cement paste. Air voids could be distinguished by painting the polished surface black and forcing white powder into the depressions.

8.7.4 Section analysis used to determine air content

The section analysis, even though it was developed theoretically much earlier than the linear traverse method and the point counting method, has been used only recently using image analysis equipment. The delayed use of section analysis happened because it was very difficult, if not impossible, to measure areas of circles on a section of air entrained concrete with the stereoscopic microscope.

G. Verbeck, 1947, used the camera lucida method on polished hardened concrete to determine air content [8.69]. The prepared surface may be observable directly or treated with pigment to fill all observable air cells, which is more convenient. The pigmentation procedure renders the air voids more easily discernible and of sharper outline, and avoids confusion with the small dark grains frequently found in cement paste. The magnification used during his measurements was in the range of about 50× to 90× depending upon the specimen. The parameters measured were the total area of the air voids in the selected field, the total area of the aggregate particles, and the total number of discrete air voids in the selected field. Verbeck found that the air content determined using the section analysis method on hardened concrete compared well to the air content determined in the plastic concrete after mixing.

In a recent study, K. Aligizaki and P. Cady, 1999, compared the air content in air-entrained cement pastes with low air contents and different water–cement ratios using the point count method and the section analysis method on photographs taken from a polished surface [8.76]. The results of their study are presented in Figure 8.10. In a similar study, R. Pleau *et al.*, 2001, used image analysis employing the section analysis method on air-entrained concrete and compared the air content values to the values obtained using the linear traverse method (see Figure 8.11) [8.77]. Both studies indicate that the section analysis method underestimates the air content; however, further research is needed to systematically study the application of the section analysis technique on the determination of air content in concrete. The spacing factor and the specific surface did not seem to be affected by the method used for their determination. The good estimation of the specific surface can be explained by the fact that the very small air voids, which are detected by the section analysis, but are more likely to be missed by the linear traverse method, have only a small influence on the computation of the specific surface of the air voids.

W. Mullen and C. Waggoner, 1980, have compared the air contents determined on a polished concrete surface using the point count method, and on a fracture surface using the section analysis method [8.81]. Their results are shown in Figure 8.12, and even though they show a linear relationship between the air contents determined by the two methods, they indicate that the

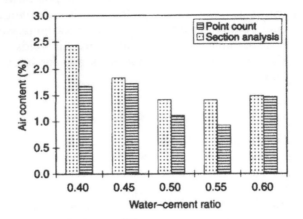

Figure 8.10 Comparison of the air content determined using the modified point-count method and the section-analysis method for air-entrained cement pastes having different water–cement ratios.

Source: Reprinted from *Cement and Concrete Research*, Vol. 29, K.K. Aligizaki, P.D. Cady, "Air content and size distribution of air voids in hardened cement pastes using the section-analysis method," pp. 273–280, copyright 1999, with permission from Elsevier Science, Amsterdam, The Netherlands.

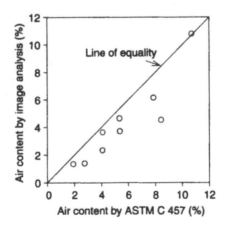

Figure 8.11 Relationship between the air content determined using the linear traverse method (ASTM C 457), and by image analysis using the section analysis method.

Source: Reprinted from *Cement and Concrete Composites*, Vol. 23, R. Pleau, M. Pigeon, J.-L. Laurencot, "Some findings on the usefulness of image analysis for determining the characteristics of the air-void system on hardened concrete," pp. 237–246, copyright 2001, with permission from Elsevier Science, Amsterdam, The Netherlands.

values obtained on the fracture surface are almost double compared to the values obtained on a polished surface. It should be noted, however, that the fracture surface is very likely to represent the path followed during fracture through the weakest areas in the solid and therefore provide a biased view

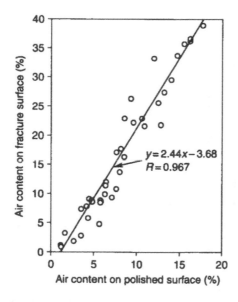

Figure 8.12 Relationship between air contents measured on a polished surface using the point count method, and on a fracture surface using the area fraction method (section analysis).

Source: Reprinted from *Cement, Concrete and Aggregates*, Vol. 2, W.G. Mullen, C.K. Waggoner, "Air content of hardened concrete from comparison electron microscope photographs," pp. 43–49, copyright 1980, with permission from the American Society for Testing and Materials, West Conshohocken, PA.

of the air voids, including more voids than what a random section through the solid would provide.

8.7.5 Air void distribution

The specific surface and the spacing factor used in air void analysis are two parameters that provide an indication of the size and distance between the air voids, but do not give information on the void size distribution. Several attempts have been made in the past to determine the size distributions of air voids, using either the lineal analysis technique or the section analysis technique.

G. Lord and T. Willis, 1950, applied the principles of mathematical probability and derived equations relating the distribution of chord lengths to the distribution of actual diameters of the spherical particles, i.e. the air voids of air-entrained concrete [8.48]. The arithmetic calculation of the sphere-diameter distribution from the observed chord length distribution is relatively simple. Other researchers have carried out similar work, using the chord analysis method to determine histograms of air voids [8.45] or to provide a continuous function for the distribution of air voids [8.82]. More recently, the void size distributions were determined using different calculation approaches

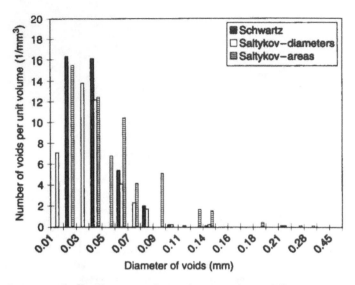

Figure 8.13 Air void distributions obtained using three different section-analysis procedures on a cement paste with w/c = 0.40. The Schwarz analysis procedure is based on diameter distributions of the circular diameters on the plane section.

Source: Reprinted from *Cement and Concrete Research*, Vol. 29, K.K. Aligizaki, P.D. Cady, "Air content and size distribution of air voids in hardened cement pastes using the section-analysis method," pp. 273–280, copyright 1999, with permission from Elsevier Science, Amsterdam, The Netherlands.

based on the section analysis method [8.76]. Examples of the distributions of air voids obtained in a cement paste using the distribution of the areas or the diameters of the circular cross sections are shown in Figure 8.13. It is worth noting the log-normal distribution of air voids that is expected in cement-based materials (see also Section 1.4.1.5). A log-normal distribution has also been obtained from other experiments using the linear traverse method on air-entrained concrete [8.77]. Attention must be drawn to the fact that the distribution of circular cross sections of air voids on the plane surface obtained using the section analysis are related, but are not identical to the air void distributions in the volume of the solid. The corresponding air void distribution can be obtained either using tables with coefficients by some of the analysis methods mentioned in Table 8.1, as it was done for Figure 8.13, or by direct integration of the distribution of the circular sections.

8.8 Advantages and limitations

The OM is cheap equipment, easily accessible and easy to use. The observation of microstructural porosity by optical techniques is limited by the resolution of the OM; typically, these analyses are carried out at low magnifications up to 400×. The fine pores are difficult to detect due to inadequate resolution on polished samples and to limitations on the number of pixels that are possible to display when an image analyzer is used.

There are statistical limitations that are associated with the magnification and the size of the field needed for examination: the higher the magnification needed to see a given pore size class, the smaller the field observed will be, and a large number of fields must be observed in order to obtain the necessary statistical validity to observations. Thus current practice in image analysis is to image at about 400×, which however, severely limits the smallest pore size that can be tallied.

Difficulties arise during sectioning, and preparation of the specimens. Extreme care is needed when using epoxy to impregnate the pores. Counting requires a finely polished concrete surface of substantial area and a traversing microscope stage on which to mount the specimen. Both specimen preparation and counting are time-consuming and require a fair amount of technical skill.

Air void analysis does not distinguish among entrapped air voids, entrained air voids, and water voids. To determine the air content, a statistical interpretation is made of the presence of air voids that are intercepted or counted and compared to areas where air voids are not intercepted or counted.

SEM can be used to study hardened cement paste and concrete in a manner similar to optical microscopy. SEM has a number of distinct advantages over optical microscopy in imaging and quantitative analysis of features. Some of these advantages include an extended range of magnification, the ability to produce high resolution images with sufficient contrast and a relatively large depth of field. SEM images have an inherent three-dimensional appearance, which gives realism to surfaces with a rough or porous structure. Furthermore, the instruments normally accept large specimens, which can be easily handled, and their preparation is usually simple and quick. Some of the limitations of the traditional SEM, like drying of the specimens and application of a conductive coating, have been eliminated with the use of the ESEM.

Image analysis can facilitate the work required by the ASTM C 457 method and provide an easier and more reliable method to assess parameters of the air void system. The method eliminates the subjectivity of the operator, which is often perceived as a major drawback of ASTM C 457 and can reduce the accuracy and reproducibility of normal measurements.

Image analysis offers cost, time, and accuracy advantages over methods currently in use. The time required for complete image analysis can be about 1 hour per sample, compared to about 3–6 hours by the standard linear traverse method. Porous regions appear as the darkest areas of the digital image of microstructure. To obtain better quality data, a larger sample area and more samples must be analyzed, and in this way, image analysis can provide better statistical accuracy and more meaningful results.

The ability to prepare the sample properly is often the most critical and most difficult factor in image analysis and much more care must be exercised in sample preparation, which can become time-consuming. The structure must exhibit adequate contrast to allow the analyzer to distinguish its various components and the cement paste content cannot be obtained easily by image analysis. The human eye is more qualified to separate the air voids from surface defects than the computer, and for this reason, a semi-automatic image analysis procedure is better than a fully automatic procedure.

Measurements by image analysis greatly overestimate the number of voids with diameters in the range in the 10–90 μm. This can be explained by the surface defects and variations in color, and by the overlapping of air voids. The separation of overlapping air voids is easy for the human eye, but difficult by image analysis.

The field examined is a two-dimensional cut of a three-dimensional reality. While the plane passes through and exposes many pores, it does not usually show them at their maximum (or characteristic) cross sections but mostly at smaller cross sections. Even though stereological equations are available to interpret the observed size distribution information in terms of the solid structure, their validity for cement paste is uncertain, because of the varying shapes of cement paste pores. Attention should be paid to the different assumptions inherent with different stereological analysis procedures in order to obtain meaningful results. For example, the linear traverse analysis can provide very useful results, but during the analysis of air voids, small voids might be difficult to be intercepted by the line of traverse. It can be shown that the traverse line will hit a great many more voids, if the voids are large than it will hit if they are small. It is obvious then that the data available for calculation cannot reflect the true relative sizes of the voids.

BSE images can provide quantitative analysis and valuable information about cement paste microstructure, but there are some serious limitations. The main problems concern sample preparation and segmentation of the image. With regard to segmentation, the question arises as to which bins of the grayscale histogram should be assigned as porosity.

References

8.1   Le Chatelier H. "Recherches experimentales sur la constitution des ciments et la theorie de leur prise," *Comptes rendus hebdomadaires des séances de l'Académie des Sciences de Paris*, Vol. 94, 1882, pp. 867–869.

8.2   De Broglie L. "A tentative theory of light quanta," *Philosophical Magazine*, Vol. 47, 1924, pp. 446–458.

8.3   Davisson C., Germer L.H. "Diffraction of electrons by a crystal of nickel," *Physical Review*, Vol. 30, 1927, pp. 705–740.

8.4   Richardson I.G. "Chapter 22: Electron microscopy of Cements," in *Structure and Performance of Cements*, 2nd Edition, J. Bensted and P. Barnes (eds), Spon Press, 2002, pp. 500–556.

8.5   Radczewski O.E., Müller H.O., Eitel W. "Zur Hydratation des Trikalziumsilikats," *Die Naturwissenschaften*, Vol. 27, 1939, p. 807.

8.6   Radczewski O.E., Müller H.O., Eitel W. "Zur Hydratation des Trikalziumaluminats," *Die Naturwissenschaften*, Vol. 27, 1939, pp. 837–838.

8.7   Chatterji S., Jeffery J.W. "Three dimensional arrangement of hydration products in set cement paste," *Nature*, Vol. 209, No. 5029, 1966, pp. 1233–1234.

8.8   Powers T.C. "Void spacing as a basis for producing air-entrained concrete," *Research Bulletin 49 of the Portland Cement Association*, March 1955, Reprint from the

Journal of the American Concrete Institute, May 1954, in Proceedings Vol. 50, pp. 741–760.

8.9 Sorby H.C. "On the microscopical structure of crystals indicating the origin of minerals and rocks," *Quarterly Journal of the Geological Society of London*, Vol. 14, 1858, pp. 453–500.

8.10 Richardson J.H. *Optical Microscopy for the Materials Sciences*, Marcel Dekker, Inc., New York, 1971.

8.11 Bradbury S. *An Introduction to the Optical Microscope*, Royal Microscopical Society Microscopy Handbooks Vol. 1, Oxford University Press, 1989, p. 86.

8.12 Thordal Andersen K., Thaulow N. "The study of alkali–silica reactions in concrete by the use of fluorescent thin-sections," in *Petrography Applied to Concrete and Concrete Aggregates*, ASTM STP 1061, B. Erlin, D. Stark (eds), American Society for Testing and Materials, West Conshohocken, PA, 1990, pp. 71–89.

8.13 St John D.A., Poole A.B, Sims I. *Concrete Petrography; A Handbook of Investigative Techniques*, Arnold, London, 1998, p. 474.

8.14 Robinson P.C., Bradbury S. *Qualitative Polarized-Light Microscopy*, Royal Microscopical Society Microscopy Handbooks, No. 09, Oxford University Press, 1992.

8.15 Bowen D.K., Hall C.R. "Modern imaging methods using electron, X-ray and ion beams," *Microscopy of Materials*, The Macmillan Press Ltd, 1975.

8.16 Shindo D., Oikawa T. *Analytical Electron Microscopy for Materials Science*, Springer Verlag, Tokyo, 2002.

8.17 Lawes G. *Scanning Electron Microscopy and X-Ray Microanalysis*, Analytical Chemistry by Open Learning (ACOL), Thames Polytechnic London, 1987.

8.18 Goldstein J., Newbury D., Joy D., Lyman C., Echlin P., Lifshin E., Sawyer L., Michael J. *Scanning Electron Microscopy and X-ray Microanalysis*, 3rd Edition, Kluwer Academic/Plenum Publishers, New York, 2003.

8.19 Stutzman P.E. "Applications of scanning electron microscopy in cement and concrete petrography," in *Petrography of Cementitious Materials*, ASTM STP 1215, S.M. DeHayes, D. Stark (eds), American Society for Testing and Materials, West Conshohocken, PA, 1994, pp. 74–90.

8.20 Scrivener K.L. "The use of backscattered electron microscopy and image analysis to study the porosity of cement paste," in *Symposium Proceedings Vol. 137: Pore Structure and Permeability of Cementitious Materials*, L.R. Roberts, J.P. Skalny (eds), Materials Research Society, Warrendale, PA, 1989, pp. 129–140.

8.21 Danilatos G.D., Robinson V.N.E. "Principles of scanning electron microscopy at high specimen pressures," *Scanning*, Vol. 2, No. 2, 1979, pp. 72–82.

8.22 Danilatos G.D. "The examination of fresh or living plant material in an environmental scanning electron microscope," *Journal of Microscopy*, Vol. 121, Part 2, 1981, pp. 235–238.

8.23 Briggs A. "An introduction to scanning acoustic microscopy," in *Royal Microscopical Society Microscopy Handbooks*, No. 12, Oxford University Press, 1985.

8.24 Briggs A. "Acoustic microscopy," in *Monographs on the Physics and Chemistry of Materials*, Vol. 47, Oxford Science Publications, Oxford University Press, New York, 1992.

8.25 Aligizaki K.K., Tittmann B.R., Gordon G.A. "Comparison between optical microscopy and scanning acoustic microscopy for detecting flaws in concrete," *Experimental Techniques*, Vol. 18, No. 5, 1994, pp. 24–28.

8.26 Livingston R.A., Manghnani M., Prasad M. "Characterization of Portland cement concrete microstructure using the scanning acoustic microscope," *Cement and Concrete Research*, Vol. 29, No. 2, 1999, pp. 287–291.

8.27 Chermant J.-L. "Why automatic image analysis? An introduction to this issue," *Cement and Concrete Composites*, Vol. 23, Nos 2–3, 2001, pp. 127–131.

8.28   *Practical Guide to Image Analysis*, ASM International, Materials Park, OH, 2000, p. 290.

8.29   Russ J.C. *The Image Processing Handbook*, 4th Edition, CRC Press, Boca Raton, FL, 2002.

8.30   Cole G.S. "Theory and concepts for computer aided microscopy," in *Quantitative Microscopy and Image Analysis, International Conference on Quantitative Microscopy and Image Analysis*, Charleston, South Carolina, 19–21 July 1993, D.J. Diaz (ed.), ASM International, Materials Park, OH, 1994, pp. 1–9.

8.31   Lange D.A., Jennings H.M., Shah S.P., Quenard D. "A fractal approach to understanding cement paste microstructure," in *Ceramic Transactions Vol. 16: Advances in Cementitious Materials*, S. Mindess (ed.), The American Ceramic Society, 1991, pp. 347–362.

8.32   Petrov I., Schlegel E. "Application of automatic image analysis for the investigation of autoclaved aerated concrete structure," *Cement and Concrete Research*, Vol. 24, No. 5, 1994, pp. 830–840.

8.33   Exner H.E., Hougardy H.P. *Quantitative Image Analysis of Microstructures*, Informationsgesellschaft Verlag, 1988.

8.34   DeHoff R.T. "Problem solving using quantitative stereology," in *Applied Metallography*, G.F. Vander Voort (ed.), Van Nostrand Reinhold Company Inc., 1986, pp. 89–99.

8.35   James N.T. "Stereology," in *Analytical and Quantitative Methods in Microscopy*, G.A. Meek, H.Y. Elder (eds), Cambridge University Press, 1977, pp. 9–28.

8.36   Hilliard J.E. "Chapter 3: Measurement of volume in volume," in *Quantitative Microscopy*, R.T. DeHoff, F.N. Rhines (eds), McGraw-Hill Book Company, 1968, pp. 45–76.

8.37   Underwood E.E. "Chapter 6: Particle size distribution," in *Quantitative Microscopy*, R.T. DeHoff, F.N. Rhines (eds), McGraw-Hill Book Company, 1968, pp. 149–200.

8.38   DeHoff R.T. "Chapter 5: Measurement of number and average size in volume," in *Quantitative Microscopy*, R.T. DeHoff, F.N. Rhines (eds), McGraw-Hill Book Company, 1968, pp. 128–148.

8.39   Weibel E.R. *Stereological Methods Vol. 2: Theoretical Foundations*, Academic Press, 1980, pp. 340.

8.40   Russ J.C. *Practical Stereology*, Plenum, New York, 1986, p. 124.

8.41   Thomson E. "Quantitative microscopic analysis," *The Journal of Geology*, Vol. 38, No. 3, 1930, pp. 193–222.

8.42   Chayes F. "A simple point counter for thin section analysis," *American Mineralogist*, Vol. 34, Nos 1–2, 1949, pp. 1–11.

8.43   Rosiwal A. "Uber geometrische gesteinsanalysen. Ein einfacher weg zur ziffemässigen festellung des quantitätsverhältnisses des mineralbestandteile gemengter gesteine," Verhandlungen der Kaiserlich-Königlichen Geologischen Reichsanstalt, Vienna, 1898, Nos 5–6, pp. 143–175.

8.44   Delesse A., "Procédé mécanique pour déterminer la composition des roches," *Annales des Mines*, Vol. 13, 1848, pp. 379–388.

8.45   Brown L.S., Pierson C.U. "Linear transverse technique for measurement of air in hardened concrete," *Journal Proceedings of the American Concrete Institute*, Vol. 47, No. 10, 1950, pp. 117–123.

8.46   Dullien F.A.L., Dhawan G.K. "Characterization of pore structure by a combination of quantitative photomicrography and mercury porosimetry," *Journal of Colloid and Interface Science*, Vol. 47, No. 2, 1974, pp. 337–349.

8.47   Dullien F.A.L., Rhodes E., Schroeter S.R. "Comparative testing of some statistical methods for obtaining particle size distributions," *Powder Technology*, Vol. 3, No. 1, 1969/1970, pp. 124–135.

8.48 Lord G.W., Willis T.F. "Calculation of air bubble size distribution from results of a Rosiwal traverse of aerated concrete," *ASTM Bulletin*, Vol. 177, 1951, pp. 56–61.

8.49 Cahn J.W., Fullman R.L. "On the use of lineal analysis for obtaining particle-size distribution functions in opaque samples," *Transactions of the American Institute of Mining and Metallurgical Engineers (AIME)*, Vol. 206, 1956, pp. 610–612.

8.50 Wicksell S.D. "The corpuscle problem-1: a mathematical study of a biometric problem," *Biometrika*, Vol. 17, 1925, pp. 84–99.

8.51 Hilliard J.E., Cahn J.W. "An evaluation of procedures in quantitative metallography for volume-fraction analysis," *Transactions of the Metallurgical Society of AIME*, Vol. 221, No. 2, 1961, pp. 344–352.

8.52 DeHoff R.T. "Sampling of material and statistical analysis in quantitative stereology," in *Stereology, Proceedings of the Second International Congress for Stereology*, H. Elias (ed.), Springer Verlag, New York, 1967, pp. 119–130.

8.53 Lange D.A., Jennings H.M., Shah S.P. "The influence of pore structure on the properties of cement paste: initial observations about research-in-progress," in *Symposium Proceedings Vol. 137: Pore Structure and Permeability of Cementitious Materials*, L.R. Roberts, J.P. Skalny (eds), Materials Research Society, Warrendale, PA, 1989, pp. 47–54.

8.54 Scrivener K.L., Pratt P.L. "The characterization and quantification of cement and concrete microstructures," in *From Materials Science to Construction Materials Engineering, Vol 1: Pore Structure and Materials Properties, Proceedings of the First International RILEM Congress*, Versailles, France, September 7–11, 1987, J.C. Maso (ed.), Chapman and Hall, pp. 61–68.

8.55 Scrivener K.L., Patel H.H., Pratt P.L., Parrott L.J. "Analysis of phases in cement paste using backscattered electron images, methanol adsorption and thermogravimetric analysis," in *Symposium Proceedings Vol. 85: Microstructural Development During Hydration of Cement*, L.J. Struble, P.W. Brown (eds), Materials Research Society, Warrendale, PA, 1987, pp. 67–76.

8.56 Zhao H., Darwin D. "Quantitative backscattered electron analysis of cement paste," *Cement and Concrete Research*, Vol. 22, No. 4, 1992, pp. 695–706.

8.57 Kjeilsen K.O., Detwiler R.J., Gjorv O.E. "Backscattered electron image analysis of cement paste specimens: specimen preparation and analytical methods," *Cement and Concrete Research*, Vol. 21, 1991, pp. 388–390.

8.58 Lange D.A., Jennings H.M., Shah S.P. "Image analysis techniques for characterization of pore structure of cement-based materials," *Cement and Concrete Research*, Vol. 24, No. 5, 1994, pp. 841–853.

8.59 Wang V., Diamond S. "An approach to quantitative image analysis for cement pastes," in *Symposium Proceedings Vol. 370: Microstructure of Cement-based Systems/Bonding and Interfaces in Cementitious Materials*, S. Diamond, S. Mindess, F.P. Glasser, L.W. Roberts, J.P. Skalny, L.D. Wakeley (eds), Materials Research Society, Warrendale, PA, 1995, pp. 23–32.

8.60 Diamond S., Leeman M.E. "Pore size distributions in hardened cement paste by SEM image analysis," in *Symposium Proceedings Vol. 370: Microstructure of Cement-Based Systems/Bonding and Interfaces in Cementitious Materials*, S. Diamond, S. Mindess, F.P. Glasser, L.W. Roberts, J.P. Skalny, L.D. Wakeley (eds), Materials Research Society, Warrendale, PA, 1995, pp. 217–226.

8.61 Willis K.L., Abell A.B., Lange D.A. "Image-based characterization of cement pore structure using Wood's metal intrusion," *Cement and Concrete Research*, Vol. 28, No. 12, 1998, pp. 1695–1705.

8.62 Kjellsen K.O., Detwiler R.J., Gjørv O.E. "Pore structure of plain cement pastes hydrated at different temperatures," *Cement and Concrete Research*, Vol. 20, No. 6, 1990, pp. 927–933.

8.63   Mouret M., Ringot E., Bascoul A. "Image analysis: a tool for the characterisation of hydration of cement in concrete – metrological aspects of magnification on measurement," *Cement and Concrete Composites*, Vol. 23, Nos 2–3, 2001, pp. 201–206.

8.64   Dullien F.A.L. "Wood's metal porosimetry and its relation to mercury porosimetry," *Powder Technology*, Vol. 29, 1981, pp. 109–116.

8.65   Tanaka K., Kurumisawa K. "Development of technique for observing pores in hardened cement paste," *Cement and Concrete Research*, Vol. 32, No. 9, 2002, pp. 1435–1441.

8.66   Richardson I.G., Groves G.W., Rodger S.A. "The porosity and pore structure of hydrated cement pastes as revealed by electron microscopy techniques," in *Symposium Proceedings Vol. 137: Pore Structure and Permeability of Cementitious Materials*, 1988, L.R. Roberts, J.P. Skalny (eds), Materials Research Society, Warrendale, PA, 1989, pp. 313–318.

8.67   Roberts L.W., Gaynor R.D. "Discussion on the paper by C. Ozyildirim 'Comparison of the air contents of freshly mixed and hardened concretes', *Cement, Concrete, and Aggregates*, Vol. 13, No. 1, 1991, pp. 11–16," *Cement, Concrete, and Aggregates*, Vol. 13, No. 1, 1991, pp. 16–17.

8.68   Sommer H. "The precision of the microscopical determination of the air-void system in hardened concrete," *Cement, Concrete, and Aggregates*, Vol. 1, No. 2, 1979, pp. 44–55.

8.69   Verbeck G.J. "The camera Lucida method for measuring air voids in hardened concrete," *Journal of the American Concrete Institute*, Vol. 43, 1947, pp. 1025–1040.

8.70   Pleau R., Pigeon M. "The use of the flow length concept to assess the efficiency of air entrapment with regards to frost durability. Part I – Description of the test method," *Cement, Concrete, and Aggregates*, Vol. 18, No. 1, 1996, pp. 19–29.

8.71   Aarre T. "Influence of measurement technique on the air-void structure of hardened concrete," *ACI Materials Journal*, Vol. 92, No. 6, 1995, pp. 599–604.

8.72   Philleo R.E. "A method for analyzing void distribution in air-entrained concrete," *Cement, Concrete, and Aggregates*, Vol. 5, No. 2, 1983, pp. 128–130.

8.73   Attiogbe E.K. "Mean spacing of air voids in hardened concrete," *ACI Materials Journal*, Vol. 90, No. 2, 1993, pp. 174–181.

8.74   Cahill J., Dolan J.C., Inward P.W. "The identification and measurement of entrained air in concrete using image analysis," in *Petrography of Cementitious Materials, ASTM STP 1215*, S.M. DeHayes, D. Stark (eds), American Society for Testing and Materials, West Conshohocken, PA, 1994, pp. 111–124.

8.75   Ohta T., Ohashi T., Konagai N., Nemoto T. "Measurement of air void parameters in hardened concrete using automatic image analysing system," *Transactions of the Japan Concrete Institute*, Vol. 8, 1986, pp. 183–190.

8.76   Aligizaki K.K., Cady P.D. "Air content and size distribution of air voids in hardened cement pastes using the section-analysis method," *Cement and Concrete Research*, Vol. 29, No. 2, 1999, pp. 273–280.

8.77   Pleau R., Pigeon M., Laurencot J.-L. "Some findings on the usefulness of image analysis for determining the characteristics of the air-void system on hardened concrete," *Cement and Concrete Composites*, Vol. 23, Nos 2–3, 2001, pp. 237–246.

8.78   Chatterji S., Gudmundsson H. "Characterization of entrained air bubble systems in concretes by mean of an image analysing microscope," *Cement and Concrete Research*, Vol. 7, No. 4, 1977, pp. 423–428.

8.79   Elsen J. "Automated air void analysis on hardened concrete. Results of a European intercomparison testing program," *Cement and Concrete Research*, Vol. 31, 2001, pp. 1027–1031.

8.80 Peterson K.W., Swartz R.A., Sutter L.L., Van Dam T.J. "Hardened concrete air void analysis with a flatbed scanner," *Transportation Research Record*, No. 1775, 2001, pp. 36–43.

8.81 Mullen W.G., Waggoner C.K. "Air content of hardened concrete from comparison electron microscope photographs," *Cement, Concrete and Aggregates*, Vol. 2, No. 1, 1980, pp. 43–49.

8.82 Larson T.D., Cady P.D., Malloy J.J. "The protected paste volume concept using new air-void measurement and distribution techniques," *Journal of Materials*, Vol. 2, No. 1, 1967, pp. 202–226.

9 Comparison of results by various methods

MIP and gas adsorption, most commonly nitrogen adsorption, are two of the most widely used methods for obtaining information about the pore structure of cement-based materials. For this reason, when a new method is introduced or developed to determine the pore structure characteristics of cement-based materials, the results are usually compared to MIP and/or to nitrogen adsorption results.

It is apparently difficult to determine the pore size distribution in cement pastes over the whole range of pore sizes by using one single technique. The MIP method is unsuitable to measure the micropore size distributions of hydrated cement paste because a large amount of micropores in cement pastes are unintrudable by mercury.

An indicative range of sizes where each technique is applicable has also been presented in Table 1.2. It is obvious that for some techniques there is no sufficient overlap over the range of pore sizes and therefore comparison is impossible. Even for techniques that can be used to determine the pore structure over the same range, results can differ significantly and the observed differences can be due to a number of reasons. Reasons for differences can be assumptions inherent with the application of each technique, and modification of the microstructure either before the experiment due to pretreatment or during the experiment due to interaction with the material used. It should be noted that disagreement of results from two methods does not imply the theoretical failure of either method. However, comparison of such techniques is sometimes still attempted in order to explain the observed differences in relation to the microstructure of cement pastes or the applicability and assumptions of the method used. In such cases, application of more than one technique on the same specimen can give better determination of the pore structure.

In the following sections, some examples are given for comparison of results obtained in the past by researchers using various methods. Comparisons of porosity, pore size distribution, and surface area are mentioned, where available. It should be noted that the results presented do not cover all the studies in the technical literature, but rather a small selection in order to highlight some aspects of the tests. In addition, different analysis techniques applied for the same testing method have already been discussed in the corresponding chapter, e.g. cryoporometry and relaxometry in NMR analysis.

9.1 Comparison with MIP results

Since the MIP has been used extensively, and for a long time on a wide variety of compositions of cement pastes and covers a wide range of pore sizes, almost all methods used in cement paste have been compared with MIP results.

9.1.1 Nitrogen adsorption

Both nitrogen adsorption and mercury intrusion provide a direct measure of the total pore volume and the pore size distribution. MIP and nitrogen adsorption are model-dependent techniques and are constrained in their applicability to particular size ranges. Nonetheless, they are considered relatively reliable for materials whose pore structures are chemically stable. Neither technique can be used alone to study the entire pore structure in well-hydrated cement due to the wide range in pore sizes existing in cement pastes.

Most MIP commercial instruments have a pressure range reaching 200 MPa although some porosimeters are capable of working to pressures slightly in excess of 400 MPa, i.e. are capable of detecting pore diameters of approximately 2 nm. The upper diameter determined may be extended in theory to approximately 1 mm. Gas adsorption results, using nitrogen in particular, are believed not to be trustworthy below 8 nm in diameter. A direct comparison of data obtained by mercury porosimetry and by methods based on capillary condensation of vapors in the pore system is possible only for pore fractions, where both methods yield meaningful results, i.e. for pore radii between 5 and 50 nm. The comparable parameters are total pore volume, pore size distribution, and specific surface area. Several researchers have carried out comparisons between MIP and nitrogen studies to determine their effectiveness in evaluating the porosity and pore size distribution [e.g. 9.1,9.2,9.3,9.4,9.5].

Because MIP covers a wider range of pore sizes compared to nitrogen adsorption, it is obvious that the *total porosity* determined by MIP is greater than the total porosity measured using nitrogen adsorption. S. Diamond, 1971, observed that the cumulative volume determined by mercury intrusion was not too different from the volume determined by nitrogen adsorption for pores with diameter between 10 and 40 nm; however this observation was not valid for pores smaller than 10 nm [9.2].

In a later study, R. Day and B. Marsh, 1988, compared the porosities measured by nitrogen adsorption and MIP and limited the comparison to sizes of pores smaller than 20 nm [9.6]. The volume of pores accessible to mercury increases with increasing the water–cement ratio and the volume of pores inaccessible to mercury varies in different hydrated materials. Pore size distributions obtained by nitrogen adsorption and mercury porosimetry did not agree well [9.5]. The volume of pores filled with nitrogen was lower than the total pore volume of the material and even lower than the volume of pores filled with mercury. In spite of an increasing total porosity, the volume of adsorbed nitrogen changed only little with increasing the water–cement ratio of the cement pastes. Consequently, the volume of pores that were not filled with nitrogen increased distinctly under these conditions.

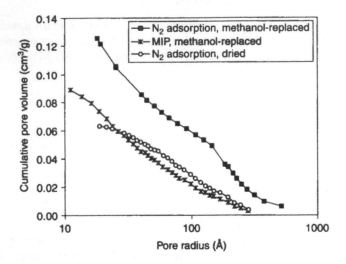

Figure 9.1 Cumulative pore size distributions obtained by nitrogen adsorption and MIP on cement pastes with w/c = 0.40, in the range of pore sizes that overlap.

Source: Reprinted from *Pore Structure and Permeability of Cementitious Materials*, L.R. Roberts, J.P. Skalny (eds), *MRS Symposium Proceedings Vol. 137*, Y. Abdel-Jawad, W. Hansen, "Pore structure of hydrated cement determined by mercury porosimetry and nitrogen sorption techniques," pp. 105–118, copyright 1989, with permission from the Materials Research Society, Warrendale, PA.

An example of cumulative pore volume curves as they have been obtained by nitrogen adsorption and MIP are shown in Figure 9.1 in the range of overlap of MIP and nitrogen sorption for a Portland cement paste [9.1]. Similar results to Figure 9.1 have been presented in other research studies [9.5,9.6,9.7].

Total pore volume by MIP is virtually unaffected by the drying method (see also Figure 2.3). However nitrogen sorption shows a substantial decrease in the volume of capillaries smaller than about 30 nm due to drying at 0% RH from the saturated surface dry state [9.7]. Therefore, a specimen treatment procedure that is expected to cause the least damage to the microstructure, such as *methanol replacement*, is necessary in order to measure the micropores. G. Litvan, 1976, used solvent replacement with methanol on bottle hydrated cements prior to nitrogen sorption measurement, and found that the surface areas determined were similar to the values obtained using water sorption [9.8]. His findings indicate that the original pore structure may have been preserved with solvent replacement.

A comparison of *pore size distributions* of hydrated cement pastes obtained using MIP and capillary condensation measurements reveals discrepancies between the results. The MIP data indicate considerably coarser size distributions with mean diameters of the order of several hundred Å. The capillary condensation data yield mean pore diameters between 50 and 100 Å, and indicate that most of the space tallied is in pores of diameters less than 100 Å [9.2]. R. Gerhard, 1989 presented pore size distributions that are not overlapping [9.9]. However, H. Midgley and J. Illston, 1983 have presented results showing some similarities (see Figure 9.2) [9.4].

It has also been reported that the curve for pore size distribution obtained using MIP is closer to the curve obtained by nitrogen desorption, than to the curve obtained by nitrogen adsorption. This observation seems reasonable, since the desorption curve is more sensitive to pore entrance diameter than the adsorption curve, a fact that also holds true for MIP [9.1].

Since *surface areas* from large pores contribute little to the total measured surface area relative to the volume they represent, it is possible that the surface area determined by MIP may agree well with that from gas adsorption, despite the disagreement observed in pore volumes. Surface areas have been determined on a large number of samples by MIP and adsorption methods with excellent agreement. Deviation between the two methods usually is observed in samples with high surface areas, where the porosimeter has inadequate pressure to fill the smallest pores. The nitrogen molecule is only 0.4 nm in size and will thus penetrate pores about 3% of the size that mercury at 103 MPa will penetrate [9.3]. Therefore, the surface area measured by MIP would be approximately $13.3 \, m^2$ less than that measured by the BET-method [9.3].

I. Odler and Y. Chen, 1995, determined the surface area using nitrogen adsorption, water absorption, and MIP [9.10]. As an example, for a cement paste with w/c = 0.60, the surface area using water was determined to be $140 \, m^2/g$, $50 \, m^2/g$ when using nitrogen adsorption, and $20 \, m^2/g$ when determined by MIP.

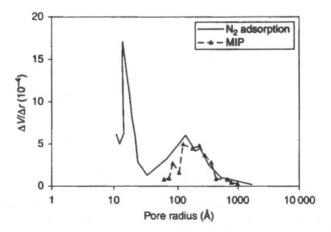

Figure 9.2 Comparison of pore size distributions obtained using nitrogen adsorption and MIP for cement paste with w/c = 0.71.

Source: Reprinted from *Cement and Concrete Research*, Vol. 13, H.G. Midgley, J.M. Illston, "Some comments on the microstructure of hardened cement pastes," pp. 197–206, copyright 1983, with permission from Elsevier Science, Amsterdam, The Netherlands.

Studies have clearly shown that MIP is intrinsically limited and that even if high pressures are applied to the system, mercury is unable to intrude the entire hydrated cement paste pore volume. It is also believed that when the surface areas measured by MIP are very high and comparable to the surface areas from nitrogen adsorption, it could be a result from ink-bottle pores. Intrusion of mercury into an ink-bottle pore will not occur until sufficient pressure is applied to force the pressure into the narrow entrance. It will therefore appear as if a large volume intruded into narrow pores, generating an excessively high calculated surface area.

One might assume that the MIP results in the high-pressure region are in error, due perhaps to crushing of the sample, or to a possible change in contact angle at high pressures. On the other hand, one might assume that the capillary condensation results are in error, especially at the finer end of the distribution, since the Kelvin equation is not applicable to pores of the order of 30–40 Å in diameter. Differences could be attributed to a variety of factors or combinations of factors, such as the existence of a wide range of pore sizes and shapes in cement pastes, the collapsing of the pore structure depending on the drying technique used, and damage of the microstructure during testing.

It is believed that capillary condensation results using nitrogen in particular are not trustworthy much below 80 Å in diameter, instead of approximately 30 Å that is commonly mentioned [9.2]. Therefore, substantial agreement of all the pore size distributions might have been achieved if the calculations had been started at 70–80 Å diameters, instead of at approximately 30 Å. Another likely source for error in the capillary condensation measurements involving condensates other than nitrogen may lie in the *t*-curves used by various researchers in their calculations (see Section 4.2.2) [9.2]. A *t*-curve is designed to represent the statistical thickness of the adsorbed film lining the pore walls at each stage of the isotherm. Any error in this curve will have significant effects on the pore size distribution in the smallest-diameter region of pore sizes. A *t*-curve is obtained from an adsorption isotherm of a non-porous substance having the same surface characteristics as the porous substance under study. However, it is questionable whether the *t*-curves used which are invariably derived from measurements on solids of compositions other than that of hydrated cement paste are appropriate for the pore size distribution calculations on cement pastes.

D. Winslow and S. Diamond, 1970, have suggested that other source of error might be attributed to the "lost porosity," that is the fraction of the total pore space unintruded by mercury [9.24]. The researchers have attributed the lost porosity to ultrafine pores, which were too fine to be measured by MIP. It has also been suggested that the lost porosity might not be in the ultrafine pores, but in "encapsulated pockets of gel," which would be unintrudable to mercury at any pressure. J. Beaudoin, 1979, suggested the existence of microspace between aggregations of C–S–H sheets accessible to helium but not to mercury [9.11].

Another factor which can lead to surface areas from nitrogen adsorption to be slightly higher than the surface areas from mercury porosimetry is pore wall roughness. Slight surface roughness will not alter the porosimetry surface area, since it is calculated from the pore volume, while the same roughness will be measured by gas adsorption. Experimental determinations of the surface-to-volume ratios have demonstrated that the surface area appears to decrease as a result of sample drying. In effect, there is a coarsening of the pore space and as a result, higher permeabilities are measured in samples that have been dried out.

9.1.2   Helium pycnometry

J. Beaudoin, 1979, carried out experiments measuring the total porosity in cement pastes using helium pycnometry and MIP up to 400 MPa pressure [9.11]. The results of his study indicate that porosity measured by MIP can be significantly different from porosity measured by helium pycnometry, either greater, or less than helium porosity depending on the cementitious material studied. Beaudoin concluded that when the MIP porosity is less than helium porosity, mercury cannot enter all the pore space available to helium. It is possible that trapped microspace exists between aggregations of C–S–H sheets (extra layer space), which is accessible to helium but not to mercury. This space possibly forms as a result of deposition and consolidation of aggregations of C–S–H layers in confined space such as would be found in cement pastes with low water–cement ratios. The geometrical constraints on the deposition of these products is such that they cannot be deposited in a regular manner. It is postulated that many surfaces would come together or "mate" within distances of molecular dimension, while other surfaces would provide the boundaries of "trapped space." However, this conclusion would not explain why mercury and helium give comparable values at higher water–cement ratios when the volume concentration of pores with diameters less than 200 Å increases with the water–cement ratio. The trapped space, if present, appears to be present only for pastes with low water–cement ratios. The solid matrix apparently has sufficient strength to resist the hydrostatic stresses imposed on it by mercury at high pressure [9.11].

B. Marsh *et al.*, 1985, based on their experimental results pointed out that differences between porosity measured by MIP and helium pycnometry only occurred, when there was evidence of significant pozzolanic reaction (see Figure 9.3) [9.12]. Porosity determined by helium pycnometry was found to measure only the immediately accessible porosity. In their studies they attributed the differences in porosity from the two methods to a discontinuous pore structure [9.6]. For blended cement pastes, which exhibit significant pozzolanic reaction, mercury and helium porosities have different meanings [9.12]. Similar results have been reported from other research [9.11].

By observing the differences in Figure 9.3a and 9.3b, some evidence exists that solvent replacement results in less damage to the pore structure during drying. With the microstructure more intact in the solvent-replaced samples, mercury produces more damage and thus the deviation from equality of the values by helium and by MIP in the fly ash pastes is much greater [9.12]. R. Feldman, 1984, has also attributed differences in MIP and helium porosity to fracture of pore walls occurring at pressures greater than 70 MPa [9.13]. However, his observations were based on cement pastes with extremely low porosities.

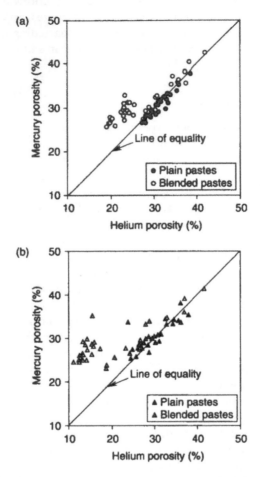

Figure 9.3 Comparison of helium and mercury porosities for cement pastes cured at temperatures from 20°C to 80°C and hydrated for periods up to 1 year. The pastes have been dried by (a) oven drying at 105°C, and (b) by solvent replacement.

Source: Reprinted from *Very High Strength Cement-Based Materials*, J. Francis Young (ed.), *MRS Symposium Proceedings Vol. 42*, B.K. Marsh, R.L. Day, "Some difficulties in the assessment of pore-structure of high performance blended cement pastes," pp. 113–121, copyright 1985, with permission from the Materials Research Society, Warrendale, PA.

M. Krus *et al.*, 1997, carried out experiments to compare the porosity measured with helium pycnometry, MIP, and water saturation for different cement pastes [9.14]. The porosities measured with the water saturation method in a wet state are significantly higher than the porosities measured by helium pycnometry or by MIP in a dry state. The porosities measured by MIP correspond overall very well to the porosities obtained by helium pycnometry. For cement pastes containing 30% by mass blast furnace slag, the helium porosity was 12%, higher compared to a porosity of 9% determined by MIP.

The researchers attributed the observed differences to the different accessibility of mercury and helium in the pores. The lowest pore diameter that could be registered by the MIP equipment used in their experiments was about 0.4 nm and the lowest value was 0.22 nm in the case of helium pycnometry. Similar differences in the measured porosities have been reported by other studies [9.15].

Gel pores in the hardened cement paste have a great affinity to water and take in water during the water absorption process, which leads to changes in the microstructure by the hygric expansion of these pores [9.16]. The swelling of the gel pores results in a reduction of the cross sections of the capillary-active larger pores and this effect can alter the results by various techniques.

9.1.3 Alcohol exchange and water absorption

A comparison of porosity determined by MIP and water absorption for plain and blended cement pastes is shown in Figure 9.4 [9.17]. The results indicate that the porosity measured using water is higher than the porosity measured using MIP. It has been reported by several researchers that water used as a displacement fluid always gives a higher porosity for D-dried hydrated Portland cement than other fluids such as mercury, nitrogen, methanol, propanol, pentane, and helium [9.14,9.15,9.18]. A possible reason for the increased porosity may be due to alteration in the microstructure upon drying. Many of the properties of hydrated Portland cement are measured after

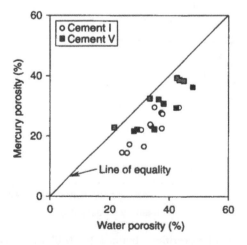

Figure 9.4 Relationship between porosity determined using MIP and water absorption on plain (I) and blended (V) cement pastes with different water–cement ratios.

Source: Reprinted from *Cement and Concrete Research, Vol. 31,* C. Gallé, "Effect of drying on cement-based materials pore structure as identified by mercury intrusion porosimetry: a comparative study between oven-, vacuum-, and freeze-drying," pp. 1467–1477, copyright 2001, with permission from Elsevier Science, Amsterdam, The Netherlands.

drying, which leads to decomposition of the hydrates. In addition, changes in porosity, surface area, and other physical properties will occur when the system is dried; if water is used for measurement, rehydration of the composed hydrates may make results difficult to interpret [9.18,9.19]. The higher porosity using water has also been attributed to water being a strong polar liquid, which can slip in between the mineral layers of the material, widen up the distances between them, and create new pore spaces. The possibility of "open porosity" mentioned by researchers by which some pores are more open to water access seems to be unrealistic, because the diameter of the helium atom is smaller than the water molecule. Therefore, the open porosities measured by helium pycnometry should be the same or even a bit greater than the porosities measured by water saturation [9.14].

Figure 9.5 shows the relationship between the porosities obtained using alcohol exchange and MIP. In plain pastes, the porosity determined by mercury porosimetry is comparable to the porosity determined using resaturation by alcohol. However for the cement pastes that show pozzolanic reaction, i.e. those containing silica fume and fly ash, the porosity determined by MIP is higher than the porosity obtained by alcohol resaturation [9.6]. The higher porosity of MIP has been explained by greater accessibility of mercury to closed pores after damage to the microstructure. The pozzolanic reaction blocks pores or encapsulates areas of porosity; these blocked areas are not accessible to intrusion by alcohols or low pressure helium. However, the

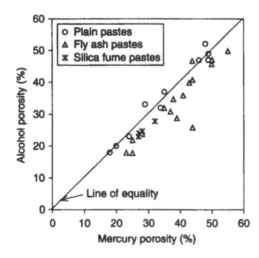

Figure 9.5 Comparison of porosities obtained by MIP (mercury porosity) and by alcohol exchange (alcohol porosity) on plain cement pastes and on cement pastes containing mineral admixtures with $w/c = 0.30$, 0.47, and 0.71, at 7 days, 28 days and 3 months. In half the specimens the alcohol porosity was determined using methanol and in the other half using isopropanol.

Source: Reprinted from *Cement and Concrete Research*, Vol. 18, R.L. Day, B.K. Marsh, "Measurement of porosity in blended cement pastes," pp. 63–73, copyright 1988, with permission from Elsevier Science, Amsterdam, The Netherlands.

intrusion of mercury at high pressures breaks through the blocked regions. Pores which exist behind the blockages are then filled and are correspondingly assigned to a smaller pore radius than they may actually possess. As a result an inaccurate assessment of total porosity and pore size distribution is obtained [9.6].

9.1.4 NMR vs. MIP and nitrogen sorption

Comparison between conventional techniques, such as nitrogen sorption and MIP and the results of NMR relaxation analysis can become complicated by the fact that different pore structure parameters are measured with each method. MIP measures the pore size of the entrance of the pores and NMR measures the true pore volume-to-surface area ratio. A research study for comparison of pore volumes measured by MIP, NMR, and nitrogen sorption techniques has been carried out by R. Valckenborg *et al.*, 2001 and is shown in Figure 9.6 [9.20]. The size of the large pores obtained from NMR is relatively close to that obtained from MIP. The NMR relaxation analysis of wet cement pastes is observed to be consistent with the existence of the capillary and gel pores. There is a peak at 20 nm by both NMR and MIP, and a peak close to 0.8 nm by gas adsorption, and 5 nm by NMR. S. Bhattacharja *et al.*, 1993, obtained similar NMR results, which showed a main peak at 10 nm, while the distribution from MIP shows only a small peak at 17 nm and a larger peak at 70 nm [9.21]. The researchers have attributed the discrepancy to the inability of mercury to intrude into the small pores, which NMR can detect.

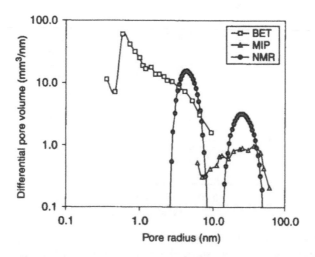

Figure 9.6 Pore size distributions determined by NMR, MIP, and water vapor adsorption using the BET-method.

Source: Reprinted from *Materials and Structures Vol. 34*, R.M.E. Valckenborg, L. Pel, K. Hazrati, K. Kopinga, J. Marchand, "Pore water distribution in mortar during drying as determined by NMR," pp. 599–604, copyright 2001, with permission from RILEM.

9.2   Comparison with nitrogen adsorption

The nitrogen adsorption technique is used predominantly to determine the surface area and compare the results to those by different techniques. However, it also determines the micropore volume and is used in conjunction with the MIP technique to provide a better view of the pore structure of cement pastes.

9.2.1   Water sorption

An example of experimental studies carried out in the past on cement pastes with different water–cement ratios using water absorption, nitrogen adsorption, and MIP to determine pore volume is shown in Figure 9.7 [9.7]. The results shown in Figure 9.7 indicate that the pore volume determined using water absorption is higher than the pore volumes determined by MIP and nitrogen adsorption combined. The difference in the sum of volumes tends to be reduced as the water–cement ratio increases.

Possible reasons of differences include rehydration of cement pastes during water absorption after drying pretreatment needed for MIP and nitrogen adsorption, and these have already been mentioned in Section 9.1.3.

9.2.2   SAXS and SANS

In two thorough studies and reviews on the surface area of cement pastes, J. Thomas *et al.*, 1998, 1999, pointed out the similarities and differences of

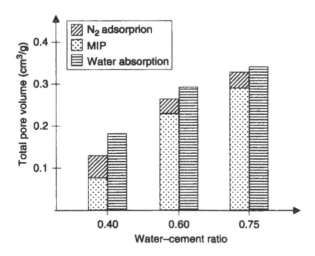

Figure 9.7 Porosity for well-hydrated Portland cement pastes, determined by nitrogen adsorption, MIP, and water absorption for solvent-replaced specimens. The pore volume by nitrogen is of pore diameters smaller than 4 nm, and the pore volume by MIP is for pore diameters greater than 4 nm; both expressed as cm³/g of dry paste. The mass loss using water is expressed as g/g of dry paste.

Source: Drawn from experimental results of W. Hansen, J. Almudaiheem, 1987 [9.7].

values obtained from SAXS and SANS [9.22,9.23]. The researchers particularly emphasized the need to normalize the reported values obtained by the various techniques for meaningful comparisons. For example, the specific surface of the saturated cement paste was found to be approximately 700 m²/g of ignited cement paste; the equivalent value expressed on a basis of D-dried weight is about 580 m²/g. Drying treatments significantly reduce the measured surface area with the reduction being related to the severity of drying. The surface areas were found to range from 272 m²/g (ignited) for P-dried paste down to 138 m²/g (ignited) for paste dried at 105°C under vacuum. The BET adsorption and the SANS/SAXS technique can be used to obtain surface areas, but SANS/SAXS also measure the surface areas of totally enclosed pores that cannot be reached by gas adsorption. Similarly, both MIP and SANS/SAXS can produce pore size distributions, but the results will not be commensurate because SANS/SAXS measures enclosed pores. Moreover, SANS/SAXS can also provide information on angular distributions of pores (anisotropy), which is not possible with MIP.

From their studies, the researchers concluded that nitrogen adsorption measurements fail to register most of the surface that is present in cement paste. Water vapor adsorption measurements are comparable in magnitude to those obtained by X-ray scattering for dried samples. However, the surface area of wet cement paste is very much higher. As an example, the surface areas for cement pastes using SAXS have been 516–618 m²/g [9.24].

9.3 Comparison with replacement techniques

9.3.1 *Helium pycnometry vs. methanol sorption*

Experiments have been carried out in the past by researchers to compare the porosity determined by helium pycnometry, water absorption, and solvent replacement for cement pastes with different water–cement ratios, and pretreated by different techniques [9.25,9.14]. The results of these studies are presented in Figure 9.8. The results indicate that the water porosity values compare well for cement pastes that have been D-dried or pretreated with solvent replacement. For D-dried cement pastes, the results indicate that the porosities obtained with helium pycnometry and methanol replacement are very similar. The difference between methanol and water porosity of D-dried cement pastes is about 8%, and helium porosity is very close to the porosity as measured by methanol replacement.

By comparing porosities of cement pastes determined by the same method, but pretreated by different techniques, one can observe that the porosities determined using water absorption compare very well for D-dried and solvent replaced pastes. The helium porosities of D-dried cement pastes seem to be higher than the porosities determined using solvent replacement.

The methanol replacement and helium pycnometry porosities compare well with each other. In a similar study, porosity values from helium pycnometry have been found to compare well to the porosity values from alcohol resaturation [9.6].

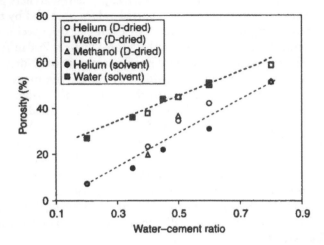

Figure 9.8 Comparison of porosities determined using helium pycnometry, water absorption, and methanol replacement for cement pastes that have been pretreated by D-drying or solvent replacement.

Source: Drawn from experimental results from R. Feldman, 1972 [9.25] and M. Krus *et al.*, 1997 [9.14].

9.3.2   Alcohol exchange vs. water saturation

Methanol and propanol exchange porosities come close to measuring the same porosity as water absorption, both for plain and for blended hardened cement pastes [9.6]. In addition, when the porosity is determined by alcohol saturation after the specimens have been dried, the porosity obtained by alcohol is consistently lower than the porosity obtained by water resaturation. This observation is usually attributed to the inability of alcohols to enter the interlayer spaces, which collapse during drying. One would expect that the plain cement pastes that have higher quantities of the C–S–H gel also have a greater amount of interlayer space compared to blended cement pastes, and the difference between alcohol and water porosity should be more profound [9.6]. Similar results have also been found in another research study [9.5].

9.3.3   SANS vs. water absorption

Experimental studies carried out by D. Harris *et al.*, 1974, to demonstrate that oven-drying removes some bound water, and their results have been discussed by R. Feldman, 1989 [9.26,9.18]. A comparison of the volume of free water calculated by neutron scattering experiments is significantly lower than the free water determined by oven-drying (see Figure 9.9). The free water determined by neutron scattering is comparable to the capillary water determined by oven-drying; this observation indicates that oven drying removes not only the capillary water, but also some of the bound water from cement pastes.

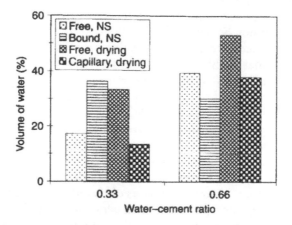

Figure 9.9 Free and bound water in cement pastes with different water–cement ratios determined using quasi-elastic neutron scattering (NS) and oven-drying. The pastes were 2 years old. The volume of water represents the volume fraction of water in the total volume of cement paste, which includes the volume of water and the hydrated cement.

Source: Drawn from experimental results of D. Harris *et al.*, 1974 [9.26].

9.4 Comparison with microscopy techniques

Several researchers have used MIP and microscopy techniques to obtain total porosity and pore size distributions of cement pastes. Such research has become more popular in recent years due to the facilitation of measurements using image analysis techniques. The total volume can be determined using the volume fraction of the surfaces (cross section of pores) (see Section 8.5.3) and the pore size distribution.

9.4.1 MIP vs. OM

A combination of MIP and OM accounts for what seems to be nearly all the pore volume, even allowing for overlap between OM and MIP. N. Alford and A. Rahman, 1981, used both MIP and OM on oven-dried cement pastes, and found that the pore volumes measured by OM and MIP methods combine to give a value in excess of the total pore volume of the sample (can be explained by certain overlap) [9.19].

Even though optical microscopy is simple and readily accessible, most researchers prefer SEM to study the pore structure of cement pastes as it is discussed in the following section. This is due to the high resolving power, the field of depth, and the great versatility of the SEM compared to the optical microscope.

9.4.2 MIP vs. SEM

Several researchers have carried out research to compare porosities obtained by MIP and SEM. K. Kjelsen *et al.*, 1990, discovered great variance to porosities obtained by SEM compared to MIP even though coarse porosity was observed by SEM [9.27]. They attributed the differences to the presence of pores that are closed and cannot be accessed by MIP such as the hollow-shell pores, which however become accessible and visible on a plane section. From their study it appears that within the resolution of the method, BSE images of cement pastes with w/c = 0.50 that were freeze-dried and subsequently oven-dried, examined at 500× show pore sizes to be roughly five times greater than the sizes indicated by MIP. For example, the main difference in pore radius distribution between the specimens shown in the images appeared in the range 0.250–1.25 μm, rather than the 20–100 nm indicated by MIP. The researchers attributed the differences to the "ink bottle effect."

In a similar study, D. Lange *et al.*, 1994, carried out experiments to characterize the nature of pore structure as observed in BSE images of polished sections [9.28]. They used plain cement pastes, pastes with silica fume, and mortars and calculated total porosity, and pore size distributions. They used image analysis of BSE and compared the pore size distributions they received to MIP. With MIP, the total porosity measured is almost twice the porosity tallied using the image analysis method. Similar results have been reported by other researchers [9.29]. The researchers constructed a pore size distribution from the results of the sizing analysis by sorting the pore clusters by size and plotting the cumulative pore area vs. pore size. This type of distribution is analogous to the distribution measured by MIP, and the image based curves bear a resemblance to typical MIP curves as shown in Figure 9.10 for cement paste and for cement mortar. Although the curves are similar in

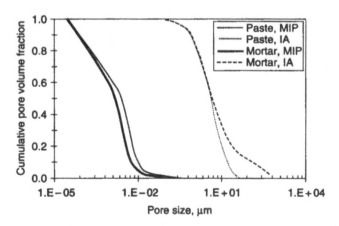

Figure 9.10 Pore size distributions determined by mercury intrusion porosimetry (MIP) and image analysis (IA) on a cement paste with w/c = 0.40, and on a cement mortar with w/c = 0.40 and sand/cement = 1. For the comparison, the researchers converted the MIP results to area assuming circular cross section of the pores.

shape, they are offset by about 3 orders of magnitude in scale. The MIP curve covers a wider range than the image based curve. For mortars, images have the ability to capture more information about large pores than MIP. MIP measures the large pores inside the specimen by intruding through smaller bottlenecks and thus, the technique systematically misrepresents the large pores' size. The specific surface areas derived from these experiments are equivalent to 0.04–$0.08\,m^2/g$, only $1/1000$ the value of specific surface area determined by adsorption methods.

Image analysis is limited to only the upper part of the size range known to the present. The small pore size observed in Lange's work, which is about $0.8\,\mu m$, can be improved using equipment with high image resolution. Image analysis is not likely to permit the whole range of sizes of pores in hardened cement paste to be assessed and replace other techniques used. However, image analysis appears to provide a reasonably accurate assessment of the actual sizes of the pores tallied, including both capillary pores and air voids. In contrast, pore size distribution measurements using MIP grossly underestimate the sizes of pores tallied and do not provide even a rough approximation to the actual pore size distribution [9.29].

Similar studies have been carried out by K. Willis *et al.*, 1998, using Wood's metal that fills the pores and makes them easy to identify in order to enhance the contrast in BSE images, and by K. Tanaka and K. Kurumisawa, 2002, using gallium as it was mentioned in Section 8.6 [9.30,9.31]. A good agreement was found between the volume of intruding mercury and the volume of molten Wood's metal that infiltrated the sample at each pressure, and the threshold diameters agreed quite well. For the image analysis on samples containing Wood's metal (WMIP), the amount of porosity for a given range is determined by computing the sum of the areas of individual pores within the pore size range, and dividing by the total intruded volume of Wood's metal. For example, 26% of the pores in the WMIP analysis were between 1 and $2\,\mu m$ average pore diameter. For the MIP analysis, the amount of porosity for each range is determined directly from the intrusion curve; and the volume of porosity within each pore size range is divided by the total intruded mercury. Comparison between the cumulative intrusion curves for MIP at 35 MPa, and WMIP at 35 and 3.5 MPa, revealed that much higher percentages of intrusion were occurring at smaller pore sizes using MIP than when using WMIP. The limited intrusion observed in the 3.5 MPa occurs at pore sizes between 1000 and $5000\,\mu m^2$, whereas the majority of the intrusion in the 35 MPa occurred between 0.01 and $30\,\mu m^2$. In their study, the pore size distributions obtained using MIP were shifted more towards smaller pores compared to the pore size distributions obtained using WMIP [9.30].

Comparison of porosity determined using gallium intrusion and mass change in cement pastes before and after intrusion, and image analysis using SEM is shown in Figure 9.11 [9.31]. The results show that for higher porosities (at w/c = 0.60), image analysis tends to determine a smaller porosity than MIP. At the early ages of cement pastes, there is no specific trend in the differences observed for porosities determined using image analysis and MIP.

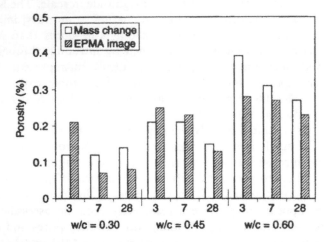

Figure 9.11 Comparison of porosity determined using gallium intrusion and mass change in cement pastes before and after intrusion, and image analysis using Electron Probe Micro-Analyzer (EPMA). The cement pastes had different water–cement ratios, and were tested at ages of 3, 7, and 28 days each, indicated in the figure.

Source: Reprinted from *Cement and Concrete Research*, Vol. 32, K. Tanaka, K. Kurumisawa, "Development of technique for observing pores in hardened cement paste," pp. 1435–1441, copyright 2002, with permission from Elsevier Science, Amsterdam, The Netherlands.

References

9.1　Abdel-Jawad Y., Hansen W. "Pore structure of hydrated cement determined by various porosimetry and nitrogen sorption techniques," in *Symposium Proceedings Vol. 137: Pore Structure and Permeability of Cementitious Materials*, L.R. Roberts, J.P. Skalny (eds), Materials Research Society, Warrendale, PA, 1989, pp. 105–118.

9.2　Diamond S. "A critical comparison of mercury porosimetry and capillary condensation pore size distributions of Portland cement pastes," *Cement and Concrete Research*, Vol. 1, No. 5, 1971, pp. 531–545.

9.3　Mikijelj B., Varela J.A., Whittemore O.J. "Equivalence of surface areas determined by nitrogen adsorption and by mercury porosimetry," *American Ceramic Society Bulletin*, Vol. 70, No. 5, 1991, pp. 829–831.

9.4　Midgley H.G., Illston J.M. "Some comments on the microstructure of hardened cement pastes," *Cement and Concrete Research*, Vol. 13, No. 2, 1983, pp. 197–206.

9.5　Odler I., Köster H. "Investigations on the structure of fully hydrated Portland cement and tricalcium silicate pastes. II. Total porosity and pore size distribution," *Cement and Concrete Research*, Vol. 16, No. 6, 1986, pp. 893–901.

9.6　Day R.L., Marsh B.K. "Measurement of porosity in blended cement pastes," *Cement and Concrete Research*, Vol. 18, No. 1, 1988, pp. 63–73.

9.7　Hansen W., Almudaiheem J. "Pore structure of hydrated Portland cement measured by nitrogen sorption and mercury intrusion porosimetry," in *Symposium Proceedings Vol. 85: Microstructural Development During Hydration of Cement*, L.J. Struble, P.W. Brown (eds), Materials Research Society, Warrendale, PA, 1987, pp. 105–114.

9.8　Litvan G.G. "Variability of the nitrogen surface area of hydrated cement paste," *Cement and Concrete Research*, Vol. 6, No. 1, 1976, pp. 139–143.

9.9 Gerhard R. "A review of conventional and non-conventional pore characterization techniques," in *Symposium Proceedings Vol. 137: Pore Structure and Permeability of Cementitious Materials*, L.R. Roberts, J.P. Skalny (eds), Materials Research Society, Warrendale, PA, 1989, pp. 75–82.

9.10 Odler I., Chen Y. "Investigations on the aging of hydrated tricalcium silicate and Portland cement pastes," *Cement and Concrete Research*, Vol. 25, No. 5, 1995, pp. 919–923.

9.11 Beaudoin J.J. "Porosity measurement of some hydrated cementitious systems by high pressure mercury intrusion-microstructural limitations," *Cement and Concrete Research*, Vol. 9, No. 6, 1979, pp. 771–781.

9.12 Marsh B.K., Day R.L., Bonner D.G. "Pore structure characteristics affecting the permeability of cement paste containing fly ash," *Cement and Concrete Research*, Vol. 15, No. 6, 1985, pp. 1027–1038.

9.13 Feldman R.F. "Pore structure damage in blended cements caused by mercury intrusion," *Journal of the American Ceramic Society*, Vol. 67, No. 1, 1984, pp. 30–33.

9.14 Krus M., Hansen K.K., Künzel H.M. "Porosity and liquid absorption of cement paste," *Materials and Structures*, Vol. 30, No. 201, 1997, pp. 394–398.

9.15 Marsh B.K., Day R.L., Bonner D.G., Illston J.M. "The effect of solvent replacement upon the pore structure characterization of Portland cement paste," in *Proceedings of the Principles and Applications of Pore Structural Characterization*, Milan, Italy, 1983.

9.16 Hall C., Hoff W.D., Taylor S.C., Wilson M.A., Yoon B.-G., Reinhardt H.W., Sosoro M., Meredith P., Donald A.M. "Water anomaly in capillary liquid absorption by cement-based materials," *Journal of Materials Science Letters*, Vol. 14, No. 17, 1995, pp. 1178–1181.

9.17 Gallé C. "Effect of drying on cement-based materials pore structure as identified by mercury intrusion porosimetry. A comparative study between oven-, vacuum-, and freeze-drying," *Cement and Concrete Research*, Vol. 31, No. 10, 2001, pp. 1467–1477.

9.18 Feldman R.F. "The porosity and pore structure of hydrated Portland cement paste," in *Symposium Proceedings Vol. 137: Pore Structure and Permeability of Cementitious Materials*, L.R. Roberts, J.P. Skalny (eds), Materials Research Society, Warrendale, PA, 1989, pp. 59–73.

9.19 Alford N.McN., Rahman A.A. "An assessment of porosity and pore sizes in hardened cement pastes," *Journal of Materials Science*, Vol. 16, 1981, pp. 3105–3114.

9.20 Valckenborg R.M.E., Pel, L., Hazrati K., Kopinga K., Marchand J. "Pore water distribution in mortar during drying as determined by NMR," *Materials and Structures*, Vol. 34, 2001, pp. 599–604.

9.21 Bhattacharja S., Moukwa M., D'Orazio F., Jehng J.Y., Halperin W.P. "Microstructure determination of cement pastes by NMR and conventional techniques," *Advanced Cement Based Materials*, Vol. 1, 1993, pp. 67–76.

9.22 Thomas J.J., Jennings H.M., Allen A.J. "The surface area of cement paste as measured by neutron scattering – evidence for two C-S-H morphologies," *Cement and Concrete Research*, Vol. 28, No. 6, 1998, pp. 897–905.

9.23 Thomas J.J., Jennings H.M., Allen A.J. "The surface area of hardened cement paste as measured by various techniques," *Concrete Science and Engineering*, Vol. 1, 1999, pp. 45–64.

9.24 Winslow D.N., Diamond S. "Specific surface of hardened Portland cement paste as determined by small-angle X-ray scattering," *Journal of the American Ceramic Society*, Vol. 57, No. 5, 1974, pp. 193–197.

9.25 Feldman R.F. "Density and porosity studies of hydrated Portland cement," *Cement Technology*, January–February 1972, pp. 5–14.

9.26 Harris D.H.C., Windsor C.G., Lawrence C.D. "Free and bound water in cement pastes," *Magazine of Concrete Research*, Vol. 26, No. 87, 1974, pp. 65–72.

9.27   Kjellsen K.O., Detwiler R.J., Gjørv O.E. "Pore structure of plain cement pastes hydrated at different temperatures," *Cement and Concrete Research*, Vol. 20, No. 6, 1990, pp. 927–933.

9.28   Lange D.A., Jennings H.M., Shah S.P. "Image analysis techniques for characterization of pore structure of cement-based materials," *Cement and Concrete Research*, Vol. 24, No. 5, 1994, pp. 841–853.

9.29   Diamond S., Leeman M.E. "Pore size distributions in hardened cement paste by SEM image analysis," in *Symposium Proceedings Vol. 370: Microstructure of Cement-based Systems/Bonding and Interfaces in Cementitious Materials*, S. Diamond, S. Mindess, F.P. Glasser, L.W. Roberts, J.P. Skalny, L.D. Wakeley (eds), Materials Research Society, Warrendale, PA, 1995, pp. 217–226.

9.30   Willis K.L., Abell A.B., Lange D.A. "Image-based characterization of cement pore structure using Wood's metal intrusion," *Cement and Concrete Research*, Vol. 28, No. 12, 1998, pp. 1695–1705.

9.31   Tanaka K., Kurumisawa K. "Development of technique for observing pores in hardened cement paste," *Cement and Concrete Research*, Vol. 32, No. 9, 2002, pp. 1435–1441.

Summary and conclusions

Pore structure influences the physical and mechanical properties of cement-based materials. Characterization of the pore structure is complicated by the existence of pores having different shapes and sizes, and by their connectivity. The range of pore sizes in hardened cement paste is very wide, and pores and voids in concrete can range in size from 1 nm to 1 cm. A variety of techniques have been developed to characterize the pore structure of the principal binding material, the hydrated Portland cement paste. Pores in the hardened cement paste include gel pores, capillary pores, hollow shell pores, and air voids. The gel pores have a size of about 0.5 nm and are the inherent pores in the calcium silicate hydrate (C–S–H) gel. Capillary pores are cavities of irregular shape of the initially water-filled spaces, and their size ranges from 2 nm to 10 µm. Hollow-shell pores are closed, very distinct pores, and form from the original cement grains boundaries; they range in size from 1 to 15 µm. Entrained air voids are introduced intentionally during production of concrete by using an air-entraining chemical admixture. Entrapped air voids occur inadvertently during mixing and placing of concrete.

A wide variety of techniques have been used for the characterization of the pore structure of cement pastes. Some methods have only access to open pores, while others have access to both open and closed pores. The techniques used to a great extent in cement paste and discussed in this book are: mercury intrusion porosimetry (MIP), gas adsorption, displacement methods, thermoporometry, nuclear magnetic resonance (NMR), small-angle scattering (SAS), optical microscopy (OM), and scanning electron microscopy (SEM).

Various pore structure parameters can be used to describe the properties of hardened cement pastes. The parameters determined for general pores using various techniques include porosity, hydraulic radius, specific surface area, threshold diameter, and pore size distribution. Other parameters are the shape factor or parameters defined for microscopic analysis. Cement paste is a disordered system and has a fractal nature that is characterized by the fractal dimension. Parameters used to characterize air voids are total air content, specific surface, and spacing factor, while other parameters have also been proposed by researchers.

Many of the techniques used to characterize the pore structure require proper pretreatment of the specimen, most commonly drying. Specimen preparation and pretreatment are very important, because the pore structure parameters determined are affected by the pretreatment technique. Water removal is done using a drying technique or using a solvent replacement technique. Drying techniques commonly used include oven-drying, vacuum-drying, P-drying, D-drying, and freeze-drying. Solvent replacement is considered to be gentle to the cement paste microstructure when used as an alternate drying technique prior to porosity measurement. The ease of penetration of a solvent in cement pastes depends on the density of the material tested and its drying history, but also on the type of the solvent used. Most solvents are physically adsorbed on the pore walls of cement pastes, while it is believed that some solvents interact chemically with the hydration products of cement during solvent replacement. The selection of the appropriate technique for water removal depends on a number of factors, such as the time available, the property of interest of the material examined, the equipment available, etc.

Specimen preparation for analysis using microscopical techniques is very important because it facilitates examination and interpretation of microstructural features. Analysis using OM or SEM requires a highly polished surface for optimum imaging. The specimens used for microscopical observation can have polished or fracture surfaces, or be thin sections. The quality of a polished surface will affect the image: a low quality polish can produce pitting and scarring that are interpreted as porous regions introducing in this way error into the analysis. Contrast during microscopical analysis may be enhanced by adding a colored or fluorescent dye to the resin, or by using a fusible metal or alloy, such as gallium or Wood's metal. Cement-based materials may be impregnated by epoxy resin in two ways: by dry vacuum impregnation, and by solvent replacement.

Thin sections may be prepared in various ways and can be particularly problematic in their manufacture, because of the frail and brittle nature of the concrete microstructure when reduced to fine dimensions. The thickness required for analysis of cement-based materials is approximately 25–30 μm. A thin section consists of a thin slice of concrete usually impregnated with fluorescent epoxy glued to an objective glass and protected by a cover glass.

MIP has been used for several decades in characterizing many porous materials. The technique has become popular because it is applicable to a broad range of pore sizes, more than any other method, and determines a variety of pore parameters. Mercury has a high surface tension and therefore, non-wetting properties. Some parameters of particular interest are the total porosity, the critical pore diameter, and the threshold diameter. Two of the most commonly used curves are the cumulative intrusion curve, and the differential distribution curve. MIP can measure pores over a wide range of diameters, from 1.5 nm up to 1 mm. Mercury is unable to intrude all pores in the system and shows the problem of lost porosity, which is believed to be due to pores that are too fine to be intruded by mercury, or completely isolated from the exterior.

MIP shows the phenomenon of hysteresis and mercury entrapment. Total mercury intruded into the sample is irreversibly trapped in the pores and never extrudes, and mercury cycles in and out of the pores as the applied pressure is increased and reduced. Several theories have been proposed to explain hysteresis, and the most commonly used theories are the presence of ink-bottle pores and differences in contact angle. Additional explanations include pore potential, network effects, surface roughness of the pore walls, and contamination of the material's surface by mercury.

The second intrusion method has been used to differentiate ink-bottle pores from the uniform pores of capillary shape, i.e. pores that contain small contractions and enlargements along their lengths. It has also been used to determine the porosity of air-entrained pores. With second intrusion, the pore size distribution can be divided into reversible and irreversible sub-distributions. The reversible sub-distribution is believed to provide the best correlation with water permabilility. The retention factor is used to quantify the fraction of mercury retained in the specimen, after the pressure is reduced to atmospheric pressure. Factors that affect the results are the pretreatment technique, the specimen size, the rate of pressure build-up, the contact angle, surface tension of mercury, and alteration to the pore structure. Possible alterations of the pore structure are checked using the second intrusion of specimens and by microscopic examination of specimens intruded by mercury.

The contact angle between mercury and cement paste can range widely, from 117° to 175°. Several parameters can affect the contact angle, such as the composition and age of the cement paste, the size of the pores, surface roughness of the pore walls, and the mercury purity. The contact angle can be measured using the sessile drop method and the mercury contact anglometer.

Despite a number of limitations, MIP continues to be regarded as a common pore structure characterization technique for cement pastes. The technique is conceptually simpler, experimentally much faster, and has the ability to evaluate a much wider range of pore sizes than any alternative method practiced currently. Extreme care and precaution is needed during the test, because mercury is a hazardous substance, and precautions are needed to avoid contact with skin and to minimize possible exposure to mercury vapor.

Compression of the specimen leads to false values for the pore volume and pore size. Due to high pressure applied, permanent structural changes could occur by breakdown of the porous structure during mercury intrusion. It should also be kept in mind that the pore size distributions obtained by MIP are based on numerous assumptions.

Gas adsorption has been one of the most popular techniques used for the study of pore structure in materials that contain micropores and mesopores. Methods developed in the past for analysis of experimental adsorption results, usually involve models for condensation of gas in capillaries, such as the Kelvin equation, and models for adsorption of gas molecules on free surfaces such as the Brunauer–Emmitt–Teller (BET) equation. The Kelvin equation and the BET equation still hold their place as the basis of many calculations in gas adsorption because of their simplicity. The vapor most commonly used for adsorption experiments is nitrogen, at its boiling point, 77.3 K, because it has low reactivity with most solids, and is readily available as pure gas or as liquid.

During the experiment, the amount of gas adsorbed and the corresponding pressure are recorded, both during adsorption and desorption. The graph of volume of adsorbed gas vs. the corresponding adsorption pressure at constant temperature is called an adsorption isotherm. The isotherm obtained from a cement paste is Type IV isotherm, which has been studied and analyzed extensively. The thickness of adsorbed film of gas on the pore walls can be calculated either theoretically assuming a packing for the gas molecules, or empirically from experimental curves. The Kelvin equation determines the Kelvin (or critical) radius of a cylindrical pore into which capillary condensation occurs at a specific relative pressure. The Kelvin equation ignores the fact that adsorption inside the pores is not caused only by capillary condensation but by adsorption on the pore walls as well. Several procedures are used to obtain pore size distributions. The Barrett–Joyner–Halenda (BJH) method is most applicable to a wide range of pore sizes and uses the desorption branch of the isotherm, and ignores pores with a radius greater than 30 nm. A complete pore structure analysis can be performed in two parts using a combination of the micropore (MP) method for micropores, and the corrected modeless method for wide pores (mesopores and macropores).

The specific surface can be determined from gas adsorption measurements using a variety of methods such as the BET equation, the Dubinin–Kaganer (DK) equation, the Harkins–Jura method, the t-plot method and the α_s-plot method. The BET method continues to be extensively used for gas adsorption analysis because of its simplicity, but can only be used in the relative pressure range from 0.05 to 0.35.

A characteristic feature of the Type IV isotherm is the hysteresis loop. Several types of hysteresis can be observed for the Type IV isotherm and can be classified into different types. Several theories have been formulated in order to explain the difference between the states of the adsorbate during adsorption and desorption that leads to hysteresis. These theories include the difference in contact angle between adsorbent and adsorbate during adsorption and desorption and the presence of ink-bottle pores. Another theory proposes that the process of capillary condensation and evaporation inside the pores do not take place as exact reverses of each other, and capillary condensation in pores of a certain size during adsorption does not occur exactly at the same relative pressure as capillary evaporation from the same pores during desorption. The hysteresis loop introduces a considerable concern of whether to adopt the adsorption or desorption branch of the hysteresis loop for pore analysis. The desorption branch of the hysteresis loop corresponds to a more stable adsorbate condition, and it has been suggested that the desorption isotherm be used for pore structure analysis.

The pretreatment method, type of adsorbate and analysis procedure followed can affect the results. Nitrogen is commonly used for routine gas adsorption measurements, but also water vapor has been used for pore structure analysis in hardened cement pastes. The surface area determined using nitrogen as adsorbate range from 10 to 200 m²/g of dried paste, while the values for surface areas using water vapor as adsorbate are typically around 140–200 m²/g of dried paste. The differences vanish, if special techniques are employed for drying the specimens before the experiment.

Pycnometry and thermoporometry are simple and relatively fast techniques to use. Liquid and gas pycnometry provide information mainly on total pore volume. Pycnometry determines the density of a material by measuring the difference between the real and total volumes of a specimen. A liquid or gas may be used for pycnometry. The most commonly used immersion liquids for cement-based materials are water and alcohols. Helium is commonly used for gas pycnometry. Helium flow is measured as the volume of helium with time. Based on helium flow, the "missing porosity" has been identified as the inter-layer space occupied by approximately one layer of water when the material is in the wet condition.

Thermoporometry provides a simple method for pore size distributions in a porous material that is fully or partially saturated with a liquid. The method is based on the thermodynamic conditions of the liquid–solid transformation (melting–solidification) of a capillary condensate inside a porous body. Thermoporometry makes use of the triple point depression of a liquid in contact with its solid phase within the porous matrix of a material. The pore radius is determined as the sum of the curvature of the solid–liquid interface and a parameter that relates to the layer of nonfreezable pore water. When a sample saturated with a liquid such as water or some organic solvent is cooled, solidification can occur progressively in the smaller pores when the size of the critical nuclei at a given temperature is the same as the size of the pores. The freezing point depression and the melting point depression can be used to determine the pore radius, but the melting point depression is more repro-ducible and is most often used in experiments. The melting curves obtained during increasing the temperature can be used to determine the pore size dis-tribution, while the existence or absence of hysteresis between melting and solidification diagrams can indicate the presence of cylindrical or spherical pores. Thermoporometry is applicable to pores with radius ranging between 2 and 200 nm. The technique has the potential for further development and application more extensively to cement-based materials. A combination of the thermoporometry and NMR techniques is the cryoporometry technique.

NMR uses the phenomenon of nuclear magnetism, i.e. compounds develop a macroscopic magnetic moment when placed in a magnetic field. NMR is fast, nondestructive, and does not require drying of the sample. During an NMR experiment, a specimen is placed in a strong magnetic field and irradi-ated with intense radio frequency pulses typically of duration 1–10 μs over a frequency range required to excite specific atomic nuclei from a low energy state to a high energy state. Resonance radiation from a sample continues for a brief but electronically measurable time after the incident radiation is removed. A free induction decay (FID) signal can be transformed into a spec-trum or an image. Many magnetic resonance experiments involve repeated excitation of the nuclear spins and manipulation of the effective spin interactions in different ways. A variety of pulses can be formed by adding several pulses, varying the pulse angle, and inserting time delays.

After the application of a RF pulse, the free precession signal is gradually attenuated due to relaxation and dephasing. Spin relaxation can be due to interactions between nuclei (spin–spin relaxation) or due to interactions with the solid matrix (spin–lattice relaxation). The time constants that describe how the magnetization returns to its equilibrium value, relate to parameters that characterize the pore structure of the material. In cement pastes, the spin–lattice relaxation time T_1 is <10 milliseconds, and the spin–spin relaxation time is

$T_2 < 0.5$ milliseconds. Paramagnetic effects in cement pastes due to the presence of iron oxides complicate NMR measurements by reducing the T_2 time and by inducing local field gradients which limit spatial resolution.

The most commonly employed methods for determining pore structure parameters are the Magnetic Resonance Relaxation Analysis (MRRA), the Cryoporometry, and the Magnetic Resonance Imaging (MRI). MRRA can be performed on a liquid that is inside the pores of a material and is only applicable to materials with an interconnected pore structure. The pore size analysis is based on the difference between the relaxation times of protons in bulk water in the middle of the pore and the more constrained protons in the near surface water at the pore boundary. Discrete pore models and diffusion cell models have been developed and are used to extract pore size distributions from relaxation data. NMR cryoporometry employs the freezing point depression in small pores with magnetic resonance to measure pore size distributions. It is suitable for spatial imaging of pore diameters in the range of few to 100 nm. MRI is a topographic imaging technique which produces an image from the NMR signal. MRI can probe structural features in porous solids containing water by providing images of the spatial distributions of the mobile fluid.

Small angle scattering of both X-rays (SAXS) and neutrons (SANS) occurs at interfaces between two phases and is used to characterize complex microstructures and to study porous materials. After a beam of neutrons is incident on the sample, the neutrons scattered are detected, and a number of standard approximations and numerical methods are used to derive microstructural parameters of interest. The most commonly used approximations are the Guinier approximation for well-defined discrete inhomogeneities, and Porod's approximation for determining the surface area between two phases. The Guinier approximation applies at low scattering angles and at small values of the scattering vector. The pore size can be estimated from the measurement of the radius of gyration. At large scattering angles or at the so-called Porod region, the shape of the curve is useful in obtaining information on the surface-to-volume ratio of the scattering objects, and determine the surface area. Modern approaches for pore structure analysis include the use of fractals to characterize the cement paste. Small angle scattering of X-rays and neutrons is a useful technique for the range of sizes 1–100 nm. SANS has been used to study the microstructural development of cement pastes. These studies have indicated the existence of globules consisting the microstructure of the C–S–H gel, and that fractal gel microstructures form in cementitious systems, with a fractal dimension ranging with age from 1.9 to 2.8. Factors that affect the SANS and SAXS data are the sample thickness, the addition of mineral admixtures, and the extent of drying. SAS can be used to detect blocked or closed pores inaccessible to adsorptives. SAXS can be carried out with conventional X-ray sources; SANS is expensive, non-portable, and accessibility is not immediate.

Microscopy is the most definitive method to observe and measure particles or features with dimensions smaller than the resolving power of the human eye, which is approximately 0.1 mm. Reflected light microscopy is used on highly polished surfaces, fracture surfaces, or thin sections. Transmitted light microscopy is performed on thin sections only. Fluorescent light microscopy is carried out after filling the voids and pore spaces in concrete with resin that has been mixed with fluorescent dye, which fluoresces under ultraviolet illumination.

The SEM is designed primarily for studying the surfaces of solids at high magnification. The SEM is similar to the optical microscope but provides greater resolution and depth of field. Secondary electron (SE) images, backscattered electron (BSE) images, and elemental X-ray maps are used as principal SEM images. The contrast in BSE images is more intense than in SE images and they are preferred. BSE images are well-suited for microstructure imaging, because contrast in BSE images is produced by variations in atomic number within the specimen. Analysis of BSE images has often been used to quantify the microstructure of cement pastes, including the determination of total porosity and pore size distribution. The selection of magnification is an important factor that affects accuracy of results and time. The use of an intruding liquid that solidifies at room temperature and provides better contrast has been used for easier identification of pore features.

Scanning acoustic microscopy SAM is a high resolution, high frequency ultrasonic imaging technique, most commonly used at a high magnification mode, to see fine detail at and below the surface of the sample. The technique can provide images very fast and does not require special preparation techniques. It involves the use of trained personnel for its operation and expensive equipment.

Image analysis has become popular since 1970 and in connection with stereology represents a method for describing the size distribution of pores in cement-based materials. An image analyzer can be interfaced with an optical microscope, SEM or SAM to permit direct analysis of the image captured. The main objective of image analysis is to aid quantifying the parameters of interest using contrast differences in 2-D images. Semi-automatic and fully automatic instruments can be used. Image analysis consists of several steps including area selection, image acquisition, and digitization, image processing, feature recognition, and data analysis and output.

Stereological relationships relate the parameters observed on the image to the parameters observed in the three-dimensional material. Stereological techniques used are the point counting, the lineal analysis, and the section analysis. Point counting can estimate total volume only without giving information on pore size distribution. Lineal analysis and section analysis are used to measure both pore volume and pore size distribution. The modified point count method and the lineal transverse method are used to analyze the air void system in concrete. Operator objectivity can affect the results, inclusion or exclusion of entrapped air voids, method of measurement, and the use of thin or polished sections can affect the results. Different mathematical parameters have been proposed to characterize the air void system, instead of the Powers spacing factor and specific surface. The section analysis in combination with image analysis is a promising technique for providing size distribution of the air voids.

Comparison of results obtained by various techniques needs to be done carefully. Since different techniques have different pore size ranges of applicability, comparison is impossible when there is not sufficient overlap over the range of pore sizes. Even for techniques that can be applied over similar ranges, differences in results can be due to several factors: assumptions inherent with the application of each technique, modification of the microstructure before the experiment due to pretreatment, or modification and alteration of the microstructure during the experiment due to interaction with the material used. Explanation of the differences in results obtained by different techniques can further advance our understanding of the cement paste microstructure, and help us refine the assumptions made during the application of an experimental technique.

List of related International Standards

In the following are listed some of the international standards that describe or closely relate to procedures followed by the testing techniques presented in Chapters 2–8. In some European countries, the national standards follow the generalized European standards (EN), with the same numbering. Other codes of practice or standards by Organizations or Societies are also listed; however, it should be noted that the list is not complete and that procedures are standardized continuously. The values in parentheses correspond to the years. This list includes only standards that are currently active.

Method	Organization	Standard (Year)	Titles
Mercury intrusion porosimetry	ASTM	D 4284 (2003)	Standard Test Method for Determining Pore Volume Distribution of Catalysts by Mercury Intrusion Porosimetry
		D 4404 (R 2004)	Standard Test Method for Determination of Pore Volume and Pore Volume Distribution of Soil and Rock by Mercury Intrusion Porosimetry
	BSI	BS 7591 Part 1 (1992, R 1998)	Porosity and Pore Size Distribution of Materials Part 1: Method of Evaluation by Mercury Porosimetry
	DIN	DIN 66133 (1993)	Determination of Pore Volume Distribution and Specific Surface Area of Solids by Mercury Intrusion [Bestimmung der Porenvolumenverteilung und der spezifischen Oberfläche von Feststoffen durch Quecksilberintrusion]
	UOP	UOP 578 (2002)	Automated Pore Volume and Pore Size Distribution of Porous Substances by Mercury Porosimetry

(continued)

List of related International Standards Continued

Method	Organization	Standard (Year)	Titles
Nitrogen adsorption	ASTM	B 922	Standard Test Method for Metal Powder Specific Surface Area by Physical Adsorption
		C 1069 (1986, R 2004)	Standard Test Method for Specific Surface Area of Alumina or Quartz by Nitrogen Adsorption
		C 1274 (2000)	Standard Test Method for Advanced Ceramic Specific Surface Area by Physical Adsorption
		D 1993 (2003)	Standard Test Method for Precipitated Silica-Surface Area by Multipoint BET Nitrogen Adsorption
		D 3663 (2003)	Standard Test Method for Surface Area of Catalysts and Catalyst Carriers
		D 4222 (2003)	Standard Test Method for Determination of Nitrogen Adsorption and Desorption Isotherms of Catalysts and Catalyst Carriers by Static Volumetric Measurements
		D 4365 (1995, R 2001)	Standard Test Method for Determining Micropore Volume and Zeolite Area of a Catalyst
		D 4567 (2003)	Standard Test Method for Single-Point Determination of Specific Surface Area of Catalysts and Catalyst Carriers Using Nitrogen Adsorption by Continuous Flow Method
		D 4641 (1994, R 1999)	Standard Practice for Calculation of Pore Size Distributions of Catalysts from Nitrogen Desorption Isotherms
		D 4824 (2003)	Standard Test Method for Determination of Catalyst Acidity by Ammonia Chemisorption
		D 5604 (1996, R 2001)	Standard Test Methods for Precipitated Silica – Surface Area by Single Point BET Nitrogen Adsorption
		D 6556 (2004)	Standard Test Method for Carbon Black – Total and External Surface Area by Nitrogen Adsorption
	BSI	BS 4359-1:1996, ISO 9277:1995	Determination of the specific surface area of powders. BET-method of gas adsorption for solids (including porous materials)
		BS 5293-11:1995, ISO 4652-1:1994	Sampling and testing carbon black for use in the rubber industry. Method for determination of specific surface area by nitrogen adsorption methods using single-point procedures

BS 7591-2:1992	Porosity and pore size distribution of materials. Method of evaluation by gas adsorption
COE	
CRD-C 268-93 (1997)	Handbook for Concrete and Cement Standard Test Method for Specific Surface Area of Alumina or Quartz by Nitrogen Adsorption [Use: ASTM C 1069]
DIN	
DIN 66132 (1975)	Determination of specific surface area of solids by adsorption of nitrogen; single-point differential method according to Haul and Dümbgen [Bestimmung der spezifischen Oberfläche von Feststoffen durch Stickstoffadsorption; Einpunkt-Differenzverfahren nach Haul und Dümbgen]
DIN 66134 (1998)	Determination of the pore size distribution and the specific surface area of mesoporous solids by means of nitrogen sorption – Method of Barrett, Joyner, and Halenda (BJH). [Bestimmuug der Porengrößenverteilung und der spezifischen Oberfläche mesoporöser Feststoffe durch Stickstoffsorption. Verfahren nach Barrett, Joyner, und Halenda (BJH)].
DIN 66135-1 (2001)	Particle characterization – Micropore analysis by gas adsorption – Part 1: Fundamentals and testing procedure. [Partikelmesstechnik – Mikroporenanalyse mittels Gasadsorption – Teil 1: Grundlagen und Messverfahren]
DIN 66135-2 (2001)	Particle characterization – Micropore analysis by gas adsorption – Part 2: Evaluation by isotherms comparison. [Partikelmesstechnik – Mikroporenanalyse mittels Gasadsorption – Teil 2: Bestimmung des Mikroporenvolumens und der spezifischen Oberfläche durch Isothermenvergleich]
DIN 66135-3 (2001)	Particle characterization – Micropore analysis by gas adsorption – Part 3: Determination of the micropore volume according to Dubinin and Radushkevich [Partikelmesstechnik – Mikroporenanalyse mittels Gasadsorption – Teil 3: Bestimmung des Mikroporenvolumens nach Dubinin und Radushkevich]
DIN 66135-4 (2004)	Particle size analysis – Micropore analysis by gas adsorption – Part 4: Determination of pore distribution according to Horvath-Kawazoe and Saito-Foley [Partikelmesstechnik – Mikroporenanalyse mittels Gasadsorption – Teil 4: Bestimmung der Porenverteilung nach Horvath-Kawazoe und Saito-Foley]

(continued)

List of related International Standards Continued

Method	Organization	Standard (Year)	Titles
	ISO	ISO 9277 (1995)	Determination of the Specific Surface Area of Solids by Gas Adsorption Using the BET-method
		DIN ISO 9277 (2003)	[Bestimmung der spezifischen Oberfläche von Feststoffen durch Gasadsorption nach dem BET-Verfahren]
		NF X11-620 (1996)	[Détermination de l'aire massique (surface spécifique) des solides par adsorption de gaz à l'aide de la méthode BET (ISO 9277 (1996))]
		ISO 4652-1 (1994)	Rubber Compounding Ingredients – Carbon Black – Determination of Specific Surface Area by Nitrogen Adsorption Methods – Part 1: Single-Point Procedures
		ISO 8008 (1986)	Aluminium Oxide Primarily Used for the Production of Aluminium – Determination of Specific Surface Area by Nitrogen Adsorption – Single-Point Method
	ASI	AS 2879.4 (1991)	Alumina Part 4: Determination of Specific Surface Area by Nitrogen Adsorption
	UOP	UOP 425 (1986)	Surface Area, Pore Volume and Pore Diameter of Porous Substances by Nitrogen Adsorption
		UOP 821 (1981)	Automated Micro Pore Size Distribution of Porous Substances by Nitrogen Adsorption and/or Desorption Using a Micromeritics Analyzer
		UOP 874 (1988)	Pore Size Distribution of Porous Substances by Nitrogen Adsorption Using a Quantachrome Analyzer
Water absorption/ helium pycnometry	ASTM	C 20 (2000)	Standard Test Methods for Apparent Porosity, Water Absorption, Apparent Specific Gravity, and Bulk Density of Burned Refractory Brick and Shapes by Boiling Water
		C 373 (1988, R 1999)	Standard Test Method for Water Absorption, Bulk Density, Apparent Porosity, and Apparent Specific Gravity of Fired Whiteware Products
		C 830 (2000)	Standard Test Methods for Apparent Porosity, Liquid Absorption, Apparent Specific Gravity, and Bulk Density of Refractory Shapes by Vacuum Pressure
		C 948 (1981, R 2001)	Standard Test Method for Dry and Wet Bulk Density, Water Absorption, and Apparent Porosity of Thin Sections of Glass-Fiber Reinforced Concrete
		B 923 (2002)	Standard Test Method for Metal Powder Skeletal Density by Helium or Nitrogen Pycnometry

		D 6093 (2003)	Standard Test Method for Percent Volume Nonvolatile Matter in Clear or Pigmented Coatings Using a Helium Gas Pycnometer
	BSI	BS/EN 1936 (1999)	Natural Stone Test Methods – Determination of Real Density and Apparent Density, and of Total and Open Porosity
		BS 1881-5 (1970)	Testing concrete. Methods of testing hardened concrete for other than strength
		BS 1881-122 (1983)	Testing concrete. Method for determination of water absorption
	ISO	ISO 10545-3 (1997) BS 10545-3 DIN 10545-3 NF 10545-3 SFS 10545-3 NS 10545-3 EN 10545-3 AS 4459.3	Ceramic Tiles Part 3: Determination of Water Absorption, Apparent Porosity, Apparent Relative Density and Bulk Density [Including Technical Corrigendum 1]
		DIN 52102 (2004)	Determination of absolute density, dry density, compactness, and porosity of natural stone and mineral aggregates (draft standard, 2004) [Prüfverfahren für Gesteinskörnungen – Bestimmung der Trockenrohdichte mit dem Messzylinderverfahren und Berechnung des Dichtigkeitsgrades]
	RILEM	CPC11.3 (1984)	Absorption of water by concrete by immersion under vacuum, 1984
	JSA	JIS R 2205 (1992, R 2003)	Testing Method for Apparent Porosity, Water Absorption, and Specific Gravity of Refractory Bricks
	ASI	AS 1774.5 (2001)	Refractories and Refractory Materials – Physical Test Methods – Part 5: The Determination of Density, Porosity, and Water Absorption
NMR	ASTM	E 386 (1990, R 2004)	Standard Practice for Data Presentation Relating to High-Resolution Nuclear Magnetic Resonance (NMR) Spectroscopy
SAXS	ISO	TS 13762 (2001)	Particle Size Analysis – Small Angle X-ray Scattering Method

(continued)

List of related International Standards Continued

Method	Organization	Standard (Year)	Titles
Microscopy	ASTM	C 457 (1998)	Standard Test Method for Microscopical Determination of Parameters of the Air-Void System in Hardened Concrete
		C 856 (2004)	Standard Practice for Petrographic Examination of Hardened Concrete
		E 175 (1982, R 1999)	Standard Terminology of Microscopy
		E 211 (1982)	Standard Specification for Cover Glasses and Glass Slides for Use in Microscopy
		E 1382-97 (2004)	Standard Test Methods for Determining Average Grain Size Using Semiautomatic and Automatic Image Analysis
	ISO	ISO 10934-1 (2002) BS ISO 10934-1	Optics and optical instruments – Vocabulary for microscopy – Part 1: Light microscopy
		ISO 8576 (1996) DIN ISO 8576	Optics and optical instruments – Microscopes – Reference system of polarized light microscopy
	JSA	JIS K 0132 (1997)	General Rules for Scanning Electron Microscopy
	EN	EN 480-11 (1999) NF EN 480-11 DS 480-11 SFS 480-11 NS 480-11 DIN EN 480-11 BS 480-11 SN EN 480-11 OENORM EN 480-11	Admixtures for Concrete, Mortar and Grout – Test Methods – Part 11: Determination of Air Void Characteristics in Hardened Concrete
	COE	CRD-C 42-92 (1997)	Handbook for Concrete and Cement Standard Practice for Microscopical Determination of Air-Void Content and Parameters of The Air-Void System in Hardened Concrete [Use: ASTM C 457]

Glossary

Some of the terms mentioned in the book, which are not thoroughly discussed are explained briefly here.

Absorption   The process by which a liquid is drawn into and tends to fill permeable pores in a porous material.

Admixture   A material other than water, aggregates, hydraulic cement, and fiber reinforcement, used as an ingredient of concrete or mortar, which is added to the batch immediately before or during its mixing.

Adsorption   A process by which gas molecules impinge upon the surface of a solid; the process can be physical or chemical in nature.

Alcohols   Compounds in which a hydroxy group, $-OH$, is attached to a saturated carbon atom R_3COH.

Atomic number (proton number)   The number of protons in the atomic nucleus.

Bimodal distribution   The occurrence of two maxima in a frequency distribution.

Binary image   A digital image having only two gray levels (usually zero and one, black and white).

Blaine apparatus   Air-permeability apparatus used for measuring the surface area of a finely ground cement, or other product.

Blast-furnace slag   The nonmetallic by-product, that is developed in a molten condition simultaneously with iron in a blast furnace. It consists essentially of silicates and aluminosilicates of calcium and other bases.

Calorimetry   A general term used to describe any experiment in which heat is measured as a chemical reaction or a physical process occurs.

Carbonation   Reaction between the carbon dioxide of the atmosphere and a hydroxide or oxide in cement paste, mortar, or concrete to form a carbonate.

Carbon black   An industrially manufactured colloidal carbon material in the form of spheres and of their fused aggregates with sizes below 1000 nm.

Cementitious materials   Cements and pozzolans which have cementing properties and are used in concrete and masonry construction.

Chemisorption (chemical adsorption)   Adsorption that results from chemical bond formation between the adsorbent and the adsorbate in a monolayer on the surface.

Coherent scattering   Scattering is coherent whenever the phases of the signals arising from different scattering centers are correlated, and incoherent whenever these phases are uncorrelated.

Colloid   A mixture with properties between those of a solution and a fine suspension.

Condensation   The physical process of converting a material from a gaseous or vapor phase to a liquid or solid phase. The process commonly results when the temperature is lowered and/or the vapor pressure of the material is increased.

Covalent bond   A region of relatively high electron density between nuclei which arises at least partly from sharing of electrons and gives rise to an attractive force and characteristic internuclear distance.

Critical point   The temperature and pressure at which the liquid and vapor intensive properties (density, heat capacity, etc.) become equal. It is the highest temperature (critical temperature) and pressure (critical pressure) at which both a gaseous and a liquid phase of a given compound can coexist.

Crystallization   The formation of a crystalline solid from a solution, melt vapor, or a different solid phase, generally by lowering the temperature or by evaporation of a solvent.

Differential scanning calorimetry (DSC)   A technique in which the difference in energy inputs into a substance and a reference material is measured as a function of temperature while the substance and reference material are subjected to a controlled temperature program.

Diffraction   A modification which light (or electron, neutron, X-ray beams, etc.) undergoes in passing through opaque bodies or through narrow slits or in being reflected from ruled surfaces (or crystalline materials).

Elastic scattering   Radiation may be scattered by its transmission through a medium containing particles. If the scatter results in no significant change in the wavelength relative to the primary radiation it is called elastic scattering.

Electron probe microanalysis (EPMA)   General term for methods using bombardment of a solid specimen by electrons which generate a variety of signals providing the basis for a number of different analytical techniques.

Enthalpy   Internal energy of a system plus the product of pressure and volume. Its change in a system is equal to the heat brought to the system at constant pressure.

Evaporation   The physical process by which a liquid substance is converted to a gas or vapor. This may occur at or below the normal boiling point of the liquid (the temperature at which a liquid boils at 1 atmosphere pressure) and the process is endothermic.

Excited state   State of a system with energy higher than that of the ground state. This term is most commonly used to characterize a molecule in one of its electronically excited states, but can also refer to vibrational and/or rotational excitation in the electronic ground state.

Feature selection   A step in the pattern recognition system development process during image analysis, in which measurements or observations are studied to identify those that can be used to assign objects to classes.

Ground state   The state of lowest Gibbs energy of a system. See also Excited state.

Heavy water Water containing a significant fraction (up to 100%) of deuterium in the form of D_2O or HDO.

Ideal gas Gas which obeys the equation of state $PV = nRT$ (the ideal gas law; P is the pressure, V the volume, n the amount of molecules, R the gas constant, and T the thermodynamic temperature).

Imbibition The uptake of a liquid by a gel or porous substance. It may or may not be accompanied by swelling.

Inelastic scattering If radiation is scattered within a medium with change in the wavelength relative to the primary radiation, there is said to be inelastic scattering.

Inert gas A non-reactive gas under particular conditions.

Interface The plane ideally marking the boundary between two phases.

Layer Any conceptual region of space restricted in one dimension, within or at the surface of a condensed phase or a film. The term doublelayer applies to layers approximated by two "distinct" sublayers.

Least-squares technique A procedure for replacing the discrete set of results obtained from an experiment by a continuous function.

Liquefaction Change of substance from the solid or gaseous state to the liquid state.

Magnetic moment Vector quantity, the vector product of which with the magnetic flux density of a homogeneous field is equal to the torque.

Magnetogyric ratio Ratio of the magnetic moment to the angular momentum. It is often misleadingly called the gyromagnetic ratio.

Material safety data sheet (MSDS) Compilation of information required under the US OSHA Hazard Communication Standard on the identity of hazardous substances, health and physical hazards, exposure limits, and precautions.

Monolayer A single, closely packed layer of atoms or molecules.

Monolayer capacity For physisorption, the amount needed to cover the surface with a complete monolayer of atoms or molecules in close-packed array. The kind of close packing needs to be stated explicitly when necessary.

Monomer An organic liquid, of relatively low molecular weight, that creates a solid polymer by reacting with itself or other compounds of low molecular weight or both.

Monomolecular It is composed of single molecules; e.g. certain chemical compounds that develop a "monomolecular layer."

Multilayer A system of adjacent layers or monolayers. The term bilayer applies to the particular case of a multilayer, two monolayers thick.

Nucleation (in colloid chemistry) The process by which nuclei are formed in solution.

Paramagnetic Substances having a magnetic susceptibility greater than 0 are paramagnetic.

Paste content Proportional volume of cement paste in concrete, mortar, or similar material of the like, expressed as volume percentage of the entire mixture.

Petrography The branch of petrology dealing with description and systematic classification of rocks mainly by laboratory methods, largely chemical and microscopical.

Physical adsorption (physisorption)   Adsorption in which the forces involved are intermolecular forces (van der Waals forces) which do not involve a significant change in the electronic orbital patterns of the species involved.

Portlandite   The mineral calcium hydroxide [$Ca(OH)_2$] which occurs naturally in Ireland and is equivalent to a common product of hydration of Portland cement.

Pozzolan   A siliceous or siliceous and aluminous material, which in itself possesses little or no cementitious value. However, in finely divided form and in the presence of moisture, it will chemically react with calcium hydroxide at ordinary temperatures to form compounds possessing cementitious properties.

Quantitative analysis   Analysis in which the amount or concentration of an analyte may be determined (estimated) and expressed as a numerical value in appropriate units. Qualitative analysis may take place without quantitative analysis, but quantitative analysis requires the identification (qualification) of the analytes for which numerical estimates are given.

Sample   A portion of material selected from a larger quantity of material. The term "sample" implies the existence of a sampling error, i.e. the results obtained on the portions taken are only estimates of the concentration of a constituent or the quantity of a property present in the parent material. If there is no or negligible sampling error, the portion removed is a test portion, or specimen. See also Specimen.

Saturated surface-dry   Condition of an aggregate particle or other porous solid when the permeable voids are filled with water and no water is on the exposed surfaces.

Saturation   (1) In general: the condition of coexistence in stable equilibrium of either a vapor and a liquid or a vapor and solid phase of the same substance at the same temperature; (2) as applied to a porous material: the condition such that no more liquid can be held or placed within it.

Saturation vapor pressure   The pressure exerted by a pure substance (at a given temperature) in a system containing only the vapor and condensed phase (liquid or solid) of the substance.

Scattering   A process in which a change in direction or energy of an incident radiation is caused by interaction with a particle, a system of particles, or a photon.

Scattering angle   The angle between the forward direction of the incident beam and a straight line connecting the scattering point and the detector.

Scattering vector   The vector difference between the wave propagation vectors of the incident and the scattered beam.

Signal (in analysis)   A representation of a quantity within an analytical instrument.

Silica fume   Very fine noncrystalline silica produced in electric arc furnaces as a byproduct of the production of elemental silicon or alloys containing silicon; also known as condensed silica fume and microsilica.

Solid angle   Of a cone, the ratio of the area cut out on a spherical surface (with its center at the apex of that cone) to the square of the radius of the sphere. It is a quantity of dimension one (dimensionless quantity) with unit steradian.

Solidification   The transition of a liquid or gas into a solid.

Specific surface   The surface area of particles or of air voids contained in a unit mass or unit volume of a material; in the case of air voids in hardened concrete, the surface area of the air void volume expressed as millimeters per cubic millimeter.

Specimen   A piece or portion of a sample used to make a test. The term has been used both as a representative unit and as a nonrepresentative (often better than most) unit of a population, in mineralogical collections.

Spectroscopy   The study of physical systems by the electromagnetic radiation with which they interact or that they produce. Spectrometry is the measurement of such radiations as a means of obtaining information about the systems and their components. In certain types of spectroscopy, the radiation originates from an external source and is modified by the system, whereas in other types, the radiation originates within the system itself.

Spectrum analysis   The interpretation of the information present in an energy spectrum in terms of radiation energy and intensity.

Standard state   State of a system chosen as standard for reference by convention. Three standard states are recognized: for a gas phase, it is the (hypothetical) state of the pure substance in the gaseous phase at the standard pressure, assuming ideal behavior. For a pure phase, or a mixture, or a solvent in the liquid or solid state it is the state of the pure substance in the liquid or solid phase at the standard pressure. For a solute in solution, it is the (hypothetical) state of solute at the standard molality, standard pressure, or standard concentration and exhibiting infinitely dilute solution behavior. For a pure substance, the concept of standard state applies to the substance in a well-defined state of aggregation at a well-defined but arbitrarily chosen standard pressure.

STP   Abbreviation for standard temperature (273.15 K or 0°C), and pressure (105 Pa); usually employed in reporting gas volumes.

Surface area   See Specific surface.

Surface layer (or interfacial layer)   The region of space comprising and adjoining the phase boundary within which the properties of matter are significantly different from the values in the adjoining bulk phases.

Synchrotron radiation   X-radiation which results from the acceleration of charged particles in circular orbits by strong electric and magnetic fields.

Thresholding   The process of producing a binary image from a gray-scale image by assigning each output pixel the value 1 if the gray level of the corresponding input pixel is at or above the specified threshold gray level, and the value 0 if the input pixel is below that level.

Triple point   The point in a one-component system at which the temperature and pressure of three phases are in equilibrium.

van der Waals forces   The attractive or repulsive forces between molecular entities (or between groups within the same molecular entity) other than those due to bond formation or to the electrostatic interaction of ions or of ionic groups with one another or with neutral molecules.

Water–cement ratio   The ratio of the amount of water, exclusive only of that absorbed by the aggregates, to the amount of cement in a concrete, mortar, grout, or cement paste mixture. It is preferably stated as a decimal by mass and abbreviated w/c.

Xerogel   A term used for the dried out open structures which have passed a gel stage during preparation (e.g. silica gel); and also for dried out compact macromolecular gels such as gelatin or rubber.

X-ray diffraction   The diffraction of X-rays by substances having a regular arrangement of atoms, and the technique used to identify substances.

X-ray intensity   The term is commonly used for X-ray measurements and is expressed as photons per unit time detected.

X-ray photoelectron spectroscopy (XPS)   Any technique in which the sample is bombarded with X-rays and photoelectrons produced by the sample are detected as a function of energy. This technique is used to identify elements, their concentrations, and their chemical state within the sample.

Name index

In the following list, the numbers in brackets correspond to the reference numbers throughout the chapters, and the numbers to the pages of the text where the reference is mentioned. The numbers in parentheses correspond to the pages where the full references are listed.

Subject index

Printed and bound by CPI Group (UK) Ltd, Croydon, CR0 4YY

01/11/2024

01782621-0011